ENDOCRINOLOGY RESEARCH AND CLINICAL DEVELOPMENTS

ANDROGENS

PRODUCTION, FUNCTIONS AND DISORDERS

ENDOCRINOLOGY RESEARCH AND CLINICAL DEVELOPMENTS

Additional books in this series can be found on Nova's website under the Series tab.

Additional e-books in this series can be found on Nova's website under the e-book tab.

ENDOCRINOLOGY RESEARCH AND CLINICAL DEVELOPMENTS

ANDROGENS

PRODUCTION, FUNCTIONS AND DISORDERS

BERKLEY F. THOMPSON
AND
DEVON J. ROBINSON
EDITORS

Nova Science Publishers, Inc.
New York

For permission to use material from this book please contact us:
Telephone 631-231-7269; Fax 631-231-8175
Web Site: http://www.novapublishers.com

Library of Congress Cataloging-in-Publication Data

Library of Congress Control Number: 2012932562

ISBN: 978-1-62081-145-0

Published by Nova Science Publishers, Inc. † New York

Contents

Preface

Androgens have diverse effects at both a physiological and behavioral level. These steroid hormones are present right from the first weeks of gestation and, in addition to playing a key role in the development of the male sex organs (both internal and external) are also involved in the development of the central nervous system in both males and females. In this book, the authors present current research in the study of androgens including androgen deprivation therapy for advanced prostate cancer; and the relationship between androgen levels and diverse cognitive abilities and social behavior in childhood.

Chapter I - Androgens and the androgen receptor (AR) have been implicated in aggressiveness of prostate cancer. Huggins and Hodges in 1941 demonstrated that bilateral orchiectomy or estrogen treatment is an effective treatment for prostate cancer. Based upon their findings, androgen deprivation therapy (ADT) has remained the main therapeutic option for patients with advanced prostate cancer for about 70 years. The patients with prostate cancer of high Gleason score have low DHT levels in the prostatic tissue. Biologically aggressive prostate cancer can occur under a low DHT level environment where the prostate cancer of a low malignancy with high DHT dependency cannot easily occur. The serum testosterone level in men being treated with ADT has never been categorically established. After ADT with castration and flutamide, the DHT level in the prostatic tissue remained at approximately 25% of the amount measured before ADT in the same patients. Conventional ADTs that we have used cannot decrease DHT levels in the prostate sufficiently to restrain the growth of the cancer in patients with aggressive prostate cancer. The androgen levels in blood were low; however, it became clear that the AR signals play important roles for progress of the cancers from the studies of the prostate cancer that became exacerbation after ADT. The androgen milieu in men with high Gleason score prostate cancer is probably less affected by conventional ADT than that in men with low Gleason score cancer, which was suggested to be associated with adrenal androgen levels. In patients treated with ADT the pituitary-adrenal axis mediated by adrenocorticotropic hormone has a central role in the regulation of androgen synthesis. Importantly, the suboptimal suppression of androgen activity in prostate cancer may account for the heterogeneity in treatment effect observed across individual patients, moreover, may contribute to the outgrowth of resistant prostate cancer clones adapted to survive in a low androgen environment, suggesting that more effective methods for suppressing the androgen axis in the tumor microenvironment are required. From the results of the high reactivity of CYP17A1 inhibitors and a new AR antagonist for patients with castration-resistant prostate cancer (CRPC), the androgen syntheses and the maintenances of the AR signal system in the prostate cancer play important roles in etiological mechanisms of CRPC. AR transcriptional activity is reactivated in CRPC and identifies the intracellular conversion of adrenal androgens to DHT as a mechanism mediating this reactivation. As for the ADT for the patients with aggressive prostate cancer, it is important to restrain productions of testosterone and DHT and to restrain AR activity together with castration from early period of treatment. New ADTs using CYP17A1 inhibitors such as abiraterone acetate and AR antagonists such as MDV3100 for patients with CRPC, which will break through the limitation of the efficacy of conventional ADTs that have been used for nearly 70 years, have been developing. Therefore, it is important to examine the status of in situ androgen metabolism and/or the AR signal system in detail in order to improve the clinical response to ADT in patients diagnosed with biologically aggressive prostate cancer from now on.

Chapter II - Androgens play an pluripotent role in males. The clinical model of its deficiency is represented by androgen deprivation therapy (ADT). Androgen deprivation therapy (ADT) is considered the standard therapy for advanced prostate cancer (PCa).

ADT significantly reduces tumor growth and improves symptoms- free survival, but is associated with several adverse effects which can affect overall survival and quality of life. These adverse effects clearly show that the physiologic effect of testosterone is pluripotent and essential for a male and will be discussed in this chapter.

The basics of physiology of testosterone and bone metabolism will be clarified as well in this chapter, which helps to elucidate the new therapeutic approaches and its importance in bone metabolism.

There is also some evidence that patients with prostate cancer (PCa) have lower baseline BMD than healthy cohort, which is further lowered by ADT. Furthermore, patients with invasive PCa have greater probability of baseline osteopenia and osteoporosis and a vice versa.

These interesting facts illustrate the role of testosterone in bone metabolism and will be mentioned in the chapter.

Metabolic changes while on ADT are gaining more importance, even though there is not so much data about this issue. More patients are treated by ADT for longer period of time which may affect cardiovascular morbidity and mortality. The development of „metabolic syndrome" (syndrome „X"), its definition and its potential importance and treatment possibilities will be disscused, too.

There are experimental studies which provide the evidence, that testosterone plays an essential role in spatial congnition in an early childhood as well as in adults. The effect of testosterone on spatial cognitive functions will be clarified in the chapter. Spatial impairment together with the loss of BMD, muscular reduction contribute to the risk of pathological fractures in patients with PCa treated with ADT.

Chapter III - Erectile dysfunction (ED) is defined as the persistent inability to attain and maintain an erection sufficient to permit satisfactory sexual performance. It is estimated that 5-34% of men aged over the age of 40 have ED and the prevalence increases with increasing age. The risk factors for ED are similar to those of cardiovascular disease and metabolic syndrome.

Chapter IV - Prostate cancer is the most commonly occurring malignancy and the second highest cause of cancer-related mortality in men in the United States. The dependence of prostate cancer on androgens has been recognized for more than 7 decades. Huggins and

Hodges provided the first solid in vivo evidence that androgen–androgen receptor (AR) signals function as a proliferation stimulator leading to prostate cancer progression. This became a commonly accepted view, supported by subsequent early success of androgen deprivation therapy (ADT) in treating patients with prostate cancer. However, although the majority of patients with advanced prostate cancer respond initially to ADT, relapse due to progress to castration resistant prostate cancer (CRPC).

Significant efforts have been focused on understanding the mechanisms involved in the development and progression of CRPC. The prostate is a glandular organ comprised of three distinct cell types (basal, luminal and neuroendocrine cells) embedded in a fibro-muscular stroma. Prostatic development, growth and secretory function requires androgens action mediated by AR. Testosterone and its more active metabolite, 5α-dihydrotestosterone (DHT) exert their biological effects through binding to the AR and inducing AR transactivation. The AR is expressed and remains functional throughout prostate cancer progression and its expression persists in the majority of patients with CRPC. Therefore, the eventual failure of ADT cannot be simply explained by loss of the AR. Recent studies have demonstrated that the AR could function as both a proliferation stimulator and a tumor suppressor depending on its expression in individual prostatic cells. This chapter discusses findings that AR plays dual yet opposite roles in individual prostatic cells to exert its diverse and differential functions for prostate development and prostate cancer progression.

Chapter V - In normal prostatic epithelial cells, androgen receptor (AR) expression and activity are regulated by androgens like dihydrotestosterone (DHT) and testosterone. These androgens are mainly produced in the testis. Prostate cancer (PCa) retains the ability to respond to androgens in the early stage of development, in contrast with advanced stages that become independent of exogenous androgens while maintaining expression of functional AR. Targeting of androgens and AR has become the foremost approach for successful PCa palliation, especially for metastatic disease, with androgen ablation inhibiting PCa growth. Chemical castration, surgery (orchidectomy/orchiectomy), or androgen deprivation therapies are effective in most cases. This outcome, however, is temporary because an important effect of androgen deprivation is that recurrent PCa evolves to a castrate-resistant phenotype. It has recently become apparent that PCa cells in advanced stages of the disease synthesize androgens from cholesterol, and adrenal precursors, *de novo* and hence survive independently of testicular androgens. A corollary is that the steroidogenic enzymes required for androgen synthesis from cholesterol need to be present in PCa for this to happen. Our recent data indicate these are present but to varying extents in different patient samples. This book chapter will discuss the current knowledge about presence and extent of steroidogenic enzymes in PCa compared with normal prostate cells and present a theory about the importance of individualising patient therapies according to the steroidogenic potential of their PCa as it develops.

Chapter VI - Androgen deprivation therapy has been the mainstay of treatment for locally advanced and metastatic prostate cancer. While most patients respond, emergence of castration resistance ultimately ensues, a condition that is ultimately fatal. Recent studies have focused on mechanisms involved in the pathways of castration resistance and have brought about new insights and development of new agents in targeting these androgen pathways.

However, the effects of treatment entail adverse effects unique to androgen deprivation. This review focuses on current thinking on novel mechanisms of castration resistance, with particular attention to intratumoral androgen synthesis and resulting therapeutic maneuvers targeting these pathways and its antecedent adverse effects.

Chapter VII - Androgens are critical steroid hormones that determine the expression of the male phenotype, including the outward development of secondary male sexual characteristics as well as the initiation and maintenance of spermatogenesis and male fertility. Their actions are mediated by the androgen receptors. Several mechanisms involved in testosterone regulation of spermatogenesis include activation of gene transcription through the translocation of the androgen receptor, phospholipase and calcium influx, as well as specific protein receptors and kinases in Sertoli cells. On the other hand, the principle of male hormonal contraception is based on the suppression of luteinizing hormone and follicle stimulating hormone and the replacement of testosterone to maintain androgenicity. Novel contraceptive methods for men are currently not available but testosterone alone or in combination with progestin shows promising leads. The reasons for slow progress in hormonal male contraception are partly biological and partly practical and economical. This book chapter evaluates the role of androgen receptors in male fertility, as well as classical and non-classical mechanisms of actions of testosterone and spermatogenesis. It will also review the main principles of hormonal male contraception, results of clinical trials and potential innovations in this field.

Chapter VIII - We currently know that androgens act in the brain during its development, affecting both its structure and its neural function. In other words, in addition to the activating effects of adolescence, these hormones also have diverse organizational effects that, since they mold the central nervous system (CNS), influence both cognitive processes and the subject's behavior. Thus, a wide variety of data gathered over recent decades show that gonadal hormones affect the cognitive abilities of both men and women. Furthermore, recent studies carried out with normal populations instead of selected groups with hormonal abnormalities, have provided evidence that associates endogen hormone levels with behavior.

This chapter reviews existing research into both the relationship between androgen levels and diverse cognitive abilities, and the relationship between androgen levels and social behavior during childhood. First of all, it presents an overview of the principal androgens and their ontogenetic development, their main action mechanisms and their influence in the sexual differentiation of the brain and behavior. Next, it reviews the existing literature on the relationship between androgen levels and diverse cognitive abilities, paying special attention to social cognition (theory of mind, etc.) in children, before reviewing those research projects focusing on the relationship between androgens and social behavior in children, paying special attention to behaviors of dominance and aggression. Finally, the principal conclusions are discussed from an evolutionist or ultimate causes perspective.

Chapter IX - Prostate cancer is the leading cause of malignancy among American men and the third leading cause of cancer deaths. Approximately 42% of men will progress following local definitive therapy, with either rising PSA or metastatic disease. Androgen-deprivation therapy (ADT) is the primary treatment for metastatic prostate cancer. Although it is increasingly used in different disease stages and patient populations, there is still controversy regarding its use in patients whose sole determinant of disease progression is rising PSA. Various treatment modalities have been studied in combination with ADT to determine their overall efficacy and benefit in terms of progression-free and overall survival, leading to the use of ADT in selected patient populations. However, ADT is not without side effects. Different dosing strategies, such as intermittent ADT and rapid hormonal cycling, are being employed to determine how to administer ADT for maximum benefit and minimal toxicity. For the many patients who will eventually develop incurable, often debilitating metastatic castrate-resistant disease, strategies for directing novel therapies at the androgen-receptor signaling pathway and apoptotic genes hold future promise. This chapter discusses the clinical applications, side effects, and overall effects of ADT, clinical studies that support or controvert its use in selected patient populations and stages of prostate cancer, and the biology of castrate-resistant prostate cancer and novel approaches to its treatment.

In: Androgens: Production, Functions and Disorders
Editors: B. F. Thompson and D. J. Robinson
ISBN: 978-1-62081-145-0

Chapter I

Androgen Deprivation Therapy for Advanced Prostate Cancer on the Ground of Androgen Milieu in the Prostate

Tsutomu Nishiyama*
Division of Urology, Department of Regenerative and Transplant Medicine, Niigata University Graduate School of Medical and Dental Sciences, Niigata, Japan

Abstract

Androgens and the androgen receptor (AR) have been implicated in aggressiveness of prostate cancer. Huggins and Hodges in 1941 demonstrated that bilateral orchiectomy or estrogen treatment is an effective treatment for prostate cancer. Based upon their findings, androgen deprivation therapy (ADT) has remained the main therapeutic option for patients with advanced prostate cancer for about 70 years. The patients with prostate cancer of high Gleason score have low DHT levels in the prostatic tissue. Biologically aggressive prostate cancer can occur under a low DHT level environment where the prostate cancer of a low malignancy with high DHT dependency cannot easily occur. The serum testosterone level in men being treated with ADT has never been categorically established. After ADT with castration and flutamide, the DHT level in the prostatic tissue remained at approximately 25% of the amount measured before ADT in the same patients. Conventional ADTs that we have used cannot decrease DHT levels in the prostate sufficiently to restrain the growth of the cancer in patients with aggressive prostate cancer. The androgen levels in blood were low; however, it became clear that the AR signals play important roles for progress of the cancers from the studies of the prostate cancer that became exacerbation after ADT. The androgen milieu in men with

* Address correspondence and reprint requests to: Tsutomu Nishiyama, MD, PhD, Division of Urology, Department of Regenerative and Transplant Medicine, Niigata University Graduate School of Medical and Dental Sciences, Asahimachi 1-757, Chuo-ku, Niigata 951-8510, Japan. Telephone: +81-25-227-2285, Facsimile: +81-25-227-0784, E-mail: nisiyama@med.niigata-u.ac.jp.

high Gleason score prostate cancer is probably less affected by conventional ADT than that in men with low Gleason score cancer, which was suggested to be associated with adrenal androgen levels. In patients treated with ADT the pituitary-adrenal axis mediated by adrenocorticotropic hormone has a central role in the regulation of androgen synthesis. Importantly, the suboptimal suppression of androgen activity in prostate cancer may account for the heterogeneity in treatment effect observed across individual patients, moreover, may contribute to the outgrowth of resistant prostate cancer clones adapted to survive in a low androgen environment, suggesting that more effective methods for suppressing the androgen axis in the tumor microenvironment are required. From the results of the high reactivity of CYP17A1 inhibitors and a new AR antagonist for patients with castration-resistant prostate cancer (CRPC), the androgen syntheses and the maintenances of the AR signal system in the prostate cancer play important roles in etiological mechanisms of CRPC. AR transcriptional activity is reactivated in CRPC and identifies the intracellular conversion of adrenal androgens to DHT as a mechanism mediating this reactivation. As for the ADT for the patients with aggressive prostate cancer, it is important to restrain productions of testosterone and DHT and to restrain AR activity together with castration from early period of treatment. New ADTs using CYP17A1 inhibitors such as abiraterone acetate and AR antagonists such as MDV3100 for patients with CRPC, which will break through the limitation of the efficacy of conventional ADTs that have been used for nearly 70 years, have been developing. Therefore, it is important to examine the status of in situ androgen metabolism and/or the AR signal system in detail in order to improve the clinical response to ADT in patients diagnosed with biologically aggressive prostate cancer from now on.

Abbreviations

ADT	androgen deprivation therapy
AKR1C3	Aldo-keto reductase family 1, member C3
AR	androgen receptor
b5	cytochrome-b5 reductase
CRPC	castration-resistant prostate cancer
CRPC	castration-resistant prostate cancer
CYP17A1	Cytochrome P450, family 17, subfamily A, polypeptide 1
DHEA	dehydroepiandrosterone
DHEA-S	DHEA sulfate
DHT	dihydrotestosterone
FSH	follicle-stimulating hormone
GnRH	gonadotropin-releasing hormone
HSD	hydroxysteroid dehydrogenase
LH	luteinizing hormone
PCPT	The Prostate Cancer Prevention Trial
REDUCE	the Reduction by Dutasteride of Prostate Cancer Events
SRD5A1	steroid-5-α-reductase, α polypeptide 1 (3-oxo-5 α-steroid δ 4-dehydrogenase α1)
SRD5A2	steroid-5-α-reductase, α polypeptide 2 (3-oxo-5 α-steroid δ4-dehydrogenase α2)
SRD5A3	steroid 5α-reductase 3

Keywords: Prostate cancer, intra-prostatic androgen metabolism, androgen deprivation therapy, androgen receptor, testosterone, dihydrotestosterone, adrenal androgen precursors, pituitary-adrenal axis, adrenocorticotropic hormone, androgen metabolic enzymes, CYP17A1 inhibitors, androgen receptor antagonists

Introduction

Androgens and the androgen receptor (AR) have been implicated in a number of human diseases and conditions, most notably prostate cancer. [1] Huggins and Hodges in 1941 demonstrated that bilateral orchiectomy or estrogen treatment is an effective treatment for prostate cancer.[2] Based upon their findings, androgen deprivation therapy (ADT) has remained the main therapeutic option for patients with advanced prostate cancer for about 70 years. Different therapeutic approaches, including surgical or medical castration with gonadotropin-releasing hormone (GnRH) agonists, and an antiandrogen therapy, have become the gold standard for prostate cancer treatment with metastases. ADT involves the use of GnRH drugs (agonists or antagonists) or orchiectomy, consistently resulting in a 90% to 95% reduction in circulating levels of testosterone. ADTs are quite successful, leading to regression of tumors in a majority of cases; however, selection pressure induced by this treatment ultimately leads to the emergence of a tumor phenotype characterized by disease progression despite castrate levels of testosterone (typically defined as 50 ng/dL, 1.73 nmol/L) , coupled with metastatic spread, which renders the disease lethal. Following 12-18 months, most prostate cancers recur in a more aggressive, castration-resistant form (castration-resistant prostate cancer (CRPC)) that is incurable. Median survival for these patients after recurrence is 24-36 months.[3, 4] New ADTs and other promising therapies for CRPC have been actively developed.[5] This chapter will provide an overview of prostate cancer aggressiveness and the androgen levels in blood and prostatic tissues, androgen milieu during ADT, androgen milieu during CRPC, and the prospects of ADT on the ground of the androgen milieu in the prostate.

Androgen Milieu in Blood

Biosynthetic pathway for androgen synthesis is Δ-5 pathway in primates like humans.[6] (Fig. 1) Cytochrome P450s play critical roles in the metabolism of various bioactive compounds. One of the crucial functions of cytochrome P450s in Chordata is in the biosynthesis of steroid hormones. Steroid 17α-hydroxylase/17,20-lyase (CYP17A1) is localized in endoplasmic reticulum membranes of steroidogenic cells. CYP17A1 catalyzes the 17α-hydroxylation reaction of Δ4-C_{21} steroids (progesterone derivatives) and Δ5-C_{21} steroids (pregnenolone derivatives) as well as the 17,20-lyase reaction producing C_{19}-steroids, a key branch point in steroid hormone biosynthesis. Depending on CYP17A1 activity, the steroid hormone biosynthesis pathway is directed to either the formation of mineralocorticoids and glucocorticoids or sex hormones. [7] The first activity gives hydroxylation of pregnenolone and progesterone at the C(17) position to generate 17α-hydroxypregnenolone and 17α-hydroxyprogesterone, while the 17,20-lyase activity cleaves the C(17)-C(20) bond of 17α-hydroxypregnenolone and 17α-hydroxyprogesterone to form DHEA and androstenedione respectively. The modulation of these two activities occurs through cytochrome b(5). [8] A subset of androgens, adrenal androgens, includes any of C_{19}-steroids synthesized by the adrenal cortex, that function as weak steroids or steroid precursors, including dehydroepiandrosterone (DHEA), its sulfate, DHEA sulfate (DHEA-S), and androstenedione. [9] As a result, a large amount of DHEA and DHEA-S are in the blood in humans and nonhuman primates. The synthesis and secretion of the adrenal androgens DHEA and DHEA-S is a phenomenon apparently unique to humans. DHEA-S is a metabolite of DHEA produced by the addition of a sulfate group, catalyzed by the sulfotransferase. DHEA-S can also be back-converted to DHEA through the action of steroid sulfatase. This layer of the adrenal cortex is thought to be the primary source of serum DHEA-S. DHEA-S levels decline as a person ages as the reticularis layer diminishes in size. In rodents, biosynthetic pathway for androgen synthesis is the Δ-4 pathway.

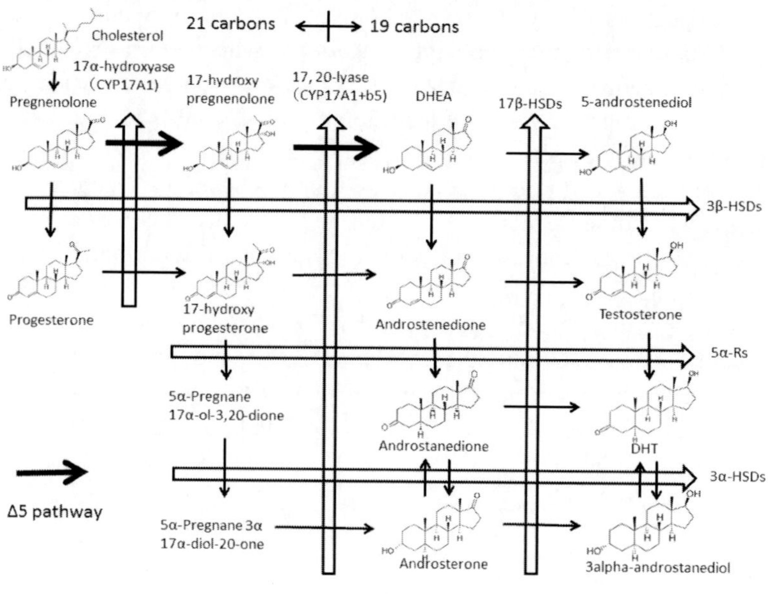

Figure 1. The androgen metabolic pathway and enzyme related products of human being.

Biosynthetic pathway for androgen synthesis is the Δ-5 pathway in primates like humans. As a result, a large amount of DHEA and DHEA-S is in the serum in humans.

b5: cytochrome-b5 reductase

CYP17A1: Cytochrome P450, family 17, subfamily A, polypeptide 1

DHEA: dehydroepiandrosterone

DHT: dihydrotestosterone

HSD: hydroxysteroid dehydrogenase

5α-Rs: 5-α-reductases.

In human males, testosterone is the major circulating androgen. More than 95% of testosterone is secreted by the Leydig cells, situated in the testicular interstitium, while the adrenal cortex also contributes to this production. [10, 11] The Leydig cells produce testosterone in response to hormonal stimulation by the pituitary gonadotropin luteinizing hormone (LH). Testosterone plays a key role in health and well-being as well as in sexual functioning. Testosterone effects can be classified as virilizing and anabolic effects. In men, testosterone plays a key role in the development of male reproductive tissues such as the testis and prostate as well as promoting secondary sexual characteristics such as increased muscle, bone mass, and the growth of body hair. [12] In addition, testosterone is essential for health and well-being as well as the prevention of osteoporosis. [13,14] Serum testosterone levels decline with advancing age. [15]

Recently, a pooled analysis of moderately sized prospective studies including 3886 cases of prostate cancer and 6438 matched controls reported no association of circulating androgens and risk of prostate cancer, either overall or for subgroups according to tumor stage or grade. [16] Prostate cancer risk and tumour aggressiveness are not related to serum levels of total and free testosterone. [17] However, this negative result does not deny the role of androgens for the prostate cancer, rather emphasizes the importance of the androgen metabolism in the prostate. It is not clear to what extent the testosterone and dihydrotestosterone (DHT) in the prostate derive from adrenal androgens or other steroid precursors.

Androgen Milieu in the Prostate

Androgens play an essential role in the development and differentiation of the prostate gland; however, their contribution to pathological conditions, such as benign prostatic hyperplasia and prostate cancer, remains unclear. The main male hormone in circulation is testosterone, which is produced mainly by the testicles. Some male hormones are produced in small amounts by the adrenal glands. However, DHT potently activating AR regulates the prostate. Originating from the gonads, testosterone is released into circulation. Estimates of serum concentrations of total testosterone are unequivocally eugonadal (12.5-40 nmol/L (350-1200 ng/dL)) or hypogonadal (<7.0 nmol/L (200 ng/dL)). [18] Testosterone and DHT can be synthesized from DHEA with various androgen synthetic enzymes in the human prostate. The conversion of testosterone from 17-ketosteroid, androstenedione, is said to be catalyzed by the 17β-hydroxysteroid dehydrogenase (HSD) type 1, 3, 5, 12. [19] 17β-HSD type3 is mainly expressed to convert into testosterone from androstenedione in the testis. Aldo-keto reductase family 1, member C3 (AKR1C3), known as 17β-HSD type5, is mainly expressed to convert into testosterone from androstenedione in the prostate. [20] Testosterone is converted to DHT by steroid-5α-reductase in the prostate. [21, 22] Humans have three isoenzymes of 5α-reductase. Both type 1 and type 2 5α-reductase catalyze the conversion of testosterone to DHT. Type 1 enzyme (encoded by SRD5A1) is expressed mostly in skin and hair, while type 2 enzyme (encoded by SRD5A2) is located primarily in androgen target tissues, including genital skin and prostate. Testosterone is reduced to DHT by both 5α-reductase type 1 and type 2 where the type 2 enzyme plays the more dominant role in the prostate [23]. Recently, the existence of 5α-reductase type 3 (encoded by SRD5A3) is identified. [24] Godoy et al. showed that expression of 5α-reductase type 3 in "classical" as well as "non-classical" androgen-regulated tissues is consistent with 5α-reductase type 3 enzyme having functions other than converting testosterone to DHT in human tissues, such as participation in the N-glycosylation process.[25] They also showed that over-expression of 5α-reductase type 3 in breast, testis, lung, thyroid, and particularly prostate cancer, compared to their benign counterparts, suggests a potential role for 5α-reductase type 3 as a biomarker of malignancy, and over-expression of 5α-reductase type 3 in androgen sensitive prostate cancer and CRPC suggests a potential role for this enzyme in synthesizing DHT in both an androgen-stimulated and an androgen-deprived human prostate microenvironment. Therefore, the prostate is not only the androgen dependent organ but also the androgen productive organ. DHT is reduced by 3α-HSD type 3 (AKR1C2) to produce 3α-diol and by 3β(20α)-HSD (AKR1C1 (98 % sequence identity with AKR1C2)) to produce 3β-diol. [26, 27] AKR1C2 is the dominant 3-ketosteroid reductase that will convert DHT to 3α-diol and will prevent DHT from binding to its receptor. AKR1C1 instead converts DHT to 3β-diol, where 3β-diol is a pro-apototic ligand for estrogen receptor β [28,29].

Recently, DHT mainly synthesizes via testosterone as stated above, but also synthesizes without testosterone biosynthesis, 'backdoor' pathway. [22, 30, 31] An alternate 'backdoor' pathway to DHT was elucidated in the tammar wallaby pouch young, and studies in knockout mice showed that this pathway uses 5-α-reductase type 1 to convert 17-hydroxyprogesterone to 5-α-reduced androgen precursors.[32] In this pathway progesterone is converted to 17α-hydroxyprogesterone by CYP17α-hydroxylase/17,20 lyase and then reduced by SRD5A1 to yield 5α-pregnane-17α-ol-3,20-dione. This is then reduced by AKR1C2 to yield 5α-pregnane-3α,17α-diol-20-one. Subsequent reaction by CYP17α-hydroxylase/17,20-lyase would produce androsterone which is then reduced by AKR1C3 to 3α-diol. Thus a new molecular switch that controls ligand access to the AR has been identified as AKR1C2 which reduces DHT to 3α-diol and oxidative 3α-HSDs which oxidizes 3α-diol back to DHT (in stromal cells). [33] Five different short-chain dehydrogenase genes (HSD17B10, HSD17B6, RDH5, DHRS9, RODH4) have been invoked as being the oxidative 3α-HSDs responsible for this back conversion. However, detailed kinetic analysis using mammalian expression systems and expression profiling have provided unequivocal evidence that the enzyme responsible is “RoDH like 3α-HSD” (HSD17B6) [34]. The role of 'backdoor' pathway to DHT in human androgen metabolism, however, is not clear yet.

The traditional belief that prostate cancer growth is dependent on the serum testosterone level has been challenged by recent negative studies in non-castrated men.[35, 36] Studies have failed to show increased risk of prostate cancer in men with higher serum testosterone, and supraphysiologic testosterone fails to increase prostate volume or prostate-specific antigen in healthy men. This apparent paradox is explained by the Saturation Model, which posits a finite capacity of androgen to stimulate prostate cancer growth. The evidence clearly indicates that there is a limit to the ability of androgens to stimulate prostate cancer growth. A Saturation Model based on androgen-AR binding provides a satisfactory conceptual framework to account for the dramatic effects seen with castration as well as the minor impact of testosterone administration in non-castrated men. Modern studies indicate no increased risk of prostate cancer among men with serum testosterone in the therapeutic range. [37, 38]

Prostate Cancer Aggressiveness and the Androgen Milieu in the Prostate and in Blood

Sex hormones in serum have been hypothesized to influence the risk of prostate cancer. A androgen milieu may be associated with tumor aggressiveness; however, large prospective study and a pooled analysis of moderately sized prospective studies did not show convincing evidence of a relationship between serum sex hormones and prostate cancer.[16, 39] Prostate cancer risk and tumor aggressiveness are not related to serum levels of total and free testosterone. [40] However, these negative results do not deny the role of androgens for the prostate cancer, rather emphasizes the importance of the androgen metabolism in the prostate. It is not clear to what extent the testosterone and DHT in the prostate derive from adrenal androgens or other steroid precursors. The 90-95% decrease in serum testosterone is observed after castration; however, the prostate efficiently transforms the adrenal androgen precursors DHEA-S, DHEA, and androstenedione into the active androgens, testosterone and DHT. All the enzymes are reported to exist in the prostate metabolizing from androgen precursors to testosterone and DHT. [41] The plasma concentration of DHEA-S is 100-500 times higher than that of testosterone. [42] High circulating levels of DHEA-S provide large amounts of the precursors required for conversion into active androgens in the prostate. The active androgens made locally in the prostate act in the prostate. In our data, the level of DHT in the prostate before ADT was not correlated with the serum level of testosterone, but was correlated with the serum level of DHEA and DHEA-S.[42] These results suggested that DHT in the prostate comes not only from testosterone produced in the testes, but also from adrenal androgens. There are still many questions about the androgen environment in the prostate in patients with prostate cancer. There have been few reports about the relations between histological aggressiveness and the DHT levels in the prostate in patients with prostate cancer. We revealed that the DHT levels in the prostate in patients with Gleason scores of 7 to 10 were significantly lower than those with Gleason scores of 6 and under, although there were no significant differences between the patients with Gleason scores of 7 to 10 and the patients with Gleason scores of 6 and under prostate cancer with respect to serum levels of androgens. [43] The result of the association between the DHT level in the prostate and the Gleason score gives useful information to predict the biological aggressiveness of prostate cancer and consider the treatment of the patients with such cancers. One hypothesis as to the correlation between clinical aggressiveness and low androgen levels in the prostate is that the hormonal milieu with low androgen levels might be varied enough to disrupt the normal growth and maintenance of the prostate tissue, and may cause the resultant compensatory hyperplasia that increases the risk of a consequent selection of androgen independent, aggressive prostate cancer formation due to low androgen levels. [44] Several studies suggested that men with smaller prostates had more high-grade cancers and more advanced diseases. [45-47] Prostate size was inversely associated with the risk of high grade cancer. The Prostate Cancer Prevention Trial (PCPT), a Phase III, randomized, double-blind, placebo-controlled trial of finasteride, a 5-α-reductase inhibitor, for the prevention of carcinoma of the prostate was held. [48] A total of 18,882 essentially healthy men, aged 55 and older, had been randomized to receive finasteride (5 mg daily) or placebo for seven years. PCPT results suggested that finasteride could reduce the frequency of prostate cancer by 25% compared with placebo. However, the patients with Gleason scores of 7 to 10 were detected significantly more frequently in the finasteride group than in the placebo group. Several questions were raised by the results of the trial that finasteride could induce the development of high grade disease. They did not find evidence that the drug caused changes in tumor composition that might contribute to aggressive cancer, though they don't entirely rule out the possibility that finasteride may have led to high-grade prostate cancer in some men in the study. One possible explanation was the occurrence of over-grading due to histological alterations. Several reports, however, showed that although finasteride use might cause some alterations in the assignment of the Gleason grade, a consistent hormonal therapy effect with finasteride treatment has not been observed. [49] Lucia et al. revealed that finasteride may have contributed to the increased rate of high-grade cancers detected in the PCPT by making the prostate smaller, helping the biopsy find the cancer. [50] Moreover, the results of the Reduction by Dutasteride of Prostate Cancer Events (REDUCE) trial using the dual 5-α-reductase isoenzyme inhibitor dutasteride, has recently been reported. [51] The REDUCE trial, contrary to the Prostate Cancer Prevention Trial (PCPT), is performed in 8,231 men ages 50-75, who would have been candidates for biopsy anyway because of PSA values of 2.5 - 10 ng/ml. Although over the course of 4 years, significantly fewer prostate cancers were detected in men taking dutasteride than in those taking placebo, representing a relative risk reduction of 22.8% (P <0.001), there were 12 tumors with a Gleason score of 8 to 10 in the dutasteride group, as compared with only 1 in the placebo group (P=0.003). These findings and our results suggest that biologically aggressive prostate cancer can exist under a low DHT environment, and may proliferate under the low DHT environment with finasteride or dutasteride medication where the prostate cancer of a low malignancy with high DHT dependency cannot easily exist.

Both AR and androgen-converting enzymes have been reported to be up-regulated in high-grade or advanced prostate cancer. [52-54] AR is a ligand-inducible transcription factor, a member of the nuclear receptor superfamily, which plays an important role in the development and progression of prostate cancer. High levels of AR are said to be associated with the increased proliferation, markers of aggressive disease, and to be predictive of decreased biochemical recurrence-free survival independently. [52] This confirms the role of AR in tumor growth and progression in hormonally naive prostate cancer. Several reports and our results revealed that 17β-HSD5 (known as AKR1C3) positive cases were significantly associated with clinical stage. [53, 54] Fung et al. revealed that elevated expression of AKR1C3 is highly associated with prostate cancer. [53] We also revealed that both AR and androgen-converting enzymes were up-regulated in the high-grade or advanced prostate cancer. [54] AKR1C3 (17β-HSD5) may be involved in increasing the local concentration of testosterone in prostate cancer tissues, resulting in the progression, invasion and further development of prostate cancer following ADT.

Intraprostatic Androgen Milieu during Androgen Deprivation Therapy

The castration level of serum testosterone is disputed; however, historically, circulating testosterone levels have been used to assess the efficacy of androgen depletion with a target total testosterone level below 50 ng/dL (1.73 nmol/L). [55-58] In men with prostate cancer, orchiectomy reduces serum testosterone to anorchid levels within 12 hours. GnRH agonists have played an important role in preventing progression of prostate cancer by lowering testosterone levels. However, initial use of agonists causes testosteronc to surge while repeated doses cause microsurges. [59] Morcover, breakthrough increases over castrate levels increase over time. [60] The clinical significance of different serum levels of testosterone yielded during ADT has not yet been well elucidated. The serum testosterone level in men being treated with ADT has never been categorically established. [61] The typical serum testosterone level in men after an orchiectomy is 14 ng/dL, which is equivalent to 0.5 nmol/L. Over the years, there have been many specialists who have argued that the appropriate serum testosterone level for a man who is being well-controlled on ADT should, in fact, be below 20 ng/dL (0.7 nmol/L). Morote et al. have shown that 25 percent of men receiving continuous ADT with a GnRH agonist have a serum testosterone level greater than 50 ng/dl.[62] Forty-four percent of the same group of men studied had a serum testosterone level that was consistently below 20 ng/dL. Also, in the same group of men, they demonstrated a direct correlation between serum testosterone levels and time to development of CRPC. Perachino et al. have also reported a significant association between serum testosterone levels 6 months after initiation of ADT and overall survival. [63] Zaitsu et al. showed that even hormone refractory prostate cancer patients are candidates for surgical castration because of endocrinological, oncological, and economical reasons. [64] Testosterone levels should be monitored routinely in patients on GnRH agonists so that the treatment can be switched or an antiandrogen agent can be added when there is a breakthrough. [65]

Several reports showed that DHT in the prostate after ADT is supplied by the precursors of androgens from the adrenal origin. We showed that after ADT with castration and flutamide, the DHT level in the prostatic tissue remained at approximately 25% of the amount measured before ADT in the same patients. [42] Previous reports revealed that the mean DHT levels in the prostate tissue treated with ADT were between 10% and 40% of those of untreated. [42, 66-70] Testosterone is converted to DHT by 5-α reductase in the prostate. Labrie et al. showed that even if testosterone levels in blood drop to low levels by castration, DHT levels in the prostate were maintained about 40% before the treatment. [67] These results showed that the considerable levels of DHT exist in the prostate after castration. We also revealed that the levels of DHT in the prostate before ADT were correlated with the serum levels of adrenal androgens. [42] We could not recognize the correlation between testosterone levels in blood and DHT levels in the prostate before the treatment; however, we recognized the correlation between DHT levels in the prostate and DHEA levels (or DHEA-S levels) in blood. The levels of DHT in the prostate after ADT using a GnRH agonist and flutamide were also correlated with the serum level of testosterone. The serum DHT levels after ADT were correlated with serum levels of adrenal androgen. The testosterone levels in blood after ADT are supplied from the adrenal androgen precursors converted in the prostate, and the adrenal androgen precursors are converted into DHT in the peripheral organs including the prostate. These results showed that DHT in the prostate can be made not only from testosterone derived from the testis but also from the adrenal precursors. These findings suggest that serum testosterone after ADT mostly comes from adrenal androgens converted in the prostatic cells. These results revealed that the prostate is the main DHT producing organ and DHT levels in the prostate are correlated with the adrenal androgen precursor levels and testosterone levels in the prostate.

Substantial levels of DHT remain in the prostate after castration; however, the influence on the change in the DHT levels in the prostate before and after ADT in connection with prostate cancer aggressiveness has not yet been elucidated. We revealed that the change in the DHT levels before and after ADT in the prostate with the prostate cancer of Gleason scores of 7 to 10 was significantly smaller than that in the prostate with the prostate cancer of Gleason scores of 6 or less. [71] Recently, 5α-reductase type 1(SRD5A1) is reported to express in aggressive and recurrent prostate cancer. [54,72, 73] Arai et al. showed that the levels of adrenal androgens in prostate cancer tissues after ADT were similar to those in untreated prostate cancer. [74] Especially, DHEA was the most existing androgen precursor in prostate cancer tissues after ADT. The levels of DHEA were high in prostate cancer tissues, irrespective of ADT. They assumed that DHEA played a significant role in the synthesis of testosterone and DHT in prostate cancer tissues after ADT. Several reports revealed that AR signaling pathway is an important mechanism in proliferation of aggressive prostate cancer during ADT. [75] Stanbrough et al. revealed that AR transcriptional activity is reactivated in androgen-independent prostate cancer and identifies the intracellular conversion of adrenal androgens to testosterone as a mechanism mediating this reactivation. [76] These findings and our results suggest that the low DHT levels in the prostate with aggressive prostate cancers are probably sufficient to propagate the growth of the tumor, and the prostate with aggressive prostate cancer can produce androgens from the adrenal precursors more autonomously than that with non-aggressive prostate cancer does under the low testosterone environment with testicular suppression. Mostaghel et al. showed that the reduction of circulating androgens to castrate levels and interruption of AR signaling by androgen deprivation are neither reliably effective nor uniform in suppressing prostatic androgen-regulated gene and protein expression. [77] Their findings suggest that medical castration based on serum testosterone levels cannot be equated with ablation of androgen levels or androgen-mediated activity in the prostate tissue microenvironment. These findings underscore the need for demonstrating suppression at the target tissue and molecular level before drawing conclusions regarding the clinical efficacy of hormonal treatment strategies. We showed that the serum PSA levels well reflect the androgen milieu in localized prostate cancer patients receiving ADT, which can be explained by the Saturation Model and disease control. [78] The androgen milieu in men with high Gleason score prostate cancer is probably less affected by conventional ADT than that in men with low Gleason score cancer, which was suggested to be associated with adrenal androgen levels. We also showed that in patients treated with ADT the pituitary-adrenal axis mediated by adrenocorticotropic hormone has a central role in the regulation of androgen synthesis.[79] Serum adrenocorticotropic hormone and adrenal androgen concentrations were correlated with the post-treatment PSA. Adrenocorticotropic hormone mediated androgen synthesis is a potential target for advanced androgen deprivation therapy.

Androgen Milieu in the Castration-Resistant Prostate Cancer

The pathogenesis of CRPC involves a mixture of multiple pathways that increase and broaden the function of the AR, and others that bypass the AR.[80] Analysis of tumor samples from patients with CRPC has revealed several mechanisms utilized by tumor cells to reactivate the AR signaling at sub-physiological serum concentration of androgens or even in the presence of AR antagonists.[81] CRPC has been used synonymously with androgen-independent prostate cancer and hormone-refractory prostate cancer but is the preferred term, as we now know that many men with CRPC respond to additional manipulations that ablate or block prostate cancer growth stimulation by androgens. CRPC is clinically detected by a rise in PSA, typically defined as three consecutive rises over nadir in the context of castrate levels of serum testosterone and after anti-androgen withdrawal for at least 4 weeks and despite secondary hormonal manipulations and/or radiologic progression. In the face of testicular androgen suppression the adrenal androgens DHEA, DHEA-S and androstenedione retain the potential to stimulate prostate cancer growth since these steroids are converted to testosterone and DHT in the prostate, which are active androgens even after prostate cancer acquires the ability to activate during ADT. Chen et al. showed that a modest increase in AR mRNA was the only change consistently associated with the development of resistance to anti-androgen therapy. [82] This increase in AR mRNA and protein was both necessary and sufficient to convert prostate cancer growth from a hormone-sensitive to a hormone-refractory stage, and was dependent on a functional ligand-binding domain. Conventional AR antagonists showed agonistic activity in cells with increased AR levels; this antagonist-agonist conversion was associated with alterations in the recruitment of coactivators and corepressors to the promoters of AR target genes. Increased levels of AR confer resistance to anti-androgens by amplifying signal output from low levels of residual ligands, and by altering the normal response to antagonists.

Mohler et al. revealed that recurrent prostate cancer may develop the capacity to biosynthesize testicular androgens from adrenal androgens or cholesterol. [83, 84] They showed that tissue levels of testosterone were similar in recurrent prostate cancer and benign prostate. They also showed that the DHT level in recurrent prostate cancer tissue was decreased to 18% of the level in benign prostate tissue. Testosterone and dihydrotestosterone occur in recurrent prostate cancer tissue at levels sufficient to activate AR. Montgomery et al. showed that soft tissue metastases in patients with anorchid serum testosterone contain levels of testosterone that are up to three times higher than those in prostate tumors in eugonadal men. [85] Although the pathway from adrenal precursor steroids to intratumoral synthesis of DHT was widely believed to require testosterone, the main route instead circumvents testosterone through the 5α-androstanedione pathway and requires expression of SRD5A1. Strategies for the development of better therapeutic approaches should account for the unanticipated dominance of the 5α-androstanedione pathway in CRPC. [86] Moreover, AR transactivation in prostate cancer cells in the presence of androstanediol resulted from the cell-specific conversion of androstanediol to DHT. [87] Transcript levels of enzymes involved in androgen synthesis were up-regulated in the same tumors, suggesting that tumoral synthesis of androgens from cholesterol might occur. [88, 89] Androgen-independent prostate cancer cell lines synthesize testosterone from radio-labeled cholesterol in vitro, and human prostate cancer xenografts are capable of synthesizing DHT from acetate and cholesterol, confirming that tumoral androgen synthesis is possible. [89] Thus, experimental and clinical evidence suggests that the enzymes of steroid metabolism are activated in the prostate cancer with CRPC state, and might provide multiple new targets for therapeutic intervention. Moreover, substantial evidences suggest that CRPC may not, in fact, be androgen independent, but occurs in a setting of continued AR-mediated signaling that is driven by the presence of residual tissue androgens. We used androgen independent and androgen sensitive cells, LNCaP cells, and examined androgen metabolism in the cells. [90] We confirmed that DHT converted from the androgen precursors is important in the cellular activity of the cancer cells. This shows that metabolism of the androgen in the prostate cancer cells plays an important role even in the exacerbation with androgen independent and androgen sensitive state after ADT. This conclusion is supported by the evidence that about 30% of recurrent prostate cancers show transient clinical responses to secondary or tertiary hormonal manipulation. [91] As a consequence, strategies to decrease adrenal cortical androgen production in patients with CRPC have been developed and are currently in widespread use. Ryan et al. revealed that higher androstenedione levels predict the likelihood of response to ketoconazole, which is used as an inhibitor of adrenal androgen synthesis, and improved survival compared with patients with lower levels. [92] These data suggest that the therapy with ketoconazole is less effective in patients with low levels of androgens at baseline. Ketoconazole therapy as secondary hormonal therapies may be less effective in the patients with low androgen levels in the prostate.

Clinical-Translational Advances of Prostate Cancer

The hypothesized dependence of prostate cancer on androgen stimulation for survival and proliferation even in the CRPC state corroborate the rationale for drug development strategies that target androgenic ligands as well as the AR itself. The promising agents of therapies that directly target androgen synthesis as well as the AR confirm that prostate cancer retains dependence on AR signaling, even in the CRPC state. Recently, encouraging GnRH antagonists and early phase trial results of promising ADT agents that interfere with AR signaling for patients with hormone-naive prostate cancer and CRPC have been reported.

Gonadotrophin-Releasing Hormone (GnRH) Antagonists: Degarelix

Medical castration using GnRH agonists currently provides the mainstay of ADT for prostate cancer. Although effective, these agents only reduce testosterone levels after a delay of 14 to 21 days. [93] The more recently developed GnRH antagonists act immediately. GnRH antagonists compete with natural GnRH for binding to GnRH receptors, thus decreasing or blocking GnRH action in the body. GnRH antagonists induce fast and profound LH and follicle-stimulating hormone (FSH) suppressions from the pituitary. In men, the reduction in LH subsequently leads to rapid suppression of testosterone release from the testes. Unlike the GnRH agonists, which cause an initial stimulation of the hypothalamic-pituitary-gonadal axis, leading to a surge in testosterone levels, and under certain circumstances, a flare-up of the tumor, GnRH antagonists do not cause a surge in testosterone or clinical flare. [94] Clinical flare is a phenomenon that occurs in patients with advanced disease, which can precipitate a range of clinical symptoms such as bone pain, ureteral obstruction, and spinal cord compression. As testosterone surge does not occur with GnRH antagonists, there is no need for patients to receive an antiandrogen as flare protection during prostate cancer treatment. GnRH agonists also induce an increase in testosterone levels after each reinjection of the drug a phenomenon that does not occur with GnRH antagonists. GnRH antagonists have an immediate onset of action leading to a fast and profound suppression of testosterone and are therefore especially valuable in the treatment of patients with prostate cancer, where fast control of disease is needed. Abarelix, initial gonadotropin-releasing hormone antagonist for prostate cancer, has been linked with immediate-onset systemic allergic reactions. [95] Degarelix is a new type of hormonal therapy for prostate cancer without stimulation of the release of histamine. [96] The reduction in LH subsequently leads to a rapid and sustained suppression of testosterone release from the testes and subsequently reduces the size and growth of the prostate cancer. This in turn results in a reduction in PSA levels in the patient's blood. Degarelix has an immediate onset of action, binding to GnRH receptors in the pituitary gland and blocking their interaction with GnRH. [97] The effectiveness of degarelix was recently evaluated in a Phase III, randomised, 12 month clinical trial (CS21) in 610 patients with prostate cancer, for whom androgen deprivation therapy was indicated. [98] Patients were randomised to receive treatment with one of two doses of degarelix or the GnRH agonist, leuprolide. Both degarelix doses were at least as effective as leuprolide at suppressing testosterone to castration levels (<or=0.5 ng/mL) from Day 28 to study end (Day 364). Testosterone levels were suppressed significantly faster with degarelix than with leuprolide, with nearly all of the degarelix patients achieving castration levels by Day 3 of treatment compared with none in the leuprolide group. No patients receiving degarelix experienced a testosterone surge compared with 81% of those patients receiving leuprolide. Furthermore, degarelix resulted in a significantly faster reduction in PSA levels compared with leuprolide indicating faster control of the prostate cancer. Recent results also suggest that degarelix therapy may result in longer control of prostate cancer compared with leuprolide. [99] As with all hormonal therapies, degarelix is commonly associated with hormonal side effects such as hot flushes and weight gain. [98] Due to its mode of administration (subcutaneous injection), degarelix is also associated with injection-site reactions such as injection-site pain, erythema or swelling. Injection-site reactions are usually mild or moderate in intensity and occur predominantly after the first dose, decreasing in frequency thereafter.

The GnRH agonists initially stimulate before suppressing testosterone and follicle-stimulating hormone (FSH) production. FSH is speculated to play a role in prostate cancer progression. [100] Compared to normal prostate cells, cancer cells contain more FSH receptors and more so in androgen refractory prostate cancer cells. FSH receptor is selectively expressed on the surface of the blood vessels of a wide range of tumors. [101]

Cytochrome P450, Family 17, Subfamily A, Polypeptide 1 (CYP17A1) Inhibitors

The multifunctional 17α-hydroxylase/17,20-lyase encoded by Cytochrome P450, family 17, subfamily A, polypeptide 1 (CYP17A1) is a cytochrome P450 enzyme localized to the endoplasmic reticulum in the adrenals, testes, placenta and ovaries, and lies at the crossroads of sex steroid and glucocorticoid synthesis. It is encoded by a single gene found on chromosome 10 in humans. The enzyme possesses both 17α-hydroxylase activity and C17,20-lyase activity, and requires P450 reductase, cytochrome-b5 reductase (b5), to transfer electrons in the presence of nicotinamide adenine dinucleotide phosphate (NADPH) for both catalytic activities. The degree of C17,20-lyase activity, modulated by cytochrome-b5, determines which metabolic pathway the substrate will follow in terms of sex steroid or glucocorticoid formation. [102, 103] Cytochrome-b5 is required for the C17,20 lyase activity, and high b5/CYP17A1 ratios are associated with the nearly exclusive production of androgens in the testes as opposed to the low ratios observed in the adrenals where glucocorticoids are produced . [104, 105] As the critical catalytic step in all androgen biosynthesis, therapeutic CYP17A1 inhibition to treat prostate cancer and other androgen dependent diseases has been envisioned for over 50 years, with recent discoveries in prostate cancer pathology driving much of the current refocus on CYP17A1 as a therapeutic target. [106, 107] Additionally, some evidence suggests that CYP17A1 mRNA and protein expression correlate with disease stage and relapse. [108,109] Despite the strong rational for targeting CYP17A1, this target has been largely unexploited with relatively few inhibitors progressing to clinical trials and only ketoconazole, an unspecific inhibitor of CYP17A1, in widespread clinical use.

Abiraterone Acetate (Zytiga, Johnson & Johnson)

One of CYP17A1 inhibitors, abiraterone acetate, the irreversible inhibitor of 17α-hydroxylase/C17,20-lyase (CYP17A1), is well tolerated and has significant and durable anti-tumor activity both in patients with chemotherapy-naive prostate cancer and in docetaxel-treated patients with CRPC. [110] In patients with CRPC, despite ADT, intra-tumoral androgen levels remain high, with continued AR signaling. The source of these androgens could be adrenal or de novo intra-tumoral synthesis. Researchers tried to inhibit CYP17A1, the key enzyme responsible for androgen synthesis, using abiraterone acetate, which is 10-fold or more potent of inhibition of CYP17A1 than ketoconazole, another inhibitor of CYP17A1. The phase II (open-label, single-arm) portion of the trial was an expansion of a promising phase I trial reported in 2008. [111] All 42 participants were castrated, and 38 had metastases, primarily in bone and/or lymph nodes. Their median age was 70. All received oral abiraterone acetate 1,000 mg daily in 28-day cycles. PSA declined at least 50% in 28 of the 42 patients (67%) ; eight patients (19%) had PSA decreases of at least 90%. Of the 24 patients who had measurable disease on CT scan, 9 (37.5%) had tumor regression that constituted a partial response. Further, 16 of these 24 patients (67%) showed no evidence of progression at 6 months. Baseline levels of DHEA, DHEA-S, androstenedione and estradiol were associated with the increased likelihood of a PSA decline of at least 50%. Adding dexamethasone 0.5 mg daily at the time of PSA progression reversed abiraterone acetate resistance in 33% of patients. This study and other phase II study revealed that abiraterone acetate has significant antitumor activity in post-docetaxel patients with CRPC. [112-114] Recent results of the phase III trial of abiraterone acetate in post-docetaxel patients has shown an overall survival benefit in advanced CRPC (COU-AA-301). [115] Abiraterone acetate is designed to treat these tumors by inhibiting the production of androgen in the testes, the adrenal glands and prostate cancer tumors themselves. The Phase-III trial included 1195 patients from 13 countries whose metastatic, CRPC had previously been treated with one of two chemotherapeutic agents that included docetaxel. Among the 398 patients randomly assigned to receive the corticosteroid prednisone plus placebo, median overall survival was 10.9 months. Among the 797 who received abiraterone acetate plus prednisone, median survival was 14.8 months. Significant differences also emerged between the placebo and treatment groups for all of the trial's secondary endpoints, including time to PSA progression, radiographic progression-free survival and PSA response rate. The benefits of abiraterone were determined during a pre-specified interim analysis of the study, prompting the trial's Independent Data Monitoring Committee to recommend unblinding the trial, and allowing anyone on the placebo arm to be offered abiraterone acetate. The US Food and Drug Administration (FDA) has approved abiraterone acetate (Zytiga, Cougar Biotechnology) in combination with prednisone for the treatment of metastatic CRPC in men who have received prior docetaxel chemotherapy. Several ongoing trials are examining the agent in earlier settings (i.e., a phase III in metastatic CRPC pre-docetaxel, and smaller studies in combination with radiation therapy or as neoadjuvant pre-surgery for localized disease). Herein, several potential strategies for abiraterone acetate are presented to clarify the clinical utilization of this agent in the future.

Abiraterone acetate is a prodrug of abiraterone. Most notably, it has a food effect greater than any other marketed drug (five- to 10-fold, depending on fat content), yet it is labeled to be taken fasting. The marked effect of a high-fat meal (10-fold increase in exposure) implies that absolute bioavailability in the fasting state is no more than 10%. This is likely the result of low absorption under fasting conditions, because 77% of the administered drug is excreted as abiraterone (or its prodrug acetate) in feces. In addition, there is marked interindividual pharmacokinetic variability, with a coefficient of variation in the plasma concentration area under the curve of 59% under labeled conditions. [116] Food can alter the bioavailability of orally administered drugs. Description of food effects in product labels and information about administration in relation to food are influenced by a variety of factors. Because food effects can change drug efficacy and toxicity, it is important that physicians and patients be aware of them. [117, 118]

Orteronel (TAK-700) [(1S)-1-(6,7-Dimethoxy-2-Naphthyl)-1-(1H-Imidazol-4-Yl)-2-Methylpropan-1-Ol]

Orteronel is a non-steroidal imidazole CYP17A1 inhibitor (IC50 = 28nM) currently in clinical trials in prostate cancer patients. The compound was found to be selective for CYP17A1 over 11β-hydroxylase by an impressive 260-fold, and exhibited suppressive effects on testosterone biosynthesis in rats and reduction in the weight of prostate and seminal vesicles in the rat. [119, 120] Orteronel was also capable of reducing serum testosterone levels down to castration level after 8 hours following a single oral administration of 1mg/kg to monkeys. A practical asymmetric synthesis of Orteronel has been established in eight steps from 2,3-dihydroxy naphthalene. This phase I/II, open-label, dose-escalation study assesses the safety and tolerability of oral, twice-daily (BID) Orteronel in patients with metastatic CRPC. Secondary objectives include assessment of Orteronel efficacy, as shown by PSA response, and Orteronel pharmacokinetics. Orteronel >= 300 mg BID appears active and well tolerated in patients with metastatic CRPC. The recommended phase II dose is 400 mg BID. [ASCO 2010 Genitourinary Cancers Symposium] Orteronel is performed on two phase 3 clinical trials: one is a randomized, double-blind, multicenter, phase 3 study evaluating

Orteronel plus prednisone compared with placebo plus prednisone in the treatment of men with progressive, chemotherapy-naive, metastatic CRPC, the other is in men with metastatic CRPC that has progressed following taxane-based therapy.

TOK-001 (VN/124-1) [3β-Hydroxy-17-(1H-Benzimidazole-1-Yl)Androsta-5,16-Diene]

The increased efficacy of TOK-001 in several prostate cancer models both in vitro and in vivo is believed to arise from its ability to down regulate the AR as well as competitively block androgen binding. [121, 122]. TOK-001 can act by several mechanisms (CYP17A1 inhibition, competitive inhibition and down-regulation of the AR). In prostate cancer cell lines, TOK-001 inhibited the growth of CRPCs, which had increased AR and were no longer sensitive to bicalutamide. Indeed, ARMOR1(Androgen Receptor Modulation Optimized for Response1) phase1/2 trials with TOK-001 was recently initiated in CRPC patients.

3β-Hydroxysteroid Dehydrogenase Inhibitors

Prostate cancer exhibits robust 3β-HSD enzymatic activity, which is required for the AR-mediated response to Δ-5 adrenal androgens and that 3β-HSD may serve as a pharmacologic target for the treatment of CRPC. [123] Pharmacologic inhibition of 3β-HSD might have therapeutic value regardless of the relative contribution of adrenal androgens as opposed to de novo steroidogenesis in the reconstitution of intratumoral testosterone and DHT. We showed that trilostane, a 3β-HSD inhibitor, actually inhibits the in situ conversion of DHEA to androstenedione in prostate cancer cells, suggesting a new therapeutic target to more efficiently suppress intracellular androgen levels.[124] However, trilostane has a direct androgenic ability in cancer cells with mutated or wild-type AR. Trilostane should be used with particular concern when treating prostate cancer, possibly suggesting the combination of trilostane and an AR antagonist such as bicalutamide or designing an inhibitor of 3β-HSD free of AR agonist activity.

Apoptone (HE3235) (17α-Ethynyl-5α-Androstan-3α, 17β-Diol)

A novel androstane steroid, Apoptone (HE3235), an orally bioavailable synthetic analogue of 3β-androstanediol, has a significant inhibitory activity for 5'-androstenediol-stimulated LNCaP proliferation. [125, 126] Apoptone exhibits a wide range of effects, including alteration of androgen receptor signaling and reductions in levels of intratumoral androgens. This inhibitory activity is accompanied by an increase in the number of apoptotic cells. Animal studies have confirmed the cytoreductive activity of apoptone on LNCaP tumors. The results suggest that this compound may be of clinical use in CRPC. From the results of phase I/II clinical trial, apoptone is a novel, orally delivered agent with clinical activity in advanced CRPC patients. It is well tolerated and has an AUC linear to dose. Based on the safety and activity observed, the trial continues to accrue an expansion cohort at 100 mg/day of patients with chemotherapy-naive CRPC. [127]

Inhibitors Targeting Androgen Receptor

A New Androgen Receptor Antagonist, MDV3100

As for the conventional anti-androgen, affinity for AR and agonistic activity for AR are matters of concern. [82] Because AR signaling appears to drive this cancer, blocking this signaling should stop its growth, the researchers theorized, and they went on to develop a drug that would do just that. Another novel ADT drug, AR antagonist MDV3100 (a diarylthiohydantoins) is a small-molecule, pure AR antagonist that inhibits AR nuclear translocation and DNA binding. [128, 129] MDV3100 is generally well tolerated, and shows encouraging anti-tumor activity in patients with CRPC. The drug was selected as a result of its activity in models of AR over-expression, in which conventional anti-androgens develop agonist properties. Currently available agents in the setting of AR over-expression actually act as agonists, rather than antagonists, so they actually make the disease worse, while MDV3100 has no known agonist activity. MDV3100 has demonstrated activity in models of bicalutamide-resistant prostate cancer. The results come from a phase 1/2 trial (NCT00510718) of 140 men, 65 of whom were chemotherapy-naive. Patients with progressive, metastatic CRPC were enrolled in dose-escalation cohorts of three to six patients and given an oral daily dose of MDV3100 from 30 mg to 600mg. The primary objective was to identify the safety and tolerability profile of MDV3100 and to establish the maximum tolerated dose. They noted antitumor effects at all doses, including decreases in serum prostate-specific antigen of 50% or more in 78 (56%) patients, responses in soft tissue in 13 (22%) of 59 patients, stabilized bone disease in 61 (56%) of 109 patients, and conversion from unfavorable to favorable circulating tumor cell counts in 25 (49%) of the 51 patients. PET imaging of 22 patients to assess androgen-receptor blockade showed decreased (18)F-fluoro-5α-dihydrotestosterone binding at doses from 60 mg to 480 mg per day (range 20-100%). The median time to progression was 47 weeks (95% CI 34-not reached) for radiological progression. The maximum tolerated dose for sustained treatment (>28 days) was 240 mg. The most common grade 3-4 adverse event was dose-dependent fatigue (16 [11%] patients), which generally resolved after dose reduction. The encouraging antitumor activity with MDV3100 in CRPC validates preclinical studies that have implicated AR signaling as a driver of this disease. MDV3100 is performed on two phase 3 clinical trials: one is a randomized, double-blind, multicenter, phase 3 study evaluating MDV3100 compared with placebo in the treatment of men with progressive, chemotherapy-naive, metastatic CRPC (PREVAIL), the other is in men with metastatic CRPC that has progressed following taxane-based therapy (AFFIRM). The company, Medivation, announced on November 3, 2011, that MDV3100 improved survival by 4.8 months, compared with placebo, in the AFFIRM trial. The AFFIRM trial is a randomized, double-blind, multinational trial comparing MDV3100 (160 mg/day) with placebo in 1199 men with advanced prostate cancer who were previously treated with docetaxel-based chemotherapy. Enrollment was completed in November 2010, and the interim analysis was triggered at 520 events. A planned interim analysis of the AFFIRM trial revealed that estimated median survival was 18.4 months for men treated with MDV3100, compared with 13.6 months for men treated with placebo (P <0.0001). This was construed that MDV3100 improved survival by 4.8 months compared with placebo. This translates into a 37% reduction in the risk for death with MDV3100 (hazard ratio, 0.631). As a result, the trial's Independent Data Monitoring Committee recommended that AFFIRM be stopped early and that men who were receiving placebo be offered MDV3100.

A New Androgen Receptor Antagonist, ARN-509

A new androgen receptor antagonist, ARN-509 is also an AR antagonist that inhibits nuclear translocation and DNA binding of the receptor, thereby modulating expression of genes that drive prostate cancer growth. Phase 1/2 clinical trial of ARN-509 in patients with CRPC was announced. (NCT01171898)

An Inhibitor of the Amino Terminal Domain (NTD) of the AR: EPI-001

AR is activated by androgen or by other factors when androgen is removed, resulting in the transactivation of the amino terminal domain (NTD) of AR and AR binding to androgen response elements (AREs) in the promoters of androgen-regulated genes. All current therapies that target the AR are dependent on the presence of its C-terminal ligand-binding domain (LBD). A library of extracts from marine sponges for an inhibitor of AR NTD transactivation were screened and the small molecule EPI-001 was identified, which inhibited transcriptional regulation of androgen-regulated genes, but not those regulated by the related progesterone and glucocorticoid steroid receptors. [130] Importantly, EPI-001 was effective at inhibiting both ligand-dependent and ligand-independent transactivation of AR, as well as inhibiting constitutively active AR splice variants that are devoid of LBD. EPI-001 has a new mode of action, as it targets the transactivation of AR regardless of the presence of androgen, and is a promising therapy to delay the progression of CRPC.

The Prospects of the Androgen Deprivation Therapy on the Ground of Androgen Milieu in the Prostate

Conventional ADTs that we have used cannot decrease DHT levels in the prostate sufficiently to restrain the growth of the cancer in patient with aggressive prostate cancer, as we mentioned above. The androgen levels in blood were low; however, it became clear that the AR signals play important roles for progress of the cancers from the studies of the prostate cancer that became exacerbation after ADT. The androgen milieu in men with high Gleason score prostate cancer is probably less affected by conventional ADT than that in men with low Gleason score cancer, which was suggested to be associated with adrenal androgen levels. In patients treated with ADT the pituitary-adrenal axis mediated by adrenocorticotropic hormone has a central role in the regulation of androgen synthesis. Adrenocorticotropic hormone mediated androgen synthesis is a potential target for advanced ADT. Importantly, the suboptimal suppression of androgen activity in the prostate cancer may account for the heterogeneity in treatment effect observed across individual patients, and, moreover, may contribute to the outgrowth of resistant prostate cancer clones adapted to survive in a low androgen environment, suggesting that more effective methods for suppressing the androgen axis in the tumoral microenvironment are required. From the results of the high reactivity of a CYP17A1 inhibitor and a new AR antagonist for patients with CRPC, the androgen syntheses and the maintenances of the AR signal system in the prostate cancer play important roles in etiological mechanisms of CRPC. As for the ADT for the patients with aggressive prostate cancer, it is important to restrain productions of testosterone and DHT and to restrain AR activity together with castration from early period of treatment. The results with new ADT agents, such as MDV3100 and abiraterone acetate seem to validate the hypothesis that androgens and the AR are vital to progression of metastatic CRPC.

Conclusion

The patients with prostate cancer of high Gleason score have low DHT levels in the prostatic tissue. Biologically aggressive prostate cancer can occur under a low DHT level environment where the prostate cancer of a low malignancy with high DHT dependency cannot easily occur. Conventional ADTs that we have used cannot decrease DHT levels in the prostate sufficiently to restrain the growth of the cancer in patients with aggressive prostate cancer. Persistent levels of DHT in the prostate after castration are derived from adrenal androgens in the prostate. The androgen milieu in men with high Gleason score prostate cancer is probably less affected by conventional ADT than that in men with low Gleason score cancer, which was suggested to be associated with adrenal androgen levels. In patients treated with ADT the pituitary-adrenal axis mediated by adrenocorticotropic hormone has a central role in the regulation of androgen synthesis. These locally produced androgens can play important roles in the pathogenesis and development of prostate cancer. AR signaling pathway is an important mechanism in proliferation of aggressive prostate cancer during ADT. AR transcriptional activity is reactivated in CRPC and identifies the intracellular conversion of adrenal androgens to DHT as a mechanism mediating this reactivation. New ADTs using CYP17A1 inhibitors such as abiraterone acetate and AR antagonists such as MDV3100 for patients with CRPC, which will break through the limitation of the efficacy of conventional ADTs that have been used for nearly 70 years, have been developing. Therefore, it is important to examine the status of in situ androgen metabolism and/or the AR signal system in detail in order to improve the clinical response to ADT in patients diagnosed with biologically aggressive prostate cancer from now on.

References

[1] Heinlein CA, Chang C. Androgen receptor in prostate cancer., *Endocr. Rev.* 2004;25(2):276-308.

[2] Huggins C, Hodges CV. Studies on prostate cancer. Effect of castration, estrogen and androgen injection on serum phosphatases in metastatic carcinoma of the prostate. *Cancer Res.* 1941; 1:293-7.

[3] McLeod DG, Crawford ED, DeAntoni EP. Combined androgen blockade: the gold standard for metastatic prostate cancer. *Eur. Urol.* 1997;32 Suppl 3:70-7.

[4] Pienta KJ, Bradley D. Mechanisms underlying the development of androgen-independent prostate cancer. *Clin. Cancer Res.* 2006; 12: 1665-71.

[5] Di Lorenzo G, Buonerba C, Autorino R, De Placido S, Sternberg CN. Castration-resistant prostate cancer: current and emerging treatment strategies. *Drugs.* 2010;70(8):983-1000.

[6] Winters SJ, Clark BJ. Testosterone Synthesis, Transport, and Metabolism. In Bagatell CJ, Bremner WJ (eds), *Androgens in Health and Disease.* pp3-22. Humana Press, 2003.

[7] Gilep AA, Sushko TA, Usanov SA. At the crossroads of steroid hormone biosynthesis: the role, substrate specificity and evolutionary development of CYP17. *Biochim. Biophys. Acta.* 2011 Jan;1814(1):200-9.

[8] Akhtar MK, Kelly SL, Kaderbhai MA. Cytochrome b(5) modulation of 17{alpha} hydroxylase and 17-20 lyase (CYP17) activities in steroidogenesis. *J. Endocrinol.* 2005 Nov;187(2):267-74.

[9] Rainey WE, Nakamura Y. Regulation of the adrenal androgen biosynthesis. *J. Steroid Biochem. Mol. Biol.* 2008 Feb;108(3-5):281-6.

[10] Rommerts FFG.. Testosterone: overview of biosynthesis, transport, metabolism and non-genomic actions. In Nieschlag E, Behre HM, Niechlag S (eds) *Testosterone.* 3rd edn, pp1-38. Cambridge, 2004.

[11] Haider SG., Cell biology of Leydig cells in the testis., *Int. Rev. Cytol.* 2004;233:181-241.

[12] Mooradian AD, Morley JE, Korenman SG. Biological actions of androgens. *Endocr. Rev.* 1987 Feb;8(1):1-28.

[13] Bassil N, Alkaade S, Morley JE. The benefits and risks of testosterone replacement therapy: a review. *Ther. Clin. Risk Manag.* 2009 Jun;5(3):427-48.

[14] Tuck SP, Francis RM. Testosterone, bone and osteoporosis. *Front Horm. Res.* 2009;37:123-32.

[15] Midzak AS, Chen H, Papadopoulos V, Zirkin BR. Leydig cell aging and the mechanisms of reduced testosterone synthesis. *Mol. Cell Endocrinol.* 2009;299(1):23-31.

[16] Endogenous Hormones and Prostate Cancer Collaborative Group, Roddam AW, Allen NE, Appleby P, Key TJ. Endogenous sex hormones and prostate cancer: a collaborative analysis of 18 prospective studies. *J. Natl. Cancer Inst.* 2008;100(3):170-83.

[17] Morote J, Ramirez C, Gomez E, Planas J, Raventos CX, de Torres IM, Catalan R. The relationship between total and free serum testosterone and the risk of prostate cancer and tumour aggressiveness. *BJU Int.* 2009;104(4):486-9.

[18] Diver MJ; Clinical Scince Reviews Committee of the Association for Clinical Biochemistry. Analytical and physiological factors affecting the interpretation of serum testosterone concentration in men. *Ann. Clin. Biochem.* 2006 Jan;43(Pt 1):3-12.

[19] Luu-The V, Belanger A, Labrie F. Androgen biosynthetic pathways in the human prostate. *Best Pract. Res. Clin. Endocrinol. Metab.* 2008;22(2):207-21.

[20] Penning TM, Byrns MC. Steroid hormone transforming aldo-keto reductases and cancer. *Ann. N. Y. Acad. Sci.* 2009;1155:33-42.

[21] Ghayee HK, Auchus RJ. Clinical implications of androgen synthesis via 5α-reduced precursors. *Endocr. Dev.* 2008;13:55-66.

[22] Zhu YS, Imperato-McGinley JL. 5α-reductase isozymes and androgen actions in the prostate. *Ann. N. Y. Acad. Sci.* 2009;1155:43-56.

[23] Thomas LN, Douglas RC, Lazier CB, Too CK, Rittmaster RS, Tindall DJ. Type 1 and type 2 5alpha-reductase expression in the development and progression of prostate cancer. *Eur. Urol.* 2008 Feb;53(2):244-52.

[24] Uemura M, Tamura K, Chung S, Honma S, Okuyama A, Nakamura Y, Nakagawa H. Novel 5α-steroid reductase (SRD5A3, type-3) is overexpressed in hormone-refractory prostate cancer. *Cancer Sci.* 2008;99(1):81-6.

[25] Godoy A, Kawinski E, Li Y, Oka D, Alexiev B, Azzouni F, Titus MA, Mohler JL. 5α-reductase type 3 expression in human benign and malignant tissues: a comparative analysis during prostate cancer progression. *Prostate.* 2011 Jul;71(10):1033-46.

[26] Steckelbroeck S, Jin Y, Gopishetty S, Oyesanmi B, Penning TM. Human cytosolic 3alpha-hydroxysteroid dehydrogenases of the aldo-keto reductase superfamily display significant 3beta-hydroxysteroid dehydrogenase activity: implications for steroid hormone metabolism and action. *J. Biol. Chem.* 2004 Mar 12;279(11):10784-95.

[27] Jin Y, Penning TM. Multiple steps determine the overall rate of the reduction of 5alpha-dihydrotestosterone catalyzed by human type 3 3alpha-hydroxysteroid dehydrogenase: implications for the elimination of androgens. *Biochemistry.* 2006 Oct 31;45(43):13054-63.

[28] Weihua Z, Lathe R, Warner M, Gustafsson JA. An endocrine pathway in the prostate, ERbeta, AR, 5alpha-androstane-3beta,17beta-diol, and CYP7B1, regulates prostate growth. *Proc. Natl. Acad. Sci. U. S. A.* 2002 Oct 15;99(21):13589-94.

[29] Weihua Z, Makela S, Andersson LC, Salmi S, Saji S, Webster JI, Jensen EV, Nilsson S, Warner M, Gustafsson JA. A role for estrogen receptor beta in the regulation of growth of the ventral prostate. *Proc. Natl. Acad. Sci. U. S. A.* 2001 May 22;98(11):6330-5.

[30] Auchus RJ., The backdoor pathway to dihydrotestosterone., *Trends Endocrinol. Metab.* 2004;15(9):432-8.

[31] Ghayee HK, Auchus RJ., Basic concepts and recent developments in human steroid hormone biosynthesis., *Rev. Endocr. Metab. Disord.* 2007;8(4):289-300.

[32] Sharifi N, McPhaul MJ, Auchus RJ. "Getting from here to there"--mechanisms and limitations to the activation of the androgen receptor in castration-resistant prostate cancer. *J. Investig. Med.* 2010;58(8):938-44.

[33] Penning TM, Jin Y, Rizner TL, Bauman DR. Pre-receptor regulation of the androgen receptor. *Mol. Cell Endocrinol.* 2008 Jan 16;281(1-2):1-8.

[34] Bauman DR, Steckelbroeck S, Williams MV, Peehl DM, Penning TM. Identification of the major oxidative 3alpha-hydroxysteroid dehydrogenase in human prostate that converts 5alpha-androstane-3alpha,17beta-diol to 5alpha-dihydrotestosterone: a potential therapeutic target for androgen-dependent disease. *Mol. Endocrinol.* 2006 Feb;20(2):444-58.

[35] Morgentaler A., Testosterone and prostate cancer: an historical perspective on a modern myth., *Eur. Urol.* 2006;50(5):935-9.

[36] Isbarn H, Pinthus JH, Marks LS, Montorsi F, Morales A, Morgentaler A, Schulman C., Testosterone and prostate cancer: revisiting old paradigms., *Eur. Urol.* 2009;56(1):48-56.

[37] Morgentaler A, Traish AM., Shifting the paradigm of testosterone and prostate cancer: the saturation model and the limits of androgen-dependent growth., *Eur. Urol.* 2009;55(2):310-20.

[38] Morgentaler A. Testosterone and prostate cancer: what are the risks for middle-aged men? *Urol. Clin. North Am.* 2011 May;38(2):119-24.

[39] Weiss JM, Huang WY, Rinaldi S, Fears TR, Chatterjee N, Hsing AW, Crawford ED, Andriole GL, Kaaks R, Hayes RB. Endogenous sex hormones and the risk of prostate cancer: a prospective study. *Int. J. Cancer.* 2008;122(10):2345-50.

[40] Morote J, Ramirez C, Gómez E, Planas J, Raventós CX, de Torres IM, Catalán R. The relationship between total and free serum testosterone and the risk of prostate cancer and tumour aggressiveness. *BJU Int.* 2009 Aug;104(4):486-9.

[41] Bartsch W, Klein H, Schiemann U, Bauer HW, Voigt KD. Enzymes of androgen formation and degradation in the human prostate. *Ann. N. Y. Acad. Sci.* 1990;595:53-66.

[42] Nishiyama T, Hashimoto Y, Takahashi K. The influence of androgen deprivation therapy on dihydrotestosterone levels in the prostatic tissue of patients with prostate cancer. *Clin. Cancer Res.* 2004;10(21):7121-6.

[43] Nishiyama T, Ikarashi T, Hashimoto Y, Suzuki K, Takahashi K. Association between the dihydrotestosterone level in the prostate and prostate cancer aggressiveness using the Gleason score. *J. Urol.* 2006;176(4 Pt 1):1387-91.

[44] Prehn RT. On the prevention and therapy of prostate cancer by androgen administration. *Cancer Res.* 1999;59(17):4161-4.

[45] Freedland SJ, Isaacs WB, Platz EA, Terris MK, Aronson WJ, Amling CL, Presti JC Jr, Kane CJ. Prostate size and risk of high-grade, advanced prostate cancer and biochemical progression after radical prostatectomy: a search database study. *J. Clin. Oncol.* 2005;23(30):7546-54.

[46] Kwon T, Jeong IG, You D, Park MC, Hong JH, Ahn H, Kim CS. Effect of prostate size on pathological outcome and biochemical recurrence after radical prostatectomy for prostate cancer: is it correlated with serum testosterone level? *BJU Int.* 2010;106(5):633-8.

[47] Newton MR, Phillips S, Chang SS, Clark PE, Cookson MS, Davis R, Fowke JH, Herrell SD, Baumgartner R, Chan R, Mishra V, Blume JD, Smith JA Jr, Barocas DA. Smaller prostate size predicts high grade prostate cancer at final pathology. *J. Urol.* 2010;184(3):930-7.

[48] Thompson IM, Goodman PJ, Tangen CM, Lucia MS, Miller GJ, Ford LG, Lieber MM, Cespedes RD, Atkins JN, Lippman SM, Carlin SM, Ryan A, Szczepanek CM, Crowley JJ, Coltman CA Jr. The influence of finasteride on the development of prostate cancer. *N. Engl. J. Med.* 2003;349(3):215-24.

[49] Rubin MA, Allory Y, Molinie V, Leroy X, Faucon H, Vacherot F, Huang W, Kuten A, Salomon L, Rebillard X, Cussenot O, Abbou C, de la Taille A. Effects of long-term finasteride treatment on prostate cancer morphology and clinical outcome. *Urology.* 2005;66(5):930-4.

[50] Lucia MS, Epstein JI, Goodman PJ, Darke AK, Reuter VE, Civantos F, Tangen CM, Parnes HL, Lippman SM, La Rosa FG, Kattan MW, Crawford ED, Ford LG, Coltman CA Jr, Thompson IM. Finasteride and high-grade prostate cancer in the Prostate Cancer Prevention Trial. *J. Natl. Cancer Inst.* 2007;99(18):1375-83.

[51] Andriole GL, Bostwick DG, Brawley OW, Gomella LG, Marberger M, Montorsi F, Pettaway CA, Tammela TL, Teloken C, Tindall DJ, Somerville MC, Wilson TH, Fowler IL, Rittmaster RS; REDUCE Study Group. Effect of dutasteride on the risk of prostate cancer. *N. Engl. J. Med.* 2010 Apr 1;362(13):1192-202.

[52] Li R, Wheeler T, Dai H, Frolov A, Thompson T, Ayala G. High level of androgen receptor is associated with aggressive clinicopathologic features and decreased biochemical recurrence-free survival in prostate: cancer patients treated with radical prostatectomy. *Am. J. Surg. Pathol.* 2004;28(7):928-34.

[53] Fung KM, Samara EN, Wong C, Metwalli A, Krlin R, Bane B, Liu CZ, Yang JT, Pitha JV, Culkin DJ, Kropp BP, Penning TM, Lin HK. Increased expression of type 2 3α-hydroxysteroid dehydrogenase/type 5 17β-hydroxysteroid dehydrogenase (AKR1C3) and its relationship with androgen receptor in prostate carcinoma. *Endocr. Relat. Cancer.* 2006;13(1):169-80.

[54] Wako K, Kawasaki T, Yamana K, Suzuki K, Jiang S, Umezu H, Nishiyama T, Takahashi K, Hamakubo T, Kodama T, Naito M. Expression of androgen receptor through androgen-converting enzymes is associated with biological aggressiveness in prostate cancer. *J. Clin. Pathol.* 2008;61(4):448-54.

[55] Maatman TJ, Gupta MK, Montie JE. Effectiveness of castration versus intravenous estrogen therapy in producing rapid endocrine control of metastatic cancer of the prostate. *J. Urol.* 1985;133:620-1.

[56] Oefelein MG, Feng A, Scolieri MJ, Ricchiutti D and Resnick MI Reassessment of the definition of castrate levels of testosterone: implications for clinical decision making. *Urology.* 2000; 56: 1021-1024.

[57] Tammela T. Endocrine treatment of prostate cancer. *J. Steroid Biochem.* Mol Biol. 2004;92:287-95.

[58] Tombal B. Appropriate Castration with Luteinising Hormone Releasing Hormone (LHRH) Agonists: What is the Optimal Level of Testosterone? *Eur. Urol. Suppl.* 2005; 4:14-9.

[59] Boccon-Gibod L, van der Meulen E, Persson BE. An update on the use of gonadotropin-releasing hormone antagonists in prostate cancer. *Ther. Adv. Urol.* 2011 Jun;3(3):127-40.

[60] Morote J, Planas J, Salvador C, Raventós CX, Catalán R, Reventós J. Individual variations of serum testosterone in patients with prostate cancer receiving androgen deprivation therapy. *BJU Int.* 2009 Feb;103(3):332-5.

[61] Crawford ED, Rove KO. Incomplete testosterone suppression in prostate cancer. *N. Engl. J. Med.* 2010;363(20):1976.

[62] Morote J, Orsola A, Planas J, Trilla E, Raventos CX, Cecchini L, Catalan R. Redefining clinically significant castration levels in patients with prostate cancer receiving continuous androgen deprivation therapy. *J. Urol.* 2007 ;178(4 Pt 1) :1290-5.

[63] Perachino M, Cavalli V, Bravi F. Testosterone levels in patients with metastatic prostate cancer treated with luteinizing hormone-releasing hormone therapy: prognostic significance? *BJU Int.* 2010 Mar;105(5):648-51.

[64] Zaitsu M, Yamanoi M, Mikami K, Takeshima Y, Okamoto N, Imao S, Tonooka A, Takeuchi T. Surgical castration in hormone-refractory metastatic prostate cancer patients can be an alternative for medical castration. *Adv. Urol.* 2012;2012:979154.

[65] Gomella LG. Effective testosterone suppression for prostate cancer: is there a best castration therapy? *Rev. Urol.* 2009 Spring;11(2):52-60.

[66] Labrie F, Belanger A, Luu-The V, Labrie C, Simard J, Cusan L, Gomez J, Candas B. Gonadotropin-releasing hormone agonists in the treatment of prostate cancer. *Endocr. Rev*. 2005; 26: 361-79.

[67] Labrie F, Belanger A, Dupont A, Luu-The V, Simard J, Labrie C. Science behind total androgen blockade: from gene to combination therapy. *Clin. Invest. Med.* 1993; 16: 475-92.

[68] Hammond GL. Endogenous steroid levels in the human prostate from birth to old age: a comparison of normal and diseased tissues. *J. Endocrinol.* 1978; 78: 7-19.

[69] Belanger B, Belanger A, Labrie F, Dupont A, Cusan L, Monfette G. Comparison of residual C-19 steroids in plasma and prostatic tissue of human, rat and guinea pig after castration: unique importance of extratesticular androgens in men. *J. Steroid Biochem.* 1989; 32: 695-8.

[70] Geller J, Albert JD. Effects of castration compared with total androgen blockade on tissue dihydrotestosterone (DHT) concentration in benign prostatic hyperplasia (BPH). *Urol. Res.* 1987; 15: 151-3.

[71] Nishiyama T, Ikarashi T, Hashimoto Y, Wako K, Takahashi K. The change in the dihydrotestosterone level in the prostate before and after androgen deprivation therapy in connection with prostate cancer aggressiveness using the Gleason score. *J. Urol.* 2007;178(4 Pt 1):1282-8.

[72] Titus MA, Gregory CW, Ford OH 3rd, Schell MJ, Maygarden SJ, Mohler JL. Steroid 5α-reductase isozymes I and II in recurrent prostate cancer. *Clin. Cancer Res.* 2005;11(12):4365-71.

[73] Thomas LN, Lazier CB, Gupta R, Norman RW, Troyer DA, O'Brien SP, Rittmaster RS. Differential alterations in 5α-reductase type 1 and type 2 levels during development and progression of prostate cancer. *Prostate.* 2005;63:231-9.

[74] Arai S, Miyashiro Y, Shibata Y, Tomaru Y, Kobayashi M, Honma S, Suzuki K. Effect of castration monotherapy on the levels of adrenal androgens in cancerous prostatic tissues. *Steroids.* 2011 Feb;76(3):301-8.

[75] Grossmann ME, Huang H, Tindall DJ. Androgen receptor signaling in androgen-refractory prostate cancer. *J. Natl. Cancer Inst.* 2001 ;93 :1687-97.

[76] Stanbrough M, Bubley GJ, Ross K, Golub TR, Rubin MA, Penning TM, Febbo PG, Balk SP. Increased expression of genes converting adrenal androgens to testosterone in androgen-independent prostate cancer. *Cancer Res.* 2006;66(5):2815-25.

[77] Mostaghel EA, Page ST, Lin DW, Fazli L, Coleman IM, True LD, Knudsen B, Hess DL, Nelson CC, Matsumoto AM, Bremner WJ, Gleave ME, Nelson PS. Intraprostatic androgens and androgen-regulated gene expression persist after testosterone suppression: therapeutic implications for castration-resistant prostate cancer. *Cancer Res.* 2007;67:5033-41.

[78] Takizawa I, Nishiyama T, Hara N, Isahaya E, Hoshii T, Takahashi K. Serum prostate-specific antigen levels reflect the androgen milieu in patients with localized prostate cancer receiving androgen deprivation therapy: Tumor malignant potential and androgen milieu., *Prostate.* 2010;70(13):1395-401.

[79] Takizawa I, Hara N, Nishiyama T, Isahaya E, Hoshii T, Takahashi K. Adrenocorticotropic hormone is involved in regulation of androgen synthesis in men

receiving androgen deprivation therapy for localized prostate cancer. *J. Urol.* 2010;184(5):1971-6.

[80] Attar RM, Takimoto CH, Gottardis MM. Castration-resistant prostate cancer: locking up the molecular escape routes. *Clin. Cancer Res.* 2009;15(10):3251-5.

[81] Yuan X, Balk SP. Mechanisms mediating androgen receptor reactivation after castration. *Urol. Oncol.* 2009;27(1):36-41.

[82] Chen CD, Welsbie DS, Tran C, Baek SH, Chen R, Vessella R, Rosenfeld MG, Sawyers CL. Molecular determinants of resistance to antiandrogen therapy. *Nat. Med.* 2004;10(1):33-9.

[83] Mohler JL, Gregory CW, Ford OH 3rd, Kim D, Weaver CM, Petrusz P, Wilson EM, French FS. The androgen axis in recurrent prostate cancer. *Clin. Cancer Res.* 2004;10(2):440-8.

[84] Titus MA, Schell MJ, Lih FB, Tomer KB, Mohler JL. Testosterone and dihydrotestosterone tissue levels in recurrent prostate cancer. *Clin. Cancer Res.* 2005;11(13):4653-7.

[85] Montgomery RB, Mostaghel EA, Vessella R, Hess DL, Kalhorn TF, Higano CS, True LD, Nelson PS. Maintenance of intratumoral androgens in metastatic prostate cancer: a mechanism for castration-resistant tumor growth. *Cancer Res.* 2008;68(11):4447-54.

[86] Chang KH, Li R, Papari-Zareei M, Watumull L, Zhao YD, Auchus RJ, Sharifi N. Dihydrotestosterone synthesis bypasses testosterone to drive castration-resistant prostate cancer. *Proc. Natl. Acad. Sci. U. S. A.* 2011 Aug 16;108(33):13728-33.

[87] Mohler JL, Titus MA, Bai S, Kennerley BJ, Lih FB, Tomer KB, Wilson EM. Activation of the androgen receptor by intratumoral bioconversion of androstanediol to dihydrotestosterone in prostate cancer. *Cancer Res.* 2011 Feb 15;71(4):1486-96.

[88] Locke JA, Guns ES, Lubik AA, Adomat HH, Hendy SC, Wood CA, Ettinger SL, Gleave ME, Nelson CC. Androgen levels increase by intratumoral de novo steroidogenesis during progression of castration-resistant prostate cancer. *Cancer Res.* 2008;68(15):6407-15.

[89] Dillard PR, Lin MF, Khan SA. Androgen-independent prostate cancer cells acquire the complete steroidogenic potential of synthesizing testosterone from cholesterol. *Mol. Cell Endocrinol.* 2008;295(1-2):115-20.

[90] Suzuki K, Nishiyama T, Hara N, Yamana K, Takahashi K, Labrie F. Importance of the intracrine metabolism of adrenal androgens in androgen-dependent prostate cancer. *Prostate Cancer Prostatic Dis.* 2007;10(3):301-6.

[91] Lam JS, Leppert JT, Vemulapalli SN, Shvarts O, Belldegrun AS. Secondary hormonal therapy for advanced prostate cancer. *J. Urol.* 2006;175(1):27-34.

[92] Ryan CJ, Halabi S, Ou SS, Vogelzang NJ, Kantoff P, Small EJ. Adrenal androgen levels as predictors of outcome in prostate cancer patients treated with ketoconazole plus antiandrogen withdrawal: results from a cancer and leukemia group B study. *Clin. Cancer Res.* 2007;13(7):2030-7.

[93] Van Poppel H. Evaluation of degarelix in the management of prostate cancer. *Cancer Manag. Res.* 2010 Jan 25;2:39-52.

[94] van Poppel H, Nilsson S. Testosterone surge: rationale for gonadotropin-releasing hormone blockers? *Urology.* 2008 Jun;71(6):1001-6.

[95] Debruyne F, Bhat G, Garnick MB. Abarelix for injectable suspension: first-in-class gonadotropin-releasing hormone antagonist for prostate cancer. *Future Oncol.* 2006 Dec;2(6):677-96.

[96] Koechling W, Hjortkjaer R, Tanko LB. Degarelix, a novel GnRH antagonist, causes minimal histamine release compared with cetrorelix, abarelix and ganirelix in an ex vivo model of human skin samples. *Br. J. Clin. Pharmacol.* 2010 Oct;70(4):580-7.

[97] Princivalle M, Broqua P, White R, Meyer J, Mayer G, Elliott L, Bjarnason K, Haigh R, Yea C. Rapid suppression of plasma testosterone levels and tumor growth in the dunning rat model treated with degarelix, a new gonadotropin-releasing hormone antagonist. *J. Pharmacol. Exp. Ther.* 2007 Mar;320(3):1113-8.

[98] Klotz L, Boccon-Gibod L, Shore ND, Andreou C, Persson BE, Cantor P, Jensen JK, Olesen TK, Schroder FH. The efficacy and safety of degarelix: a 12-month, comparative, randomized, open-label, parallel-group phase III study in patients with prostate cancer. *BJU Int.* 2008 Dec;102(11):1531-8.

[99] Tombal B, Miller K, Boccon-Gibod L, Schroder F, Shore N, Crawford ED, Moul J, Jensen JK, Kold Olesen T, Persson BE. Additional Analysis of the Secondary End Point of Biochemical Recurrence Rate in a Phase 3 Trial (CS21) Comparing Degarelix 80mg Versus Leuprolide in Prostate Cancer Patients Segmented by Baseline Characteristics. *Eur. Urol.* 2010 May;57(5):836-42.

[100] Porter AT, F A C R O, Ben-Josef E. Humoral mechanisms in prostate cancer: A role for FSH. *Urol. Oncol.* 2001 Jul;6(4):131-138.

[101] Radu A, Pichon C, Camparo P, Antoine M, Allory Y, Couvelard A, Fromont G, Hai MT, Ghinea N. Expression of follicle-stimulating hormone receptor in tumor blood vessels. *N. Engl. J. Med.* 2010 Oct 21;363(17):1621-30.

[102] Lee-Robichaud P, Shyadehi AZ, Wright JN, Akhtar ME, Akhtar M. Mechanistic kinship between hydroxylation and desaturation reactions: acyl-carbon bond cleavage promoted by pig and human CYP17 (P-450(17)α; 17 α-hydroxylase-17,20-lyase). *Biochemistry.* 1995;34(43):14104-13.

[103] Akhtar MK, Kelly SL, Kaderbhai MA. Cytochrome b(5) modulation of 17α-hydroxylase and 17-20 lyase (CYP17) activities in steroidogenesis. *J. Endocrinol.* 2005;187(2):267-74.

[104] Katagiri M, Kagawa N, Waterman MR. The role of cytochrome b5 in the biosynthesis of androgens by human P450c17. *Arch. Biochem. Biophys.* 1995;317(2):343-7.

[105] Dharia S, Slane A, Jian M, Conner M, Conley AJ, Parker CR Jr. Colocalization of P450c17 and cytochrome b5 in androgen-synthesizing tissues of the human. *Biol. Reprod.* 2004;71(1):83-8.

[106] Gaunt R, Steinetz BG, Chart JJ. Pharmacologic alteration of steroid hormone functions. *Clin. Pharmacol. Ther.* 1968;9(5):657-81.

[107] Arth GE, Patchett AA, Jefopoulus T, Bugianesi RL, Peterson LH, Ham EA, Kuehl FA Jr, Brink NG. Steroidal androgen biosynthesis inhibitors. *J. Med. Chem.* 1971;14(8):675-9.

[108] Montgomery RB, Mostaghel EA, Vessella R, Hess DL, Kalhorn TF, Higano CS, True LD, Nelson PS. Maintenance of intratumoral androgens in metastatic prostate cancer: a mechanism for castration-resistant tumor growth. *Cancer Res.* 2008;68(11):4447-54.

[109] Stigliano A, Gandini O, Cerquetti L, Gazzaniga P, Misiti S, Monti S, Gradilone A, Falasca P, Poggi M, Brunetti E, Agliano AM, Toscano V. Increased metastatic lymph node 64 and CYP17A1 expression are associated with high stage prostate cancer. *J. Endocrinol.* 2007;194(1):55-61.

[110] Attard G, Reid AH, Yap TA, Raynaud F, Dowsett M, Settatree S, Barrett M, Parker C, Martins V, Folkerd E, Clark J, Cooper CS, Kaye SB, Dearnaley D, Lee G, de Bono JS. Phase I clinical trial of a selective inhibitor of CYP17, abiraterone acetate, confirms that castration-resistant prostate cancer commonly remains hormone driven. *J. Clin. Oncol.* 2008;26(28):4563-71.

[111] Attard G, Reid AH, A'Hern R, Parker C, Oommen NB, Folkerd E, Messiou C, Molife LR, Maier G, Thompson E, Olmos D, Sinha R, Lee G, Dowsett M, Kaye SB, Dearnaley D, Kheoh T, Molina A, de Bono JS. Selective inhibition of CYP17 with abiraterone acetate is highly active in the treatment of castration-resistant prostate cancer. *J. Clin. Oncol.* 2009;27(23):3742-8.

[112] Danila DC, Morris MJ, de Bono JS, Ryan CJ, Denmeade SR, Smith MR, Taplin ME, Bubley GJ, Kheoh T, Haqq C, Molina A, Anand A, Koscuiszka M, Larson SM, Schwartz LH, Fleisher M, Scher HI. Phase II multicenter study of abiraterone acetate plus prednisone therapy in patients with docetaxel-treated castration-resistant prostate cancer. *J. Clin. Oncol.* 2010;28(9):1496-501.

[113] Reid AH, Attard G, Danila DC, Oommen NB, Olmos D, Fong PC, Molife LR, Hunt J, Messiou C, Parker C, Dearnaley D, Swennenhuis JF, Terstappen LW, Lee G, Kheoh T, Molina A, Ryan CJ, Small E, Scher HI, de Bono JS. Significant and sustained antitumor activity in post-docetaxel, castration-resistant prostate cancer with the CYP17 inhibitor abiraterone acetate. *J. Clin. Oncol.* 2010;28(9):1489-95.

[114] Ryan CJ, Smith MR, Fong L, Rosenberg JE, Kantoff P, Raynaud F, Martins V, Lee G, Kheoh T, Kim J, Molina A, Small EJ. Phase I clinical trial of the CYP17 inhibitor abiraterone acetate demonstrating clinical activity in patients with castration-resistant prostate cancer who received prior ketoconazole therapy. *J. Clin. Oncol.* 2010;28(9):1481-8.

[115] de Bono JS, Logothetis CJ, Molina A, Fizazi K, North S, Chu L, Chi KN, Jones RJ, Goodman OB Jr, Saad F, Staffurth JN, Mainwaring P, Harland S, Flaig TW, Hutson TE, Cheng T, Patterson H, Hainsworth JD, Ryan CJ, Sternberg CN, Ellard SL, Fléchon A, Saleh M, Scholz M, Efstathiou E, Zivi A, Bianchini D, Loriot Y, Chieffo N, Kheoh T, Haqq CM, Scher HI; COU-AA-301 Investigators. Abiraterone and increased survival in metastatic prostate cancer. N. *Engl. J. Med.* 2011;364(21):1995-2005.

[116] Ratain MJ. Flushing oral oncology drugs down the toilet. *J. Clin. Oncol.* 2011 Oct 20;29(30):3958-9.

[117] Kang SP, Ratain MJ. Inconsistent labeling of food effect for oral agents across therapeutic areas: differences between oncology and non-oncology products. *Clin. Cancer Res.* 2010 Sep 1;16(17):4446-51.

[118] Jain RK, Brar SS, Lesko LJ. Food and oral antineoplastics: more than meets the eye. *Clin. Cancer Res.* 2010 Sep 1;16(17):4305-7.

[119] Matsunaga N, Kaku T, Itoh F, Tanaka T, Hara T, Miki H, Iwasaki M, Aono T, Yamaoka M, Kusaka M, Tasaka A. C17,20-lyase inhibitors I. Structure-based de novo design and SAR study of C17,20-lyase inhibitors. *Bioorg. Med. Chem.* 2004;12(9):2251-73.

[120] Matsunaga N, Kaku T, Ojida A, Tanaka T, Hara T, Yamaoka M, Kusaka M, Tasaka A. C(17,20)-lyase inhibitors. Part 2: design, synthesis and structure-activity relationships of (2-naphthylmethyl)-1H-imidazoles as novel C(17,20)-lyase inhibitors. *Bioorg. Med. Chem.* 2004;12(16):4313-36.

[121] Handratta VD, Vasaitis TS, Njar VC, Gediya LK, Kataria R, Chopra P, Newman D Jr, Farquhar R, Guo Z, Qiu Y, Brodie AM. Novel C-17-heteroaryl steroidal CYP17 inhibitors/antiandrogens: synthesis, in vitro biological activity, pharmacokinetics, and antitumor activity in the LAPC4 human prostate cancer xenograft model. *J. Med. Chem.* 2005;48(8):2972-84.

[122] Vasaitis T, Belosay A, Schayowitz A, Khandelwal A, Chopra P, Gediya LK, Guo Z, Fang HB, Njar VC, Brodie AM. Androgen receptor inactivation contributes to antitumor efficacy of 17α-hydroxylase/17,20-lyase inhibitor 3β-hydroxy-17-(1H-benzimidazole-1-yl)androsta-5,16-diene in prostate cancer. *Mol. Cancer Ther.* 2008;7(8):2348-57.

[123] Evaul K, Li R, Papari-Zareei M, Auchus RJ, Sharifi N. 3β-hydroxysteroid dehydrogenase is a possible pharmacological target in the treatment of castration-resistant prostate cancer. *Endocrinology.* 2010;151(8):3514-20.

[124] Takizawa I, Nishiyama T, Hara N, Hoshii T, Ishizaki F, Miyashiro Y, Takahashi K. Trilostane, an inhibitor of 3β-hydroxysteroid dehydrogenase, has an agonistic activity on androgen receptor in human prostate cancer cells. *Cancer Lett.* 2010;297(2):226-30.

[125] Trauger R, Corey E, Bell D, White S, Garsd A, Stickney D, Reading C, Frincke J. Inhibition of androstenediol-dependent LNCaP tumour growth by 17α-ethynyl-5α-androstane-3α, 17β-diol (HE3235). *Br. J. Cancer.* 2009;100(7):1068-72.

[126] Koreckij TD, Trauger RJ, Montgomery RB, Pitts TE, Coleman I, Nguyen H, Reading CL, Nelson PS, Vessella RL, Corey E. HE3235 inhibits growth of castration-resistant prostate cancer. *Neoplasia.* 2009;11(11):1216-25.

[127] Ahlem C, Kennedy M, Page T, Bell D, Delorme E, Villegas S, Reading C, White S, Stickney D, Frincke J. 17α-Alkynyl 3α, 17β-androstanediol non-clinical and clinical pharmacology, pharmacokinetics and metabolism. *Invest. New Drugs.* 2010 Sep 3. [Epub ahead of print]

[128] Tran C, Ouk S, Clegg NJ, Chen Y, Watson PA, Arora V, Wongvipat J, Smith-Jones PM, Yoo D, Kwon A, Wasielewska T, Welsbie D, Chen CD, Higano CS, Beer TM, Hung DT, Scher HI, Jung ME, Sawyers CL. Development of a second-generation antiandrogen for treatment of advanced prostate cancer. *Science.* 2009;324(5928):787-90.

[129] Scher HI, Beer TM, Higano CS, Anand A, Taplin ME, Efstathiou E, Rathkopf D, Shelkey J, Yu EY, Alumkal J, Hung D, Hirmand M, Seely L, Morris MJ, Danila DC, Humm J, Larson S, Fleisher M, Sawyers CL; Prostate Cancer Foundation/Department of Defense Prostate Cancer Clinical Trials Consortium. Antitumour activity of MDV3100 in castration-resistant prostate cancer: a phase 1-2 study. *Lancet.* 2010;375(9724):1437-46.

[130] Andersen, R. J. et al. Regression of castrate-recurrent prostate cancer by a small-molecule inhibitor of the amino-terminus domain of the androgen receptor. *Cancer Cell.* 2010;17: 535–546.

In: Androgens: Production, Functions and Disorders
Editors: B. F. Thompson and D. J. Robinson
ISBN: 978-1-62081-145-0

Chapter II

Clinical Model of Androgen Deficiency - Androgen Deprivation Therapy and Its Effect on Bone Metabolism, Metabolism and Cognitive Functions

Stanislav Źiaran
University hospital in Bratislava, Urology department
Bratislava, Slovakia

Introduction

Androgens play an pluripotent role in males. The clinical model of its deficiency is represented by androgen deprivation therapy (ADT). Androgen deprivation therapy (ADT) is considered the standard therapy for advanced prostate cancer (PCa).

ADT significantly reduces tumor growth and improves symptoms- free survival [1,2], but is associated with several adverse effects which can affect overall survival and quality of life (see below). These adverse effects clearly show that the physiologic effect of testosterone is pluripotent and essential for a male and will be discussed in this chapter.

The basics of physiology of testosterone and bone metabolism will be clarified as well in this chapter, which helps to elucidate the new therapeutic approaches and its importance in bone metabolism.

There is also some evidence that patients with prostate cancer (PCa) have lower baseline BMD than healthy cohort, which is further lowered by ADT. Furthermore, patients with invasive PCa have greater probability of baseline osteopenia and osteoporosis and a vice versa (see below).

These interesting facts illustrate the role of testosterone in bone metabolism and will be mentioned in the chapter.

Metabolic changes while on ADT are gaining more importance, even though there is not so much data about this issue. More patients are treated by ADT for longer period of time which may affect cardiovascular morbidity and mortality. The development of „metabolic syndrome" (syndrome „X"), its definition and its potential importance and treatment possibilities will be disscused, too.

There are experimental studies which provide the evidence, that testosterone plays an essential role in spatial congnition in an early childhood as well as in adults. The effect of testosterone on spatial cognitive functions will be clarified in the chapter. Spatial impairment together with the loss of BMD, muscular reduction contribute to the risk of pathological fractures in patients with PCa treated with ADT.

Methods

Systematic review of relevant articles, authors published and so far unpublished studies, including the guidelines of AUA and EAU and the effect of possible (future) treatment modalities.

Androgen Deprivation Therapy (ADT)

Since 1941, when Huggins and Hodges proved the favourable effect of surgical castration and oestrogen administration on the progression of metastatic prostate cancer (PCa) [1,2], androgen deprivation therapy (ADT) became the mainstay of management of advanced PCa till now. They demonstrated for the first time the responsiveness of PCa to androgen deprivation.

ADT effectively palliates the symptoms of advanced disease, significantly reduces tumor growth, but there is no conclusive evidence at present that it prolongs survival. Moreover, significant amount of data report that ADT is associated with several adverse effects. The most prominent include: loss of bone mineral density (BMD), which leads into increased fracture risk [3], induction of insulin resistance [4], unfavorable changes in serum lipid profile [5], changes in body composition [6] which can lead into increased cardiovascular morbidity [7] and changes in cognitive functions [8].

The aim of ADT is to cause severe hypogonadism, and adverse effects of ADT clearly demonstrate the essential and pluripotent role of male´s most important androgen – testosterone (TST).

Testosterone: A Basal Overview of Biosynthesis, Metabolism and Its Action

In the human male, the main circulating androgen is testosterone (TST). More than 95% of circulating TST is secreted by the testis (Leydig cells) which produce aproximately 6-7 mg of TST daily [9]. The rest is secreted by the adrenal cortex, and very small quantities (especially pregnan derivatives) are formed by the cells of the brain [10].

Physiologic TST level in a male is 3-8 ng / ml. The source for the synthesis of steroids is cholesterol. This substrate may be synthetized *de novo* from acetate but it may be also taken up from plasma lipoproteins. Cleavage of the side chain of cholesterol in the mitochondria and the formation of pregnenolone (biologically inactive) is the start of steroidogenic cascade. Pregnenolone is further converted into various steroids by enzymes (cytochromes P450) in the endoplasmatic reticulum.

TST secretion is regulated by the hypothalamic-pituitary-gonadal axis. The hypothalamic luteinising hormone-releasing hormone (LHRH) stimulates the anterior pituitary gland to release luteinising hormone (LH) and follicle-stimulating hormone (FSH). The main regulator of Leydig cell function is luteinizing hormone (LH), acting through the LH receptor (LHr) in Leydig cells. Luteinizing hormone (LH) and follicle stimulating hormone (FSH) are required for the development and maintance of testicular functions.The natural ligand for the LHr is LH, but also human choriogonatropin (hCG) can equally well activate the LHr. Activated LHr stimulate adenylyl cyclase via GTP binding proteins and this results in increased production of cyclic AMP (cAMP). cAMP increases steroid production [11].

The total concentration of steroids in target tissues (central and peripheral nervous system, bone, muscle, adipose tissue, haematopoetic system and myocardium) and body fluids is dependent on the presence of binding proteins (sex hormone binding globulin, SHBG, albumin). Binding proteins represent a storage form of circulating steroids, which bind 98% of circulating TST. The rest-2% is "free testosterone" (fTST), which is biologically active. Homeostasis is achieved by “closed” steroid feedback inhibition mechanism, where the plasmatic level of steroids affects the secretion of LH from adenohypophysis.

The effect of testosterone on target tissues is modulated by metabolic pathways.

1) *Aromatisation* of TST gives rise to 17b- estradiol. When the target cell is estrogen-dependent, the aromatase activity in target cells and supply of androgen substrate (TST) are of major importance for determining the rate of synthesis of estrogens. Aromatase cytochrome P450 enzyme is expressed in many tissues including placenta, ovary, testis, fat tissue, liver, brain, hair follicles.
2) *Reduction* of TST into 5-alfa-dihydrotestosterone (DHT) is achieved by 5a-reductase. This active form of TST (5 to 10 times more biologically effective) can fully activate the androgen receptor (AR) [12]. There are 3 isoforms of 5a-reductase. Isoform 2 is clinically more important because its deficiency is associated with distinct clinical manifestations [13]. Isoform 2 predominates in cells of the prostate and external genitalia, while isoform 1 predominates in the cells of the skin (except genitals), and in liver cells in small amounts. However, in prostate cancer cells, overexpression of isoform 1 is a common finding, thus increasing its clinical significance [14]. In the total deficiency of isoform 2 (autosomal recessive) there is a serious alteration of the development of sex organs in utero (male pseudohermafroditism). Many mutations of the gene encoding isoform 2 are known and can result into a number of different clinical manifestations. Signs of its deficiency are small phallus, severe hypospadias, scrotum bifidum, residual prostate utriculus [15]. The newly discovered isoform 3 may play an important role in the development of hormone-refractory prostate cancer (HRCaP) as its overexpression is found in the HRCaP cells [16].

In addition to these metabolic pathways, the level of DHT in target tissues is affected by other enzymes (hydroxysteroid dehydrogenases), which "fine-tune" the effect of androgens in the target tissues [17]. Owing to these local conversions the peripheral plasma concentration of androgens are only a rough indicator for their biological activities [12].

The mechanism of action of androgens can be divided into *genomic* and *non-genomic* effect [12]. *Non - genomic* effects of androgens include mechanisms affecting the flow of calcium in the cells and the effect on phosphorylation cascade of Map-kinase [18, 19], or membrane effects [20]. *Genomic* effects are mediated by activation of androgen receptor (AR). AR acts as a transcription factor activated by its ligand (TST). By androgen binding, AR is translocated from the cytoplasm to the nucleus where it binds to its DNA domain and interacts as a homodimer with specific DNA sequences that are referred as androgen responsive elements (ARES) [21]. Its binding to DNA leads into interactions with transcription factors [22] and other co-factor proteins [21]. This results into the "up" or "down" regulation of transcription of target genes [23].

Androgen metabolites are excreted as free steroids or bound (conjugated). Conjugated steroids are bound to glucuronide or sulfate group. Androgens are mostly degraded in the liver (glucuronate, sulphates), but the prostate the skin also contribute significantly to the metabolism of androgens. All the steroid - metabolising enzymes constitute a network for transforming androgens into secretion products (conjugated, unconjugated) that finally leave the body via the urine or the skin. The flux though this network is great, because the half-life of TST in men is only 12 minutes [12].

Androgens and Bone Metabolism

Growth and resorption of bone tissue are mediated by *osteoblasts* and *osteoclasts*. Both types of cells exert mutual influence on each other and equilibrium between the activity of both cell lines maintains net bone mass during constant renewal and turnover. Decreased osteoblast activity and increased activity of osteoclasts leads into loss of bone mass. Androgen receptors (AR) have been located on normal human osteoblasts [24] and both aromatizable and non-aromatizable androgens can stimulate of human osteoblasts proliferation in vitro [25].

Bone deformation strain represents a stimulus for osteoblastic activity. Androgens modify the effects induced by the mechanoreception of human osteoblastic cells by affecting adhesion molecule expression, i.e. fibronectin and the fibronectin receptor. These substances facilitate the adhesion of bone cells to the extracellular matrix, which represents a crucial requirement for osteoblastic development and function [26]. In addition, the secretion of osteoprotegerin (OPG), which is unaffected by mechanical strain alone, is doubled when this stimulus occurs in the presence of androgens. OPG is a decoy receptor for RANKL (receptor activator of nuclear factor-kappaB ligand). RANKL is secreted by osteoblasts, it induces osteoclastogenesis stimulates osteoclast differentiation [27]. Thus, OPG inhibits bone resoptive effect induced by RANKL [26]. Accordingly, TST levels directly correlate with OPG concentrations in healthy men [28].

Parathyroid hormone (PTH) induces osteoclast formation a differentiation. Androgens have direct inhibiting effect on this process via osteoclasts, which express AR and these cells are also blocked form PTH effects by androgens when no conversion to estrogens occurs [29].

Androgens decrease the number of bone remodeling cycles by modifying the genesis of osteoclasts and osteoblasts from their respective progenitor cells. In addition, androgens also exert effects on the lifespan of mature bone cells: they exert pro apoptotic effects on osteoclasts and anti-apoptotic effects on osteoblasts and osteocytes. TST also modulates effects induced by other hormones and cytokines involved in bone metabolism [30].

Osteoblast activity is reflected by concentration of procollagen type 1 (carboxy-terminal: P1CP or amino-terminal: P1NP) and other non collagenous proteins secreted by osteoblasts, eg. osteocalcin and bone specific alkaline phosphatase (BSAP). Also OPG as a decoy receptor for RANKL, can serve as marker of osteoblast activity. Bone resorption, hence osteoclast activity, therefore, can be estimated by urinary excretion of degradation products of type I collagen, such as deoxypyridinoline (DPD) and collagen type I cross- linked N-telopeptide (NTX) [31].

An independent role of androgens in protecting bone mass, both by promoting bone formation and attenuating bone resorption has been demonstrated in humans. Nevertheless, the role of its metabolite estradiol is pivotal in bone metabolism [30].

Aromatization of TST to estradiol is a pivotal event concerning effects of sex steroids on bone metabolism. Estrogen receptors (ER) have been localized in human osteoblasts [32], osteoclasts [33], and osteocytes [34]. Human males with mutations of ER or aromatase genes do not achieve normal bone density, despite normal or increased levels of serum TST [35].

ADT and Its Side Effects

ADT is increasingly attained through the use of luteinizing hormone-releasing hormone (LHRH) agonists, which down-regulate anterior pituitary receptors and lead to therapeutic hypogonadism, or directly by inhibiting pituary receptors by LHRH antagonists. The standard castrate level is < 50 ng/dL. Prostate cells are physiologically dependent on androgens to stimulate growth, function and proliferation. TST, although not tumorigenic, is essential for the growth and perpetuation of tumor cells [36].

PCa displays a range of clinical behavior, from slow-growing tumors of no clinical significance to aggressively metastatic and lethal disease. Most PCa cases diagnosed with present diagnostic techniques fall into the moderately differentiated group (grade 2, Gleason 6), of which the cancer-specific 10–15-year mortality is 18–30%. This is even lower in the group of prostate cancers diagnosed with grade 1 and 2 or Gleason 4–6, and there are subgroups of patients who are not at risk of dying from prostate cancer even within 15years. Overall mortality is then determined by comorbidity [37]. While both short- and long-term ADT are effective for treating PCa, with the clinical benefits patients can often have significant side-effects. It is important that these complications are recognized and managed appropriately so that adverse effects on the patient's quality of life (QoL) are minimized. There has been an increase in recent years in the use of ADT at all stages of PCa. The extensive use of ADT is raising concerns about adverse effects – loss of BMD being the most prominent.

Bone Loss and Osteoporosis

As shown above, androgens are important for maintenance of bone tissue. Although ADT is used in medical practice for more than 60 years, it's only few years that the clinicians became aware of serious and potentially catastrophic consequences resulting form long-term ADT.

Bone density is determined both by peak bone mass achieved during skeletal development and the subsequent amount of maintenance and resorption of bone tissue. Androgens affect both processes and thus are a pivotal determinant of bone mass in men.

Osteoporosis

Osteoporosis is defined as a systemic metabolic bone disease characterized by low bone mass density (BMD) and microarchitectural deterioration of bone tissue, with a consequent increase in bone fragility and susceptibility to fracture [38]. Diagnosis of osteoporosis based on WHO definitions is developed for women originally [39]. When based on male cutoffs, 1-2 million (3 - 6%) of men have osteoporosis and 8 -13 million (28 - 47%) have osteopenia; when based on female cutoffs, 280,000 - 1 million (1 - 4%) have osteoporosis and 4 - 9 million (15 - 33%) have osteopenia [40, 41]. While this numbers may seem disturbing, it is believed that osteoporosis in men is substantially underdiagnosed and undertreated in the United States [42].

Female osteoporosis has been studied extensively and characterized due to its high prevalence [43] whereas male osteoporosis, especially that associated with ADT has gained focus only recently.

In men, osteoporosis occurs later than in women [44], but the prevalence of osteopenia does not differ significantly between men and women aged more than 50 years. Conversely, the prevalence of osteoporosis in men is lower than in women [40]. Even though it may be underestimated when standard female BMD parameters are considered suitable for normal mineralization in men [45]. Men generally have a higher BMD than women at the same age [46]. Accordingly, the prevalence of male osteoporosis is greater when male-specific ranges are used in men above fifties: ranging from 1% to 4% of elderly men when the diagnosis is based on female cut-off points vs. 3% to 6% when based on male cut-off points [40].

Men are estimated to lose bone mineral density (BMD) at a rate of up to 1% per year with advancing age [47, 48], and one in eight men over age 50 years will experience an osteoporosis-related fracture in their lifetime [49]. Of all osteoporotic fractures, hip fractures contribute to the greatest morbidity as well as mortality, both of which are much greater in men than in women [50-52].

Diagnosis of Osteoporosis

Current guidelines recommend assessment of bone mineral density (BMD) previous to ADT and yearly thereafter [53] with dual-energy x-ray absorptiometry (DXA) which is considered the standard method to measure BMD [54]. International Society for Clinical

Densitometry (ISCD) recommends that central skeleton sites (lumbar spine, total hip and femoral neck) are the most appropriate locations to asses BMD [55].

Diagnosis of osteoporosis can than be made according to WHO classification: if T- score is less than 2,5 of standard deviation. Values between (-) 1 and (-) 2.5 SD (standard deviation) is defined as osteopenia. T - score stands for the number of standard deviations (SD) from the density of young healthy individuals of the same sex [39].(Table 1)

To improve the identification of patients at highest risk of fracture, WHO has developed an algorithm to predict fractures - FRAX ™ http://www.shef.ac.uk/FRAX/

According to EAU guidelines, a precise evaluation of BMD should be performed by dual X-ray absorptiometry before starting long-term ADT. An initial low BMD (T-score below 2.5, or below 1 if other risk factors are present) indicates a high risk of subsequent non-metastatic fracture, suggesting the need for early use of preventive bisphosphonate therapy [56].

Table 1.

Clasiffication	T-score
Normal	>-1
Osteopenia	-1 - -2,5
Osteoporosis	<-2,5

ADT, PCa and Clinical Aspects of Bone Disease

At presence, significant amount of data report that ADT is associated with the loss of bone mineral density (BMD) in a time-dependent manner [57-59] which leads into increased fracture risk [60]. Skeletal fractures negatively correlate with overall survival in men with PCa [61] and maintaining skeletal health is crucial for quality of life and survival [62].

Treatment of complications of pathological fractures is complicated and expensive [63]. Moreover, the typical feature of PCa is the ability to metastasize into bone in more than 80 % of cases [64]. Most bone lesions in PCa are osteoblastic in nature [65]. However, studies show that osteoblastic lesions in PCa have excessive bone growth, but on the other hand also simultaneously increased osteolysis [66]. The new bone formed by tumor stimulated osteoblast is weak and poorly mineralized and subsequent osteopenia leads into increased osteolysis - result to the creation of bone matrix with seriously compromised integrity. The risk of developing bone complications is therefore increased.

Treatment of PCa does not focus on skeletal complications that may arise from bone metastases. The main symptom of bone metastases is severe bone pain, which often requires strong narcotic therapy or palliative radiation therapy. Other complications include spinal cord compression and pathological fractures, which may require surgery. These skeletal complications have a negative effect on quality of life [65].

Data from a double-blind, placebo-controlled studies show that approximately half of patients treated with ADT had one or more events associated with the skeleton. Most of these events required palliative radiotherapy, or were pathological fractures [67]. Skeletal complications are also associated with significant financial expenditure. A recent analysis of the costs of health insurance in the U.S. since 1994 until 2002 revealed that the total cost to treat patient with PCa who had skeletal event were 20 000 dollars higher than in patient who did not experience skeletal event [68].

Interestingly, it has been reported that hormone naive patients with advanced PCa have lower baseline BMD than healthy control, and relatively high prevalence of osteopenia and osteoporosis [69, 70]. The largest study that investigated the association of BMD measures with PCa risk in older men enrolled was the Osteoporotic Fractures in Men Study (MrOS) [71].

MrOS was prospective study conducted on 4597 men with mean follow up 5.2 years, which evaluated the association of BMD and incidental PCa in a cohort of older men with no history of PCa. Unexpectedly, the authors found that higher total body BMD was significantly related to reduced risk for PCa. This result was „unexpected" because authors presumed that the higher levels of androgens lead into higher prevalence of PCa, which positively correlates with BMD. Additionally, total body BMD was inversely associated with the development of high-grade, but not low-grade disease. A similar but weaker association was observed for total hip BMD with high-grade PCa. This study confirms the association, although still not elucidated, between low BMD and PCa.

Treatment of ADT Induced Osteoporosis

Lifestyle changes: Immobilization is an important cause of bone loss. Immobile patients lose bone mass more rapidly than mobile patients. Regular daily activities, overcoming gravity, walking, and exercise have a positive impact on bone density: it stimulates the osteoblasts to produce new bone and inhibits osteoclasts, thereby decreasing resorption of bone. It also improves physical coordination (prevention of falls). Cessation of smoking, decreased alcohol consumption and normalization of body mass index (BMI) helps to maintain BMD [72].

Ca supplementation: The ideal is to ensure that the amount of calcium is taken in the normal diet. If the patient is unable to take the recommended amount of calcium in the diet (lactose intolerance, hyperlipoproteinemia, etc) it is indicated for calcium supplementation (1000mg – 1500 mg daily) [72].

Vitamin D: Supplementation of vitamin D is indicated when its deficiency can be assumed or proven. The recommended daily dose is 400-800 IU (10-20 mg).

Bisphosphonates belong to most potent inhibitors of bone resorption. Effect on the reduction of osteoporotic fractures has been demonstrated for treatment with alendronate, risedronate, ibandronate, etidronate and zoledronic acid. The effectiveness in reducing vertebral and nonvertebral fractures were confirmed by many studies [73, 74, 75]. The optimal regimen for zoledronic acid is unclear, because one study recommends treatment every 3 weeks [76], while another trial has produced similar results with an annual injection [77], and finally, another study reports that single infusion of zoledronic acid in patients receiving ADT reduces bone mineral loss and maintains BMD at least at 12 months during ADT [78].

One of the most important and serious adverse effect of bisphoshonate administration is jaw necrosis [79] The initial BMD could be used to guide the choice of regimen [80].Thus, a

3-month injection might be given in osteoporotic patients, for whom a yearly injection is likely to provide insufficient protection [56].

Denosumab is a fully human monoclonal antibody against RANKL (see above). In the largest, well conducted study to date, denosumab was associated with 5.6% increase in the lumbar BMD versus 1% decrease in the placebo arm. There were also significant BMD increases at the total hip, femoral neck and distal third of the radius. 60 mg was delivered subcutaneously every 6 months, was not associated with any significant toxicity, or delayed healing in vertebral fractures [81].

Androgens and Cognitive Functions

It is known that during certain developmental stages- especially during the first years of life, during adolescence, girls surpass in boys several verbal skills. Males excel after about the tenth year of life and non-verbal skills in adulthood, especially in spatial orientation and manipulation [82].

Evidence of a link between sex hormones and spatial abilities came from studies of individuals with Turner syndrome (XO karyotype, no gonadal hormones) or testicular feminization syndrome- (XY karyotype, the tissues are refractory to normal levels of TST). These patients have female external genitalia, they are raised as girls. In these patients verbal skills surpass their spatial abilities, which is a typical pattern of cognitive abilities of women [83].

Studies on men with idiopathic or acquired hypogonadotrophic hypogonadism confirm the importance of TST for spatial abilities. As short-term androgen supplementation did not restore spatial function, suggesting that low levels of sex hormones during the intrauterine and neonatal period have a lifelong impact [84].

Direct manipulation of sex hormones supports the conclusion that androgens play an important role in cognition. The first experiments with direct hormonal manipulation can be traced back to 1941 when Simonson et al. [85] published their experiment using methyl testosterone that was administered to eunuchoids, castrated males, and elderly men. The result was an improved ability to perceive the flicker (critical flicker frequency), a measure of attention and alertness, as long as the androgen treatment lasted [86].

Androgen therapy was also administered to female to male transsexuals in high doses as a preparation before gender reassignment. Their spatial skills have significantly improved, while verbal skills declined considerably [87].

For ethical reasons, nowadays the manipulation of gonadal hormones is restricted to patients in clinical studies. Thus, the last such study was conducted in 1971. Klaiber et al. [88] studied the effect of infused TST on mental abilities in healthy male students. After a 4-hour infusion of TST or saline in the control group, performance of the control group (saline infusion) showed a significant decline in mental performance than TST infused group.

Potential Mechanisms of Action

TST may impact cognition through several mechanisms. For example, activation of calcium channels in the brain occurs through rapid, nongenomic methods of action on G-protein-coupled, agonist sequestrable TST membrane receptors that initiate a transcription-independent signal pathway [89]. TST also may impact cognitive performance directly through modulating neurotransmitters and stimulating neuronal connectivity, decreasing β-amyloid peptide production, and preventing N-methyl-D-aspartate excitotoxicity [90] mechanisms implicated in cognitive disorders such as Alzheimer disease or dementia. Furthermore, some estrogen studies have highlighted several possible mechanisms through which this hormone can impact cognitive functioning [91]. These include increasing cholinergic activity through its action of choline acetyltransferase, maintenance of dendritic spine density on CA1 pyramidal cells of the hippocampus and facilitating induction of long-term potentiation in the hippocampus, increasing serotonergic and cholinergic activity to maintain neural circuitry, altering lipoprotein, and decreasing risk of cerebral ischemia [92, 93].

Cognitive Functions and ADT

Green et al were the first to systematically research the impact of androgen-ablation therapy on the cognitive functioning of men with PCa. Sixty-five men (mean age, 73 years) with advanced PCa were assigned randomly to 1 of 4 groups: leuprolide (N = 19), goserelin (N = 20), cyproterone acetate (N = 11), and monitoring without hormone treatment (N = 15). All men participated in a battery of neuropsychological assessments at baseline (ie, 1 week before treatment) and then 6 months later. PSA and TST levels decreased significantly from baseline to 6 months for the 3 hormonally treated groups. Conflicting results emerged in the memory domain; men in the goserelin group surprisingly improved on 2 measures of memory (verbal [Wechsler Memory Scale-Revised] and visual [Rey-Osterrieth Complex Figure test]) but declined in another measure of verbal memory (Auditory Verbal Learning Test). The goserelin group also declined in a measure of executive functioning (Trails B). Of the 50 men on active treatments, 24 men showed a reliable decline (ie, >1 standard deviation) on at least 1 cognitive task, and 7 men showed a reliable change on 2 tasks, whereas the monitoring-only group showed no decline on any of the tasks [94].

Salminen et al researched the cognitive effects of ADT on 26 men who were diagnosed recently with PCa and who began ADT 2 months before radiotherapy. From baseline to 12 months, tests of visuomotor speed and of reaction time saw significant decreases. The decline in TST coincided with a decline in visuomotor processing (digit symbol), reaction time (10-choice reaction time), working memory speed (subtraction), sustained attention (vigilance), and recognition speed (recognition of letters) [95].

Jenkins et al assessed 32 men with standard neuropsychological assessments at 3 time intervals: at baseline, 3 months, and 9 months. The average age of these men was 67.5 years, and they used androgen-ablation therapy for 3 to 5 months. Twenty-five healthy men, similar in age, served as the control group. Although there was no overall group effect, a greater percent of men in the ablation group reported a significant cognitive decline in 1 task (47%) compared with the control group (17%; odds ratio, 4.412; $P < .05$) at the 3-month time point.

There were no significant differences between the groups at the 9-month time point. On specific domain analysis, the tasks most impacted at the 3 month time point were spatial memory and ability [96].

Joly et al compared physical and cognitive function in a cross-sectional study of 57 patients who were receiving ADT for nonmetastatic PCa and 51 healthy, age-matched controls. Thirty patients received ADT as adjuvant treatment after prostatectomy or radiotherapy, and 27 patients received ADT for increasing levels of PSA. The median duration on ablation therapy was 1.8 years (range, 0.4-7.4 years). To assess cognitive functioning, the researcher administered the Sensitivity Cognitive Screen and a self-reported assessment on cognitive deficits (the Functional Assessment of Cancer Treatment-Cognitive Scale [FACT-COG]). In contrast to other studies cited above, Joly and colleagues observed that, although men with nonmetastatic PCa who received ADT experienced more treatment-related symptoms, they demonstrated no differences in cognitive function on either the High Sensitivity Cognitive Screen or the FACT-COG. The authors suggested that the High Sensitivity Cognitive Screen may not be sensitive enough to detect the subtle cognitive changes that occur after ADT. In addition, self-report of cognitive function has not been correlated consistently with actual neuropsychological testing [97].

In conclusion, the data show that androgen ablation does have cosequences for cognitive functioning. Larger longitudinal studies are warranted.

Androgen Deprivation Therapy and Changes in Body Composition

Male hypogonadism (of any etiology) results in a decline in lean body mass (LBM) and an increase in fat mass, which is reversed with TST replacement [98]. A cross-sectional study showed that men undergoing long-term ADT (12–101 months) have increased fat mass in the trunk and all extremities- measured by dual-energy x-ray absorptiometry (DXA), compared with eugonadal men with PCa not undergoing ADT (treated with prostatectomy and/or radiation therapy) and age-matched eugonadal controls [99].

Another case control study examined the prevalence and magnitude of obesity and fat mass in a group of 62 men with PCa receiving ADT for 1–5 yr [100]. Healthy men (n = 47) with a PSA of less than 4.0 ng/ml were recruited as controls. The study showed that men with PCa had significantly higher body weight (86.5 vs. 80.6 kg) and percent body fat (30 vs. 26%) than controls.

Meta-analysis of sixteen studies showed that ADT increased % body fat by on average 7.7% (95% CI 4.3, 11.2, from seven studies, $P < 0.0001$) and decreased % lean body mass by on average -2.8% (95% CI -3.6, -2.0, from six studies, $P < 0.0001$) but for both there was marked heterogeneity between studies ($I2 = 99\%$ $I2 = 73\%$, respectively). Similarly, body weight (2.1%, $P < 0.0001$ from nine studies) and BMI (2.2%, $P < 0.0001$, from eight studies) increased significantly. More extensive changes were seen with longer duration of treatment [101].

These studies prove that ADT results in an unfavorable body composition. The increase in fat mass on ADT correlates positively with rising insulin levels [102]. Hence, this increasing adiposity may be the primary event leading to these metabolic complications (possibly via elaboration of adipokines and inflammatory cytokines). Similarly, it is possible that a decrease in muscle mass may result in decreased glucose uptake by the muscle. These changes may ultimately lead to insulin resistance and diabetes in this population, hence predisposing them to cardiovascular disease. Lifestyle changes or suitable interventions to minimize the effect of ADT on body composition need to be investigated.

ADT and Insulin Resistance

Epidemiological studies have shown that low TST levels predict the development of insulin resistance and type 2 diabetes [103, 104, 105]. Studies have also confirmed a direct relationship between serum TST and insulin sensitivity [106]. These findings are further supported by interventional studies showing an improvement in insulin sensitivity with TST replacement in hypogonadal obese men [107].

Early Metabolic Changes

There is some evidence, that the onset of insulin resistance can be detectable after 3 months of ADT [108, 109]. 3-month prospective study using combined androgen blockade with leuprolide and bicalutamide showed a 43% increase in fat mass and a 26% increase in insulin levels from baseline, again indicating development of insulin resistance with increasing adiposity [110]. Although there was no significant change in fasting glucose levels, a statistically significant increase in glycosylated hemoglobin was seen (though this increase was within the normal range from 5.46–5.62%). These observations suggest that insulin resistance develops within a few months of initiating ADT; however, this compensatory hyperinsulinemia prevents the development of diabetes.

Late Metabolic Changes

Observational study of a population-based cohort found that men undergoing ADT with GnRH agonists had a higher risk of incident diabetes (11%), coronary artery disease (25%), myocardial infarction, and sudden death [111]. Interestingly, orchiectomy was associated only with a higher risk of diabetes. In some men, this risk was evident within 4 months of starting ADT. These findings suggest that although both medical and surgical modalities of ADT result in increased metabolic burden, GnRH analogs are also associated with cardiovascular events.

After 12 months of ADT, serum fasting glucose increased significantly [112], suggesting that men with PCa who are receiving long-term ADT are at risk for developing insulin resistance and hyperglycemia, thus leading to their increased risk of cardiovascular disease [113].

A retrospective study which enrolled 396 patients with a median follow-up of 60.1 months, 36 (11.3%) patients developed new- onset diabetes mellitus (NODM). In 77 patients with pre-existing DM, there was an increase of >/=10% in serum HbA1c or fasting glucose levels in 15 (19.5%) and 22 (28.6%), respectively. On multivariate analysis, a BMI of >/=30 kg/m(2) was associated with an increased risk of developing NODM (odds ratio 4.65, P = 0.031) [114].

In conclusion, patients receiving ADT for PCa with or with no history of DM should have routine surveillance of glycaemic control, with appropriate preventive and treatment measures.

ADT and Lipid Alterations

Hyperlipidemia is a known risk factor for cardiovascular disease. Recent epidemiological research suggests that low serum TST levels in men are associated with an adverse lipid profile, especially elevated total cholesterol, LDL cholesterol, and triglycerides [115]. Furthermore, interventional studies have shown that TST replacement in hypogonadal men results in an improvement in lipid profile [116].

During long-term ADT, triglycerides rise by approximately 26% and total cholesterol approximately 10%. [117, 118, 119] In addition, high-density lipoprotein (HDL) rises approximately 8% to 11%. The net effect of these changes on cardiovascular risks is unknown. Significant changes can be observed within the first 3 months of treatment, with more modest subsequent change [110].

ADT, Metabolic Syndrome and Cardiovascular Disease

Metabolic syndrome (MS) is a known risk factor for cardiovascular disease (CVD) [120]. According to the Adult Treatment Panel III guidelines [121], a man is considered to have MS if he meets 3 of the following 5 criteria: fasting plasma glucose level >110 mg/dL, serum triglyceride level 150 mg/dL, serum high-density lipoprotein level <40 mg/dL, waist circumference >102 cm, and blood pressure 130/85 mmHg. Subjects on antihypertensive and antilipid medications are also considered positive for the respective criteria. Recently, male hypogonadism has surfaced as an independent risk factor for MS. Cross-sectional studies have shown that men with hypotestosteronemia have a higher prevalence of MS [122]. Longitudinal studies also show that lower androgen levels in men independently predict the development of MS [105].

These observations suggest that profound hypogonadism due to ADT imparts increased metabolic burden. Long-term prospective studies are needed to determine the time of onset of various metabolic alterations in these men.

Since MS is associated with CVD, large studies were conducted to assed the CV risk and ADT. A large SEER-Medicare–based analysis of 73,196 men aged 66 years and older with PCa identified significant GnRH agonist-associated elevations in risk for myocardial infarction (HR, 1.11; p= .03), sudden cardiac death (HR, 1.16; p = .004), and new diagnosis of coronary heart disease (HR, 1.16; p< .001) [123]. Similarly, a second SEER-Medicare–based study of 23,000 men with PCa found a 20% ADT-attributable rise in CV morbidity at 1 year [124].

In contrast, a recently reported matched cohort analysis of approximately 20,000 men in an On-tario database found no association between ADT and acute myocardial infarction (HR, 0.91; 95% CI, 0.84–1.00) [125].

A smaller population-based observational study of 3262 men who had undergone prostatectomy for PCa found that ADT was significantly associated with CV mortality, although only in the subset of men aged 65 years and older [126]. This analysis failed to validate baseline coronary artery disease and diabetes as risk factors for CV mortality. Finally, combined analysis of 3 randomized trials involving men with localized PCa found that in the subset of men aged 65 years and older, 6 months of treatment with a GnRH agonist led to earlier onset of fatal myocardial infarction [127].

Three large randomized, controlled trials by the Radiation Therapy Oncology Group (RTOG) have been retrospectively analyzed for an association between neoadjuvant /concomitant/adjuvant ADT and CV mortality. These analyses have not found convincing evidence of an association [128-130]. Secondary analyses of a randomized controlled trial from the EORTC found no association between ADT and CV mortality. The RTOG and EORTC trials were randomized, featured large enrollments, and had long-term follow-up.

Table 2. Summary of main organ systems affected by severe hypogonadism (by ADT)

Bone metabolism	Loss of BMD, skeletal events (fractures), increased morbidity&mortality
Bone marrow	Anaemia
Lipid profile alterations	Increased: overall cholesterol, TAG, LDL, HDL
Impaired insuline sensitivity	Hyperinsulinemia, new onset diabetes, worsening existing DM
Body composition	Increase in fat mass, decrease in muscle mass, increased BMI
Cardiovascular disease	Increased risk of CV morbidity and mortality (?)
Metabolic syndrome	See above
Cognitive functions	Impaired spatial cognition, reaction time, other (?)
Other	Mood changes, loss of libido

References

[1] Huggins C, Hodges CV. Studies on prostatic cancer. I. The effect of castration, of estrogen and of androgen injection on serum phosphatase in metastatic carcinoma of the prostate. *J. Urol.* 2002 Feb;167(2P 2):948-51, discussion 952.

[2] Huggins C, Stevens RE Jr, Hodges CV. Studies on prostate cancer. II. The effect of castration on advanced carcinoma of the prostate gland. *Arch. Surg.* 1941;43:209-23.

[3] Shahinian V, Kuo Y, JLFreeman, Goodwin J. Risk of fracture after androgen deprivation for prostate cancer. *N. Engl. J. Med.* 2005; 352:154-164.

[4] Smith MR, Lee H, Nathan DM. Insulin sensitivity during combined androgen blockade for prostate cancer. *J. Clin. Endocrinol. Metab.* 2006;91(4):1305-8.

[5] Hakimian P, Blute M, Jr, Kashanian J, et al. Metabolic and cardiovascular effects of androgen deprivation therapy. *BJU Int.* 2008;102:1509–1514

[6] Smith MR, Finkelstein JS, McGovern FJ, et al. Changes in body composition during androgen deprivation therapy for prostate cancer. *J. Clin. Endocrinol. Metab.* 2002;87:599–603

[7] Saigal CS, Gore JL, Krupski TL, et al. Androgen deprivation therapy increases cardiovascular morbidity in men with prostate cancer. *Cancer.* 2007;110:1493–1500

[8] Nelson CJ, Lee JS, Gamboa MC, Roth AJ. Cognitive effects of hormone therapy in men with prostate cancer: a review. *Cancer.* 2008;113(5):1097-106.

[9] Coffey DS (1988) Androgen action and the sex accesory tissues. In: Knobil E, Neill (eds): The physiology of reproduction. *J. Raven Press,* New York,pp 1081-1119

[10] Baulieu EE (1997) Neurosteroids: of the nervous system, by the nervous system, of the nervous system. In: *Rec. Prog. Horm. Res.* 52: pp 1-32

[11] Saez JM (1994) Leydig cells: endocrine, paracrine and autocrine regulation. *Endocr. Rev.* 16: 574-626

[12] Rommerts FFG (2004), Testosterone: Action, Deficiency, Subsitution. Third Edition In:Nieschlag E, Behre HM (Eds): *Testosterone: an overview of biosynthesis, transport, metabolism and non genomic functions.* Cambrigde University Press, pp 1-39

[13] Wilson JD, Griffin JE, Russel DW (1993) Steroid 5a-reductuase 2 deficiency. *Endocr. Rev.* 14:577-593

[14] Tindall DJ, Rittmaster RS (2008) : The rationale for inhibiting 5alpha-reductase isoenzymes in the prevention and treatment of prostate cancer. *J. Urol.* 2008 Apr;179(4):1235-42.

[15] Baskin LS (2004) Abnormalities of sexual determination &differentation In:Tanagho EA, McAninch WJ (Eds) *Smith´s general urology,* 16th Edition, Lange Medical Books, p660 (2004)

[16] Uemura M, Tamura K, Chung S, Honma S, Okuyama A, Nakamura Y, Nakagawa H. (2008): Novel 5 alpha-steroid reductase (SRD5A3, type-3) is overexpressed in hormone-refractory prostate cancer., *Cancer Sci.* 2008 Jan;99(1):81-6.

[17] Ziaran S, Goncalves FM, Štefančík J, Breza J Sn. The effect of androgens on cognitive function and bone. *Lek. Obzor* 2009; 3:116-20

[18] Guo Z, Benten WP, Krucken J, Wunderlich F (2002), Nogenomic testosterone calcium signaling. Genotropic actions in androgen receptor-free macrophages *J. Biol. Chem.* 277:29600-29607

[19] Castoria G, Lombardi M, Barone MV, Bilancio A, Di Domenico M, Bottero D, Vitale F, Migliaccio A, Auricchio F.: Androgen-stimulated DNA synthesis and cytoskeletal changes in fibroblasts by a nontranscriptional receptor action., *J. Cell Biol.* 2003 May 12;161(3):547-56.

[20] Papakonstanti E.A, Kampa M, Castanas E, and Stournaras C. A rapid, nongenomic, signaling pathway regulates the actin reorganization and PSA secretion induced by membrane testosterone receptors' activation. *Mol. Endocrinol.* 17(5), 870-881 (2003).

[21] Klocker H, Gromoll J, Cato ACB, (2004) The androgen receptor: molecular biology, Third Edition In:Nieschlag E, Behre HM (Eds) *Testosterone: Action, Deficiency, Subsitution.* Cambrigde University Press,pp 39-92

[22] Lee DK, Duan HO, Chang C.: From androgen receptor to the general transcription factor TFIIH. Identification of cdk activating kinase (CAK) as an androgen receptor NH(2)-terminal associated coactivator.,: *J. Biol. Chem.* 2000 Mar 31;275(13):9308-13

[23] *Quingley* CA, DeBellis A, Marschke KB, El-Awady MK, Wilson EM, French FS (1995). *Androgen receptor defects:* historical, clinical and molecular perspectives., *Endocr. Rev.* 16:271-321

[24] Colvard DS., Erikse EF., Keeting PE., Wilson EM., Lubahn DB., French FS., Riggs BL., Spelsberg TC., (1989) Identification of androgen receptors in normal human osteobast-like cells. *Proc. Natl. Acad. Sci. U.S.A.* 86: 854-857

[25] Kasperk CH., Wakley GK., Hierl T., Ziegler R., (1997) Gonadal and adrenal androgens are potent regulators of human bone cell metabolism in vitro. *J. Bone Miner Res.* 12: 464-471

[26] Liegibel UM., Sommer U., Tomakidi P., Hilscher U., Van den Heuvel L., Pirzer R., Hillmeier J., Nawroth P., Kasperk C., (2002) Concerted action of androgens and mechanical strain shifts bone metabolism from high turnover into an osteoanabolic mode. *J. Exp. Med.* 196: 1387- 1392

[27] Khosla S., (2001) Minireview: the OPG/RANKL/RANK system. *Endocrinology.* 142: 5050-5055

[28] Szulz P., Hofbauer LC., Heufelder AE., Roth S., Delmas PD., (2001) Osteoprotegerin serum levels in men: correlation with age, estrogen, and testosterone status. *J. Clin. Endocrinol. Metab.* 86: 3162- 3165

[29] Chen Q, Kaji H, Sugimoto T, Chihara K (2001). Testosterone inhibits osteoclast formation formation stimulated by parathyroid hormone through androgen receptor. *FEBS Lett.* 491:91-93

[30] Zitzmann M, Nieschlag E (2004), Androgens and bone metabolism. In: Nieschlag E, Behre HM (Eds): *Testosterone: Action, Deficiency, Subsitution.* Third Edition Cambrigde University Press, pp 233-254

[31] Riggs BL., Khosla S., Melton JR., 3 rd (2002) Sex steroid and the construction and conservation of the adult skeleton. *Endocr. Rev.* 23: 279-302

[32] Eriksen EF, Colvard DS, Berg NJ, Graham ML, Mann KG, Spelsberg TC, Riggs BL (1988). Evidence of estrogen receptors in normal human osteoblast-like cells. *Science.* 241:84-86

[33] Ourlsler MJ, Pederson L, Fitzpatrick L, Riggs BL (1994). Human giant cell tumors of the bone (osteoclastomas) are estrogen target cells. *Proc. Natl. Acad. Sci. U.S.A.* 96:505-510

[34] Braidman I, Baris C, Wood L, Selby P, Adams J, Feemont A, Hoyland J (2000). Preliminary evidence for impaired estrogen receptor protein expression in ostetoblast and osteocytes from men with idiopatic osteoporosis. *Bone.* 26:423-427

[35] Grumbach MM (2000) Estrogen, bone, growth and sex: a sea change in conventional wisdom. *J. Pediatr. Endocrinol. Metab.* 13:1439-1455

[36] Walsh PC. Physiologic basis for hormonal therapy in carcinoma of the prostate. *Urol. Clin. North Am.* 1975 Feb;2(1):125-40

[37] Bangma CH, Roemeling S, Schröder FH. Overdiagnosis and overtreatment of early detected prostate cancer.*World J. Urol.* 2007 Mar;25(1):3-9.

[38] WHO Assessment od fracture risk and its application to screening for postmenopausal osteoporosis. *WHO Technical report series.* Geneva, WHO, 1994.

[39] Kanis JA, Melton LJ, Christiansen C, Johnston CC, Khaltaev N. The diagnosis of osteoporosis. *J. Bone Miner Res.* 1994; 9: 1137-1141.

[40] Looker AC, Orwoll ES, Johnston Jr CC, Lindsay RL, Wahner HW, Dunn WL, Calvo MS, Harris TB, Heyse SP. Prevalence of low femoral bone density in older US adults from NHANES III. *J. Bone Miner Res.* 1997; 12: 1761-1768.

[41] Bilezikian JP. Osteoporosis in men. *J. Clin. Endocrinol. Metab.* 1999; 84: 3431-3434.

[42] Kiebzak GM, Beinart GA, Perser K, Ambrose CG, Siff SJ, Heggeness MH. Undertreatment of osteoporosis in men with hip fracture. *Arch. Intern. Med.* 2002; 162: 2217-2222

[43] Krane SM, Holik MF. Metabolic bone disease. In: Fauci AS, Braunwald E, Isselbacher KJ, et al., editors. *Harrison's principles of internal medicine.* 14th edition. Volume 2. New York: McGraw-Hill, 1998:2247–2253.

[44] Orwoll ES, Klein RF. Osteoporosis in men. *Endocrine Reviews.* 1995; 16: 87–116.

[45] Seeman E. Pathogenesis of bone fragility in women and men. *Lancet.* 2002; 359: 1841-1850.

[46] Smith MR. Diagnosis and management of treatment-related osteoporosis in men with prostate carcinoma. *Cancer.* 2003; 97: 789-795.

[47] Hannan MT, Felson DT, Dawson-Hughes B, Tucker KL, Cupples LA, Wilson PWF, Kiel DP. Risk factors for longitudinal bone loss in elderly men and women: the Framingham osteoporosis study. *J. Bone Miner. Res.* 2000; 15: 710-720.

[48] Jones G, Nguyen T, Sambrook P, Kelly PJ, Eisman JA. Progressive loss of bone in the femoral neck in elderly people: longitudinal findings from the Dubbo osteoporosis epidemiology study. *BMJ.* 1994; 309: 691–695.

[49] Melton LJ, Chrischilles EA, Cooper C, Lane AW, Riggs BL. How many women have osteoporosis. *J. Bone Miner. Res.* 1992; 7: 1005-1010.

[50] Forsen L, Sogaard AJ, Meyer HE, Edna T, Kopjar B. Survival after hip fracture: short- and long-term excess mortality according to age and gender. *Osteoporos. Int.* 1999; 10: 73-78.

[51] Center JR, Nguyen TV, Schneider D, Sambrook PN, Eisman JA. Mortality after all major types of osteoporotic fracture in men and women: an observational study. *Lancet.* 1999; 353: 878-882.

[52] Abrahamsen B, van Staa T, Ariely R, Olson M, Cooper C. Excess mortality following hip fracture: a systematic epidemiological review. *Osteoporos. Int.* 2009; 20(10): 1633-1650.

[53] Diamond T, Higano C, Smith M, Guise T, Singer F. Osteoporosis in men with prostate cancinoma receiving androgen deprivation therapy: recommendatios for diagnosis and therapies. *Cancer.* 2004; 100: 892-899.

[54] WHO: Assessment of fracture risk and its application to screening for postmenopausal osteoporosis repot of WHO Study Group. Geneva, 1994: *World Health Organization*

[55] Kanis J, Seeman E, Johnell O, Rizzoli R, Delmas P. The perspective of the International Osteoporosis Foundation on the official positions of the International Society for Clinical Densitometry. *Osteoporosis Int.* 2005; 16: 456-459.

[56] http://www.uroweb.org/guidelines/online-guidelines/

[57] Daniell HW, Dunn SR, Fergusson DW, Lomas G, Niazi Z, Srtatte PT. Progressive osteoporosis during andorgen deprivation therapy for prostate cancer. *J. Urol.* 2000; 163: 181-186.

[58] Greenspan SL, Coates P, Sereika SM, Nelson JB, Trump DL, Resnick NM. Bone loss after initiation of androgen deprivation therapy in patients with prostate cancer. *J. Clin. Endocrinol. Metab*. 2005; 12: 6410-6417.

[59] Ahlborg HG, Nguyen ND, Center JR, Eisman JA, Nguyen TV. Incidence and risk factors for low trauma fractures in men with prostate cancer. *Bone*. 2008; 43: 556-660.

[60] Oefelein MG, Ricchiuti V, Conrad W, Resnick MI. Skeletal fractures negatively correlate with overall survival in men with prostate cancer. *J. Urol*. 2002; 168 (3): 1005-1007.

[61] Inoue T, Segawa T, Kamba T, Yoshimura K, Nakamura E, Nishiyama H, Ito N, Kamoto T, Habuchi T, Ogawa O. Prevalence of skeletal complications and their impact on survival of hormone refractory prostate cancer patients in Japan. *Urology*. 2009; 73 (5): 1104-1109.

[62] Saad F, Olsson C, Schulman CC. Skeletal morbidity in men with prostate cancer: quality-of-life considerations throughout the continuum of care. *Eur. Urol*. 2004; 46 (6): 731-739.

[63] McKiernan JM, Delea TE, Liss M. et al. Impact of skeletal complications on total medical care costs in prostate cancer patients with bone metastases. *Proc. Am. Soc. Clin. Oncol*. 2004; 23: 531 (abstract no. 6057)

[64] Bubendorf L, Schopfer A, Wagner U, Sauter G, Moch H, Willi N, Gasser TC, Mihatsch MJ. Metastatic patterns of prostate cancer: an autopsy study of 1,589 patients. *Hum. Pathol*. 2000; 31:578–583.

[65] Liotta LA, Kohn E. Cancer invasion and metastases. JAMA 1990; 263: 1123–6

[66] Garnero P, Buchs N, Zekri J, Rizzoli R, Coleman RE, Delmas PD. Markers of bone turnover for the management of patients with bone metastases from prostate cancer. *Br. J. Cancer*. 2000; 82: 858–64

[67] Saad F., Gleason DM., Murray R. Long-term efficacy of zoledronic acid for the prevention of skeletal complications in patients with metastatic hormone-refractory prostate cancer. J Natl Cancer Inst 2004; 96: 879–82

[68] McKiernan JM, Delea TE, Liss M. et al. Impact of skeletal complications on total medical care costs in prostate cancer patients with bone metastases. *Proc. Am. Soc. Clin. Oncol*. 2004; 23: 531

[69] Ziaran S, Goncalves FM, Wendl J, Trebaticky B, Breza J Sn. Evaluation of bone mass density on patients with prostate cancer prior to the start of androgen deprivation therapy. *Bratisl. Med. J*. 2009; 110(9):559-562.

[70] Ziaran S, Goncalves FM, Breza J Sn. Hormone naive patients with advanced prostate cancer have lower baseline bone mass density than healthy control. *Bratisl. Med. J*. 2011; 131 (article in press)

[71] Farhat GN, Taioli E, Cauley JA, Zmuda JM, Orwoll E, Bauer DC, Wilt TJ, Hoffman AR, Beer TM, Shikany JM, Daniels N, Chan J, Fink HA, Barrett-Connor E, Parsons JK, Bunker CH. Osteoporotic Fractures in Men (MrOS) Study Group. The association of bone mineral density with prostate cancer risk in the Osteoporotic Fractures in Men (MrOS) Study. *Cancer Epidemiol. Biomarkers Prev*. 2009; 18: 148-154.

[72] Murphy et al. Management of Postmenopausal Osteoporosis. *JAOA*. 2003; 103: 6-11

[73] Smith MR, McGovern FJ, Zietman AL, Fallon MA, Hayden DL et al. Pamidronate to prevent bone loss during androgen-deprivation therapy for prostate cancer. *N. Engl. J. Med*. 2001; 13:948-955.

[74] Ishizaka K, Machida T, Kobayashi S, Kanbe N, Kitahara S et al. Preventive effect of risedronate on bone loss in men receiving androgen-deprivation therapy for prostate cancer. *Int. J. Urol.* 2007; 12:1071-1075.

[75] Planas J, Trilla E, Raventós C, Cecchini L, Orsola A, Salvador C, Placer J, Encabo G, Morote J. Alendronate decreases the fracture risk in patients with prostate cancer on androgen-deprivation therapy and with severe osteopenia or osteoporosis. *BJU Int.* 2009;104(11):1637-40.

[76] Smith MR, Eastham J, Gleason DM, et al. Randomized controlled trial of zoledronic acid to prevent bone loss in men receiving androgen deprivation therapy for nonmetastatic prostate cancer. *J. Urol.* 2003;169(6):2008-12.

[77] Michaelson MD, Kaufman DS, Lee H, et al. Randomized controlled trial of annual zoledronic acid to prevent gonadotropin-releasing hormone agonist-induced bone loss in men with prostate cancer. *J. Clin. Oncol.* 2007;25(9):1038-42.

[78] Satoh T, Kimura M, Matsumoto K, Tabata K, Okusa H, Bessho H, Iwamura M, Ishiyama H, Hayakawa K, Baba S. Single infusion of zoledronic acid to prevent androgen deprivation therapy-induced bone loss in men with hormone-naive prostate carcinoma. *Cancer.* 2009;115(15):3468-74.

[79] Migliorati CA, Siegel MA, Elting LS. Bisphosphonate-associated osteonecrosis: a long-term complication of bisphosphonate treatment. *Lancet Oncol.* 2006;7(6):508-1

[80] Wadhwa VK, Weston R, Parr NJ. Frequency of zoledronic acid to prevent further bone loss in osteoporotic patients undergoing androgen deprivation therapy for prostate cancer. *BJU Int.* 2009, 13. [Epub ahead of print]

[81] Smith MR, Egerdie B, Hernández Toriz N, et al; Denosumab HALT Prostate Cancer Study Group. Denosumab in men receiving androgen-deprivation therapy for prostate cancer. *N. Engl. J. Med.* 2009;361(8):745-55.

[82] Halpern DF (2000), *Sex differences in cognitive abilities* (3rd Ed) Lawrence Erlbaum, Hillsdale (NJ)

[83] Collaer ML, Hines M (1995) Human behavioral sex differences: A role for gonadal hormones during early development? *Psycholl. Bull.* 118:55-107

[84] O´Connor DB, Archer J, Hair WM, Wu FCW (2001) Activational effects of TST on cognitive function in men. *Neuropsychologia.* 39:1385-1394

[85] Simonson E, Kearnes Wm, Enzer N (1941) Effect of oral administration of methyltestosterone on fatigue in eunuchoids and castrates. *Endocrinilogy.* 28:506-512

[86] Duker H (1957) *Leistungfahigkeit und Keimdrusenhormone.* Barth, Munchen

[87] Slabbekoorn D, van Goozen SHM, Megens J, Gooren LJG (1999) Activating effects of cross-sex hormones on cognitive functioning: A study of short term and long term hormone effects in transsexuals. *Psychoneuroendocrinology.* 24:423-447

[88] Klaiber EL, Broverman DM, Vogel W, Abraham GE, Cone FL (1971) Effects of infused TST on mental performances and serum LH. *J. Clin. Endrocrinol. Metab.* 32:341-349

[89] Cherrier MMM. Androgens and cognitive function. *J. Endocrinologic. Invest.* 2005; 28 (3 suppl): 65–75.

[90] Yaffe K. Testosterone and the brain

[91] Yaffe K,Grady D,Pressman A,Cummings S. Serum estrogen levels, cognitive performance, and risk of cognitive decline in older community women. *J. Am. Geriatr. Soc.* 1998; 46: 816–821.

[92] Wefel JS,Kayl AE,Meyers CA. Neuropsychological dysfunction associated with cancer and cancer therapies: a conceptual review of an emerging target. *Br. J. Cancer.* 2004; 90: 1691–1696.

[93] Yaffe K,Lui LY,Zmuda J,Cauley J. Sex hormones and cognitive function in older men. *J. Am. Geriatr. Soc.* 2002; 50: 707–712.

[94] Green HJ,Pakenham KI,Headley BC, et al. Altered cognitive function in men treated for prostate cancer with luteinizing hormone-releasing hormone analogues and cyproterone acetate: a randomized controlled trial. *BJU Int.* 2002; 90: 427–432

[95] Salminen EK,Portin RI,Koskinen A,Helenius H,Nurmi M. Associations between serum testosterone fall and cognitive function in prostate cancer patients. *Clin. Care Res.* 2004; 10: 7575–7582.

[96] Jenkins VA,Bloomfield DJ,Shilling VM,Edginton TL. Does neoadjuvant hormone therapy for early prostate cancer affect cognition? Results from a pilot study. *BJU Int.* 2005; 96: 48–53.

[97] Joly F,Alibhai SM,Galica J, et al. Impact of androgen deprivation therapy on physical and cognitive function, as well as quality of life of patients with nonmetastatic prostate cancer. *J. Urol.* 2006; 176 (6 pt 1): 2443–2447

[98] Basaria S, Wahlstrom JT, Dobs AS 2001 Anabolic-androgenic steroid therapy in the treatment of chronic diseases. *J. Clin. Endocrinol. Metab.* 86:5108–5117

[99] Basaria S, Lieb 2nd J, Tang A, DeWeese T, Carducci M, Eisenberger M, Dobs AS 2002 Long-term effects of androgen deprivation therapy in prostate cancer patients. *Clin. Endocrinol.* (Oxf) 56:779–786

[100] Chen Z, Maricic M, Nguyen P, Ahmann FR, Bruhn R, Dalkin BL 2002 Low bone density and high percentage of body fat among men who were treated with androgen deprivation therapy for prostate carcinoma. *Cancer.* 95:2136–2144

[101] Haseen F, Murray LJ, Cardwell CR, O'Sullivan JM, Cantwell MM. The effect of androgen deprivation therapy on body composition in men with prostate cancer: systematic review and meta-analysis. *J. Cancer Surviv.* 2010;4(2):128-39

[102] Smith JC, Bennett S, Evans LM, Kynaston HG, Parmar M, Mason MD, Cockcroft JR, Scanlon MF, Davies JS2001 The effects of induced hypogonadism on arterial stiffness, body composition, and metabolic parameters in males with prostate cancer. *J. Clin. Endocrinol. Metab.* Sep;86(9):4261-7.

[103] Haffner SM, Shaten J, Stern MP, Smith GD, Kuller L 1996 Low levels of sex hormone-binding globulin and testosterone predict the development of non-insulin-dependent diabetes mellitus in men. MRFIT Research Group. Multiple Risk Factor Intervention Trial. *Am. J. Epidemiol.* 143:889–897

[104] Haffner SM, Valdez RA, Mykkanen L, Stern MP, Katz MS 1994 Decreased testosterone and dehydroepiandrosterone sulfate concentrations are associated with increased insulin and glucose concentrations in nondiabetic men. *Metabolism.* 43:599–603

[105] Laaksonen DE, Niskanen L, Punnonen K, Nyyssönen K, Tuomainen TP, Valkonen VP, Salonen R, Salonen JT2004 Testosterone and sex hormone-binding globulin predict the metabolic syndrome and diabetes in middle-aged men. *Diabetes Care.* 27:1036–1041

[106] Pitteloud N, Hardin M, Dwyer AA, Valassi E, Yialamas M, Elahi D, Hayes FJ 2005 Increasing insulin resistance is associated with a decrease in Leydig cell testosterone secretion in men. *J. Clin. Endocrinol. Metab.* 90:2636–2641

[107] Marin P, Holmang S, Jonsson L, Sjöström L, Kvist H, Holm G, Lindstedt G, Björntorp P 1992 The effects of testosterone treatment on body composition and metabolism in middle-aged obese men. *Int. J. Obes. Relat. Metab. Disord.* 16:991–997

[108] Smith JC, Bennett S, Evans LM, Kynaston HG, Parmar M, Mason MD, Cockcroft JR, Scanlon MF, Davies JS2001 The effects of induced hypogonadism on arterial stiffness, body composition, and metabolic parameters in males with prostate cancer. *J. Clin. Endocrinol. Metab.* 86:4261–4267

[109] Dockery F, Bulpitt CJ, Agarwal S, Donaldson M, Rajkumar C 2003 Testosterone suppression in men with prostate cancer leads to an increase in arterial stiffness and hyperinsulinaemia. *Clin. Sci.* (Lond) 104:195–201

[110] Smith MR, Lee H, Nathan DM 2006 Insulin sensitivity during combined androgen blockade for prostate cancer. *J. Clin. Endocrinol. Metab.* 91:1305–1308

[111] Keating NL, O'Malley AJ, Smith MR 2006 Diabetes and cardiovascular disease during androgen deprivation therapy for prostate cancer. *J. Clin. Oncol.* 24:4448–4456

[112] Mohamedali HZ, Breunis H, Timilshina N, Alibhai SM. Changes in blood glucose and cholesterol levels due to androgen deprivation therapy in men with non-metastatic prostate cancer. *Can. Urol. Assoc. J.* 201;5(1):28-32.

[113] Basaria S, Muller DC, Carducci MA, Egan J, Dobs AS. Hyperglycemia and insulin resistance in men with prostate carcinoma who receive androgen-deprivation therapy. *Cancer.* 2006;106(3):581-8.

[114] Derweesh IH, Diblasio CJ, Kincade MC, Malcolm JB, Lamar KD, Patterson AL, Kitabchi AE, Wake RW. Risk of new-onset diabetes mellitus and worsening glycaemic variables for established diabetes in men undergoing androgen-deprivation therapy for prostate cancer. *BJU.* Int. 2007;100(5):1060-5.

[115] Haffner SM, Mykkanen L, Valdez RA, Katz MS 1993 Relationship of sex hormones to lipids and lipoproteins in nondiabetic men. J. *Clin. Endocrinol. Metab.* 77:1610–1615

[116] Malkin CJ, Pugh PJ, Jones RD, Kapoor D, Channer KS, Jones TH 2004 The effect of testosterone replacement on endogenous inflammatory cytokines and lipid profiles in hypogonadal men. *J. Clin. Endocrinol.* Metab. 89:3313–3318

[117] Smith MR, Finkelstein JS, McGovern FJ, et al. Changes in body composition during androgen deprivation therapy for prostate cancer. *J. Clin. Endocrinol. Metab.* 2002;87:599–603.

[118] Dockery F, Bulpitt CJ, Agarwal S, et al. Testosterone suppression in men with prostate cancer leads to an increase in arterial stiffness and hyperinsulinaemia. *Clin. Sci.* (Lond) 2003;104:195–201

[119] Eri LM, Urdal P, Bechensteen AG. Effects of the luteinizing hormone-releasing hormone agonist leuprolide on lipoproteins, fibrinogen and plasminogen activator inhibitor in patients with benign prostatic hyperplasia. *J. Urol.* 1995;154:100–104

[120] Eckel RH, Grundy SM, Zimmet PZ. The metabolic syndrome. *Lancet.* 2005;365: 1415 –1428

[121] Expert Panel on Detection, Evaluation and Treatment of High Blood Cholesterol in Adults. Executive summary of the third report of the National Cholesterol Education Program (NCEP) Expert Panel on Detection, Evaluation, and Treatment of High Blood Cholesterol in Adults (Adult Treatment Panel III). *JAMA.* 2001;285: 2486 –2497

[122] Muller M, Grobbee DE, den Tonkelaar I, Lamberts SW, van der Schouw YT. Endogenous sex hormones and metabolic syndrome in aging men. *J. Clin. Endocrinol. Metab.* 2005; 90: 2618 –2623

[123] Keating NL, O'Malley AJ, Smith MR. Diabetes and cardiovascular disease during androgen deprivation therapy for prostate cancer. *J. Clin. Oncol.* 2006;24:4448–4456.

[124] Saigal CS, Gore JL, Krupski TL, et al. Androgen deprivation therapy increases cardiovascular morbidity in men with prostate cancer. *Cancer.* 2007;110:1493–1500.

[125] Alibhai SM, Duong-Hua M, Sutradhar R, et al. Impact of androgen deprivation therapy on cardiovascular disease and diabetes. *J. Cli. Oncol.* 2009;27:3452–3458

[126] Tsai HK, D'Amico AV, Sadetsky N, et al. Androgen deprivation therapy for localized prostate cancer and the risk of cardiovascular mortality. *J. Natl. Cancer Inst.* 2007;99:1516–1524

[127] D'Amico AV, Denham JW, Crook J, et al. Influence of androgen suppression therapy for prostate cancer on the frequency and timing of fatal myocardial infarctions. *J. Clin. Oncol.* 2007;25:2420–2425

[128] Efstathiou JA, Bae K, Shipley WU, et al. Cardiovascular mortality and duration of androgen deprivation for locally advanced prostate cancer: analysis of RTOG 92-02. *Eur. Urol.* 2008;54:816–823.

[129] Roach M, III, Bae K, Speight J, et al. Short-term neoadjuvant androgen deprivation therapy and external-beam radiotherapy for locally advanced prostate cancer: long-term results of RTOG 8610. *J. Clin. Oncol.* 2008;26:585–591

[130] Efstathiou JA, Bae K, Shipley WU, et al. Cardiovascular mortality after androgen deprivation therapy for locally advanced prostate cancer: RTOG 85-31. *J. Clin. Oncol.* 2009;27:92–99

In: Androgens: Production, Functions and Disorders
Editors: B. F. Thompson and D. J. Robinson
ISBN: 978-1-62081-145-0

Chapter III

The Role of Testosterone in Male Erectile Dysfunction

Ross P. Fowler, Osama S. Al-Bermani and Eric Chung
Department of Urology, Princess Alexandra Hospital, Brisbane QLD, Australia

1. Erectile Dysfunction

Erectile dysfunction (ED) is defined as the persistent inability to attain and maintain an erection sufficient to permit satisfactory sexual performance [1]. It is estimated that 5-34% of men aged over the age of 40 have ED and the prevalence increases with increasing age [2-5]. The risk factors for ED are similar to those of cardiovascular disease and metabolic syndrome [6, 7].

The pathophysiology of ED can be categorised as:

Causes of Erectile Dysfunction	
Vasculogenic	Cardiovascular disease, hypertension, diabetes, hyperlipidaemia, smoking, pelvic radiotherapy, prostatectomy
Neurogenic	Central cause – Multiple Sclerosis, Parkinson's Disease, stroke, tumours, disc disease, spinal cord disorders Peripheral cause –alcoholism, uraemia, polyneuropathy, surgery to prostate or retroperitoneum
Anatomical	Peyroine's disease, penile fracture, congenital curvature, micropenis, hypospadias
Hormonal	Hypogonadism, hyperprolactinaemia, hyper/hypothyroidism, Cushing's disease
Drug Induced	Diuretics, beta blockers, calcium channel blockers, antidepressants, antipsychotics, anti-androgens, antihistamines, metoclopramide, nicotine, alcohol, marajuana
Psychogenic	Generalised type (lack of arousability and disorders of sexual intimacy) Situational type (partner related, performance related issues or distress)

Vascular disease is the most common cause of ED while disorders of the endocrine system are the rarest; however within this category hypogonadism is the most common.

2. Definitions

2.1. Total Testosterone

The total sum of bound and unbound testosterone found in the serum. About two-thirds of testosterone circulates in the blood bound to sex-hormone binding protein and slightly less than one-third is bound to albumin and a small percentage (about 1-4%) circulates as free testosterone.

2.2. Free Testosterone

This is the 1-4% of testosterone circulating in blood that is totally unbound. It is the amount of testosterone that is in its active form.

2.3. Bio-Available/Bioactive Testosterone

This is the free testosterone plus the albumin bound testosterone. The fraction bound to albumin is only loosely bound and can dissociate easily and contribute to the active form.

3. Hypogonadism

Male hypogonadism is defined as the decreased ability to maintain adequate testosterone production and sperm production. There can be failure of one or both of these functions.

Primary hypogonadism is described as disease of the testis. The patient has primary hypogonadism if the serum testosterone concentration and/or the sperm count are below normal and the serum leutenizing hormone (LH) and/or follicle stimulating hormone (FSH) concentrations are above normal. On the other hand, *secondary hypogonadism* is related to the disease of the pituitary or hypothalamus. This occurs in the setting of subnormal serum testosterone concentration and/or the sperm count with normal or reduced serum LH and/or FSH concentrations.

The clinical features of hypogonadism, along with the demonstration of a morning serum total testosterone (TT) level below the reference range for young male adults are generally used in the diagnosis of androgen deficiency in men [8]. This will be discussed later in this chapter. Serum LH and FSH level help further delineate primary from secondary hypogonadism. The specific causes of hypogonadism are described elsewhere.

3.1. Late Onset Hypogonadism

There is a gradual age related decline in serum total and free testosterone levels in healthy men [9]. This decline in testosterone level has been previously described as *Andropause* or male menopause, however, it is now more correctly referred to as either of the following terms: Late-onset Hypogonadism (LoH); Age Associated Testosterone Syndrome (TDS); Androgen Decline in the Aging Male (ADAM) or Partial Androgen Deficiency of the Aging Male (PADAM). These terms will be used interchangeably throughout this chapter [10, 11]. ADAM is defined as *a clinical and biochemical syndrome associated with advancing age and characterized by a deficiency in serum testosterone levels (below the young healthy adult male reference range) with or without a decreased genomic sensitivity to androgens. It may result in significant alterations in the quality of life and adversely affect the function of multiple organ systems* [10, 12].There is a simultaneous decrease in the serum levels of other hormones, such as growth hormone, melatonin and dehydroepiandrosterone (DHEA) and an increase in sex hormone binding globulin (SHBG) [11]. The syndromes primary characteristics are listed in Table 1 [10]. The most common symptom associated with hypogonadism is low libido or sexual desire [13].

Table 1. Characteristics of ADAM

1	Diminished sexual libido/desire and erection quality, particularly nocturnal erections
2	Changes in mood with concomitant decreases in intellectual activity, spatial orientation ability, fatigue, depressed mood and irritability
3	Decrease in lean body mass with associated diminution in muscle volume and strength
4	Decrease in body hair and skin alterations
5	Decreased bone mineral density resulting in osteopenia and osteoporosis
6	Increase in visceral fat

All manifestations need not be present; the severity of these may not correlate with each other. One or more of these must corroborate with a low serum testosterone level.

Unlike the abrupt decline of oestrogen levels in females at menopause, healthy males experience an age related decline in plasma concentrations of bioactive androgens. Though the exact pathogenesis and mechanisms remain unclear, the following key factors play an important role:

- an age related decline in production and secretion of testosterone by the testes and decreased Leydig cell function [14]
- an increase in serum sex hormone-binding globulin (SHBG) [15, 16]
- blunted circadian pattern of non SHBG testosterone with an associated steady lower 24 hour level of androgens [17]
- decreased LH pulse amplitude, but not frequency [18, 19]

3.2. Hypogonadism Prevalence

As a rule of thumb, the mean serum testosterone decreases approximately 1% per year after the age 50 [11]. However there are significant inter-individual differences on the serum testosterone level depending on the type and method of testosterone measurement.

Biochemical hypogonadism, as measured by total serum testosterone, is detected in only 7% of men younger than 60 years but increases to 20% in those older than 60 and 35% in those older than 80 [19]. One could therefore argue that only a minority of men develop LoH, however these numbers do not take into account the increase in sex hormone-binding globulin

(SHBG) which has been shown to increase with age [15]. An increase in SHBG results in a further decrease in *bio-available testosterone* (free plus albumin bound testosterone). When the diagnosis is based on the measurement of *bio-available testosterone* the number of men over the age of 60 found to be hypogonadal increases to 70% [20].

4. Relationship between Testosterone and ED

Erectile dysfunction shares many of its risk factors with cardiovascular disease and the metabolic syndrome [6, 7]. Similarly, there is a bidirectional relationship between testosterone deficiency and the metabolic syndrome as shown in Figure 1 [21]. It can be inferred therefore that testosterone not only has a direct effect erectile mechanism, which will be demonstrated below, but also plays a complex role in the pathophysiology of the major risk factors which cause erectile dysfunction.

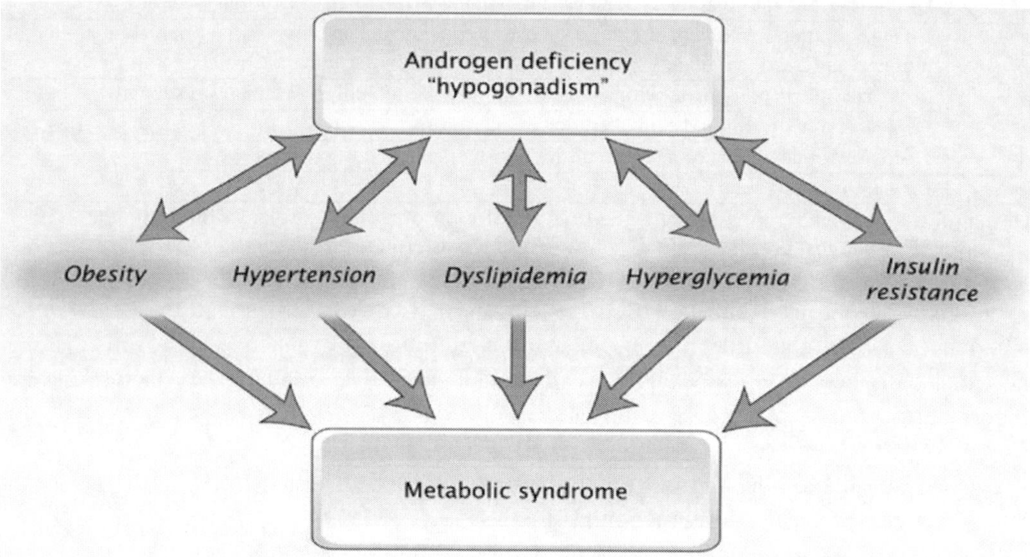

From: Abdulmaged M. Et al. The American Journal of Medicine, Vol 124, No 7, July 2011.

Figure 1. The relationship between testosterone deficiency and metabolic syndrome.

Erections are androgen dependent, evidence of this is the reduction in the frequency, amplitude and rigidity of erections and nocturnal erections in men with marked hypogonadism [22, 23].

Testosterone also influences sexual behaviour and attitudes as evidenced by decreased levels of sexual desire in men with lower testosterone levels than men with maintained sexual desire and normal testosterone levels [24]. There is also practice of autoeroticism is higher in males with higher testosterone levels than those with lower[24].

5. Physiology of Erection

Penile erection requires intact vascular, neuronal, hormonal and psychological systems. Sexual dysfunction manifests as erectile dysfunction when there is a pathology or imbalance in one or more of these interdependent systems.

Erection is triggered initially by a neurological input. This can be in the form of psychogenic or tactile reflex neural input. Psychogenic erections originate in the response to erotic visual, auditory or fantasy stimuli. They are relayed from the CNS to the thoracolumbar erection centre (T11 to L2), and corpora carvenosa. Tactile reflex erections are a simple reflex arc involving the sacral roots of S2- S4 in response to mechanical stimuli of the penis or genital area.

Acquisition and maintenance of the penis is a vascular event. Nitric oxide is responsible for the relaxation of the smooth muscle of the trabecular through increase in cavernosal blood flow and subsequent veno-occlusive mechanism by the compression of the emissary veins. This increases the intracavernosal pressure and erection of the penis occurs. However, a "normal" hormonal milieu has to be present for penile erection to occur.

6. Influence of Testosterone on Penile Erection

The mechanisms of penile erection have been widely and many potential mediators hypothesized to act either peripherally at the cavernosal tissue or centrally in the limbic system and hypothalamus to elicit and maintain an erection. The key central structures which are involved in male sexual response are the medial amygdale, medial preoptic area, paraventricular nucleus, the peri-aquaductal grey matter, and the ventral tegmentum [25, 26]. These higher centres express the androgen receptor and when stimulated increase the likelihood of sexual behaviour given favourable psychogenic and erogenic input [27]. The spinal cord then conveys these responses to the carvenosal tissues as well as having its own erectile centres which elicit erections as a tactile reflex [28]. The mechanisms of many of the involved mediators are beyond the scope of this chapter but the influences they have on penile erection are demonstrated in Figure 2. Below [26].

6.1. NOS/cGMP Pathway

The current accepted end pathway of penile erection is that of the *nitric oxide synthase/cyclic guanasine monophosphate (NOS/cGMP)* pathway. Nitric oxide (NO) mediates relaxation of the cavernosal trabecular and vascular smooth muscle of the arteries of the corpus cavernosum [29]. It is released from non-adrenergic non-cholinergic (NANC) neurons.

The NO is generated from its precursor amino acid, L-arginine by the enzyme NOS. NO diffuses locally to vascular smooth muscle where it binds to the haem moiety of guanylyl cyclase causing a conformational change and therefore activating it [30]. Activated guanylyl cyclase (GC) catalyses GTP to cGMP which then acts on protein kinases (PKG), specifically a subtype PKGIa commonly found in the walls of central and helicine arteries and trabecular smooth muscle of the carvenosal spaces, to cause intracellular calcium efflux and ultimately smooth muscle relaxation [26]. NO also works via other second messengers (such as prostaglandin E1) resulting in synthesis of cyclic adenosine monophosphate (cAMP) and an increase in calcium efflux [31].

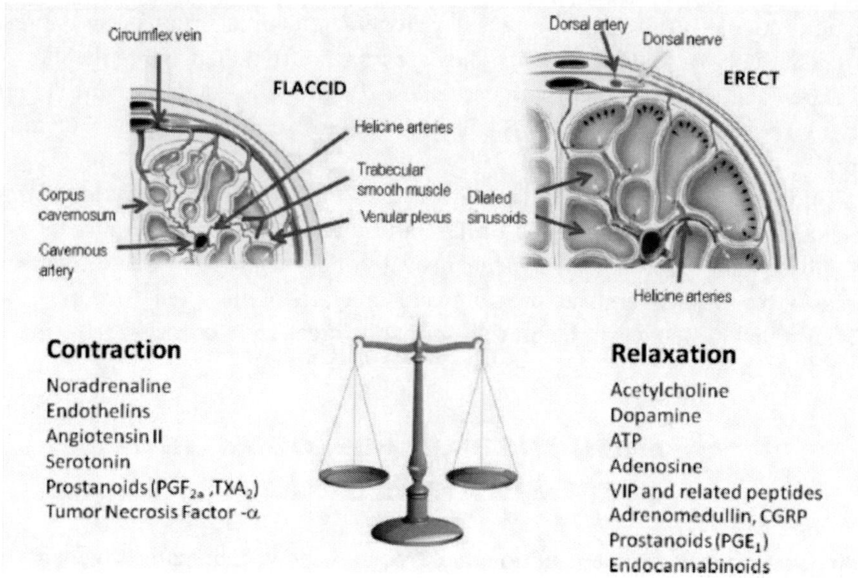

Figure 2.

6.2. Source of NO

1) Neural Nitric Oxide Synthase (nNOS) is found in high quantities in the major pelvic ganglion (inferior hypogastric plexus) and also in the penis and neurovascular structures [29]. NO is produced and released in these efferent neural pathways in response to erogenic stimuli and acts as a post-ganglionic neurotransmitter. NO locally diffuses to vascular and smooth musculature of the penis and results in vasodilation and tissue relaxation, resulting in penile tumescence. The nNOS in the penis has been found to have two distinct isoforms in rat and mouse studies, an alpha and a beta component [32]. The alpha component is more potent in NO formation at nerve terminals with the beta component produces about 10% of the normal NO levels in in-vitro studies [33].
2) Endothelial Nitric Oxide Synthase (eNOS) is an isoform that is biochemically distinct from nNOS. Endothelium of the penis liberates NO under acetylcholine stimulation. It is most likely an auxiliary source of NOS and removal of it does not inhibit relaxation of carvenosal tissue when nNOS is liberated[26].

A model of existing evidence outlining the interactions between the differing NOS isoenzymes has suggested that nNOS *initiates* the erectile response which is then *maintained* by eNOS [26]. The eNOS is activated by shear forces and second messenger pathways as shown in Figure 3.

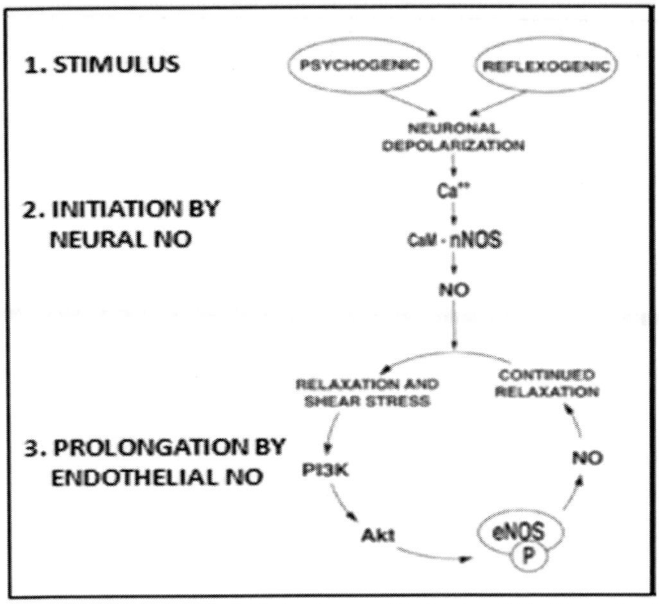

Modified from Hurt KJ, Musicki B, Palese MA, Crone JK, Becker RE, Moriarity JL, Snyder SH, and Burnett AL (2002) Akt-dependent phosphorylation of endothelial nitric-oxide synthase mediates penile erection. *Proc Natl Acad Sci USA* **99:**4061–4066.

Copyright © 2002 National Academy of Sciences, USA Treatment of Erectile Dysfunction K.-E. Andersson. *Pharmacological Reviews*. Vol. 63, No. 4 pg 811-859.

Figure 3.

6.3. Phosphodiesterases (PDE)

PDEs catalyse the hydrolysis of second messengers cAMP and cGMP, and play a central role in smooth muscle tone regulation [34]. There are more than 50 different isoforms identified all of which differ in primary structure, specificity and tissue distribution [30]. Thirteen different isoforms have been located in human cavernosal tissue including PDE type 5 which is found in high concentration and is probably the most functionally important. Only PDE5, 6 and 9 have a 100 fold substrate preference for cGMP over cAMP and are said to be cGMP-specific PDEs [26, 30].

Traish et.al 2007 outlined how androgens modulate erectile physiology at cellular, molecular and physiological levels Figure 4 [35].

- **Testosterone regulates nerve structure and function**
  Testosterone plays a role in maintaining function in pelvic ganglion neurons and maturation of terminal axon density. Testosterone therapy restores nerve fibres and myelin sheath structures to castrated rats. Stimulation of the medial preoptic area in rats elicits erection and this process is testosterone dependent. Testosterone acting on the spinal cord enhances erectile response of the cavernous nerve.

- **Testosterone regulates NOS expression and activity**
  Androgens mediate erectile function via the NO/cGMP pathway. Studies including castrated animal and those placed on androgen deprivation therapy show reduced penile NOS activity. Also older rats have reduced NOS containing nerves, NOS mRNA expression and NOS activity compared to younger rats [26]. Testosterone administration in castrated animals restores NOS expression and erectile response has been demonstrated in multiple studies. Erectile dysfunction in the setting of hypogonadism induced by diabetes and metabolic syndrome in rabbit and rat models showed decreased eNOS and nNOS expression, however this was reversed with testosterone administration [36, 37].
- **Testosterone regulates PDE5**
  Androgens up regulate both NOS and PDE5 activity paradoxically, maintaining a homeostatic environment with a relative constant ratio of critical enzymes for this pathway [38].

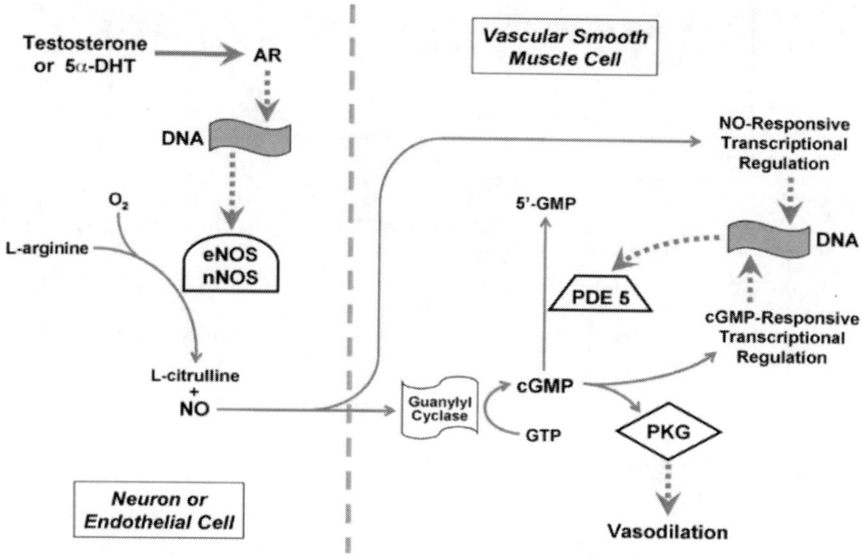

Figure 4. Potential regulation of nitric oxide synthase (NOS) and phosphodiesterase (type) 5 (PDE5) by androgens. Hypothetical mechanism by which androgens may up-regulate both NOS and PDE5 proteins.

- **Testosterone regulates cell growth and differentiation**
  Androgen deprivation results in disorganisation and reduction of trabecular smooth muscle and increased connective tissue deposition and fibrosis. These alterations in cavernosal tissue composition and structure were accompanied by reduced erectile response to nerve stimulation. Androgens seem to be critical for promoting myogenic lineage of penile pleuri-potent stem cell and inhibit their differentiation into adipocyte lineage [39]. These smooth muscle cells provide the structural and functional framework that is needed for penile erection.

7. Making the Diagnosis of Hypogonadism and Working up Erectile Dysfunction

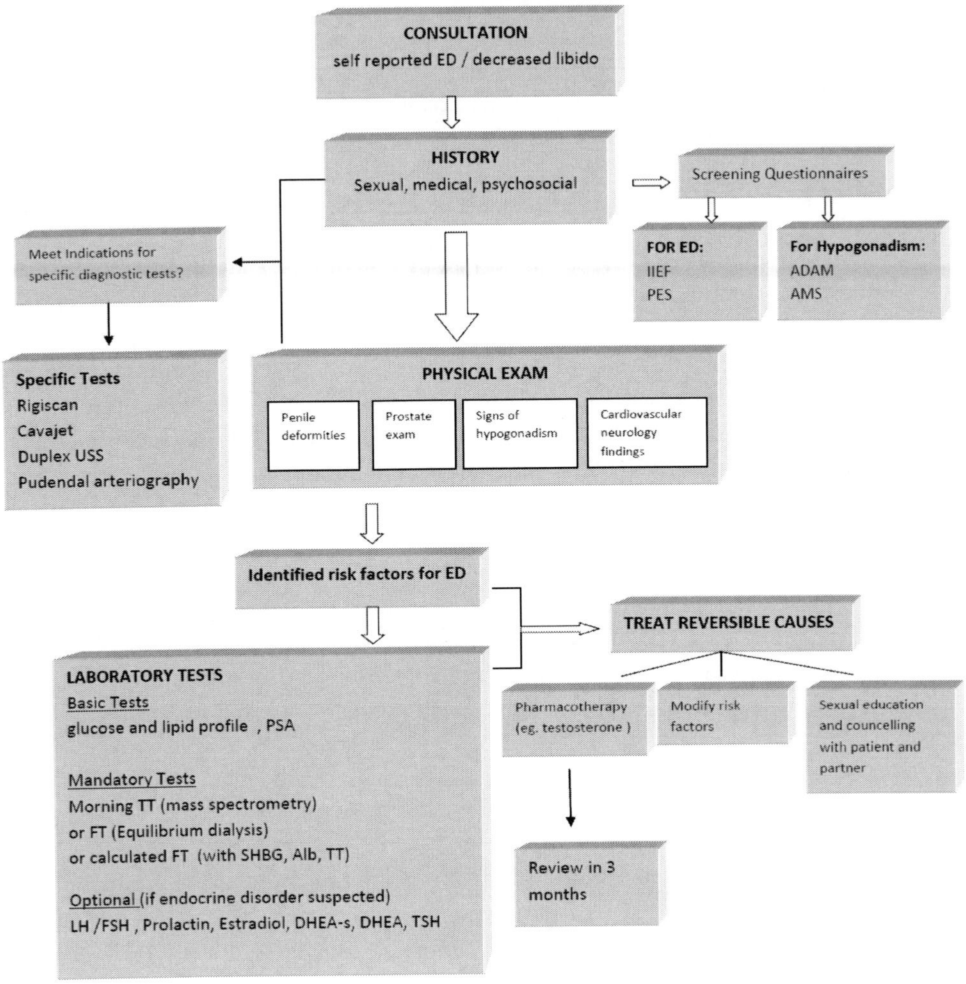

Figure 5. Diagnostic pathway for ED in hypogonadism.

For the sake of this of this chapter and the relationship between hypogonadism with erectile dysfunction the term hypogonadism will infer testosterone deficiency (TD). The current guidelines formed by the collaboration of the International Society of Andrology (ASA), International Society for the Study of Aging Male (ISSAM), European Association of Urology (EAU),European Academy of Andrology (EAA) and American Society of Andrology (ASA) published in 2008 are discussed below and summarised in Figure 5 [12].

All men with signs of hypogonadism or self reported sexual dysfunction or diminished libido should be investigated for testosterone deficiency and erectile dysfunction.

There are several tools available to screen for hypogonadism and others to assess the severity of ED. They are discussed in more detail below.

Screening Tools and Questionnaires
For Hypogonadism
• ADAM - Androgen Deficiency in Aging Males • AMS - Aging Males Symptom rating score • ANDROTEST
For Erectile Dysfunction
• IIEF - International Index of Erectile Function • SEP - Sexual Encounter Profile

Patient History

1) Take a focused history of issues important in assessing sexual dysfunction.
2) Differentiate between decreased libido and ED. Decreased libido is primary a testosterone deficiency factor whilst ED has a range of causes.
3) Inquire about a history of systemic disease like diabetes, hypertension, hypercholesterolemia, thyroid problems or problems with weight control. Tobacco and chronic alcohol intake should be assessed. History of psychiatric disorders like depression, anxiety or other mental health disorders requiring the use of antipsychotic medications. Medications such as beta blockers, over-the-counter medications like antihistamines and herbal drugs, as well as recreational drug use of marijuana, could play a part in ED.
4) Identify any social or interrelationship stressors.
5) Lastly enquire about specific sexual function like whether nocturnal erections occur. Try to quantify the rigidity and duration of erotic and morning erections and ask about problems with arousal, ejaculation and orgasm.

Physical Exam

The Examination should be focused on the genitourinary, vascular, endocrine and neurological exams. Testicular size, asymmetry and presences of varicoceles or hernias should be assessed. Penile palpation and inspection for plaques congenital deformities or curvatures is essential. A digital prostate exam in all men over 50 years old should be performed. Blood pressure and heart rate should be measured as well as palpation and auscultation of the femoral arteries for possible bruits or signs of peripheral vascular disease.

Laboratory Studies

The work up for ED and hypogonadism should include morning serum total testosterone or ideally free testosterone level. If free testosterone is not available then calculated testosterone is adequate. Fasting lipid and glucose profiles and PSA should be repeated if not done within the previous 12 months. Other endocrine markers should not be first line test unless an endocrine cause is suspected. These tests are further explored later in this chapter.

Specific Diagnostic Tests

If no reversible causes can be identified then specific diagnostic tests may be required [3].

Indications for specific diagnostic tests
Primary erectile disorder (not caused by organic disease or psychogenic disorder)
Young patients with a history of pelvic or perineal trauma who could benefit from potentially curative vascular surgery
Patients with penile deformities that might require surgical correction, e.g. Peyronie's disease, congenital curvature
Patients with complex psychiatric or psychosexual disorders
Patients with complex endocrine disorders

Once the indications for specific diagnostics tests are met there are a number of tests listed below that may lead to a diagnosis

Specific diagnostic tests are
Nocturnal penile tumescence and rigidity (NTPR) using Rigiscan®
Vascular studies Intracavernous vasoactive drug injection Duplex ultrasound of the cavernous arteries Dynamic infusion cavernosometry or cavernosography (DICC) Internal pudendal arteriography
Neurological studies, e.g. bulbocavernosus reflex latency, nerve conduction studies
Endocrinological studies
Specialised psychodiagnostic evaluation

7.1. Questionnaires

The following questionnaires have been used as screening tools for male hypogonadism (Figure 6)

1) The Androgen Deficiency in Aging Males (ADAM) questionnaire
2) Aging Males' Symptoms rating scale (AMS) questionnaire

Although the AMS questionnaire was not developed as a screening tool for androgen deficiency, its results are comparable with other screening questionnaires for androgen deficiency[16, 40].

The ADAM test has sensitivity of 88% and specificity of 60% for *bioavailable* testosterone level in males aged 40-60 yrs [20]. However other studies have found no significant association with ADAM questionnaire and hypogonadal *total* testosterone levels in men 60-80yrs of age[41].

More recently the quantitative ADAM or qADAM has been developed in an attempt to further quantify the severity of hypogonadism. It represents a viable alternative to the ADAM and AMS questionnaire in detecting androgen deficiency and assessing response to treatment[42].

The above questionnaires have previously been widely used in the literature however current guidelines mentioned above do not recommend their use to diagnose hypogonadism due to the tests low specificities [12, 41]. This emphasizes the point that if these tests are used, they are to be interpreted in context with the patients clinical presentation, physical exam and current co-morbidities which may indicate the need for testing biochemical markers.

The ANDROTEST (c) is a specifically designed interview to screen for hypogonadism in patients who already display sexual dysfunction. It has a sensitivity of 71% and specificity of 65% in screening for low free testosterone (<37pmol/L)[43].

The following questionnaires have been used to assess the severity of erectile dysfunction

1) International Index of Erectile dysfunction (IIEF)[44]
2) Sexual Encounter Profile (SEP)[45]

The IIEF is a standardised questionnaire that is used to measure ED and widely used to detect treatment related changes in symptoms of sexual dysfunction. It consists of 15 questions which provide descriptive information on the domains of erectile function, orgasm function, sexual desire, intercourse satisfaction and overall satisfaction. These domains are graded by men on a scale of 1 (never) to 5 (almost always) and the total score is then used to scale ED from normal to severe. It is validated in about 20 languages and has high sensitivity and specificity [44].

The Sexual Encounter Profile is a 5 question survey which is filled out after each sexual attempt by patients undergoing oral therapy studies. The literature is mainly concerned with answers fro question 2 and 3.

7.2. Laboratory Diagnosis

Primary or secondary hypogonadism can occur at all ages especially among the elderly. Patients at risk or those suspected of hypogonadism using the above history should be explored further with thorough physical examination and biochemical work up. Glucose and lipid profile should be determined if not done so in the last 12 months.

Risk factors for hypogonadism should be investigated and treated accordingly. These include diabetes mellitus, obesity, metabolic syndrome, haemachromatosis, chronic obstructive pulmonary disease, inflammatory arthritis, HIV and chronic renal disease.

7.3. Total Testosterone

A serum total testosterone (TT) should be obtained between 7am and 11am.

Testosterone plasma levels have a diurnal pattern with highest levels in the early morning and levels falling by 8pm[17].

Total testosterone measurements can be misleading and must be interpreted with consideration of the following:

- SHBG plasma levels increase with age - only unbound testosterone is active at a cellular level. SHBG binds with high affinity to testosterone so higher values of SHBG will lower the unbound testosterone available despite having normal total testosterone levels.
- They don't account for tissue specific androgen metabolism [12]
- There is variation in androgen sensitivity of target organs between individuals and so a certain testosterone level may be associated with erectile dysfunction in one individual but not in another[12]

Questions Used as Part of the Saint Louis University ADAM Questionnaire

1. Do you have a decrease in libido (sex drive)?
2. Do you have a lack of energy?
3. Do you have a decrease in strength and/or endurance?
4. Have you lost height?
5. Have you noticed a decreased "enjoyment of life"?
6. Are you sad and/or grumpy?
7. Are your erections less strong?
8. Have you noted a recent deterioration in your ability to play sports?
9. Are you falling asleep after dinner?
10. Has there been a recent deterioration in your work performance?

NOTE. A positive questionnaire result is defined as a "yes" answer to questions 1 or 7 or any 3 other questions.

Questions Used as Part of the qADAM Questionnaire

1. How would you rate your libido (sex drive)?
   1(terrible) 2(poor) 3(average) 4(good) 5(excellent)
2. How would you rate your energy level?
   1(terrible) 2(poor) 3(average) 4(good) 5(excellent)
3. How would you rate your strength/endurance?
   1(terrible) 2(poor) 3(average) 4(good) 5(excellent)
4. How would you rate your enjoyment of life?
   1(terrible) 2(poor) 3(average) 4(good) 5(excellent)
5. How would you rate your happiness level?
   1(terrible) 2(poor) 3(average) 4(good) 5(excellent)
6. How strong are your erections?
   (1= extremely weak 5= extremely strong)
   1 2 3 4 5
7. How would you rate your work performance over the past 4 weeks?
   1(terrible) 2(poor) 3(average) 4(good) 5(excellent)
8. How often do you fall asleep after dinner?
   1(never) 2(1-2/week) 3(3-4/week) 4(5-6/week) 5(every night)
9. How would you rate your sports ability over the past 4 weeks?
   1(terrible) 2(poor) 3(average) 4(good) 5(excellent)
10. How much height have you lost?
    1(2" or more) 2(1.5-1.9") 3(1-1.4") 4(0.5-0.9") 5(none-0.4")

Figure 6. The ADAM and qADAM questionnaires.

What is the Lower Level of Testosterone Required to Maintain Erection?

Erectile tissue may have a lower *threshold dose* of 10%of normal physiological systemic free testosterone required to maintain ordinary erectile function. Normal function is maintained over a wide range of systemic plasma concentrations as demonstrated by Armagan et.al who administered varying doses of testosterone to castrated rats [46]. They showed a reduced protein expression of nNOS and Phosphodiesterase type 5 (PDE 5) in penile tissue from orchidectomised rats, however both nNOS and PDE5 increased in testosterone infused animals. Erectile function was restored at levels above one tenth of normal physiological plasma concentrations but declined in a dose-dependent fashion below this level. Despite having this 10% threshold to work with, there has been no real consensus at what exact testosterone level this occurs at in humans [10, 12, 21, 35, 47]. These differing results highlight the difficulty in making clear diagnosis in males with Late Onset Hypogonadism which are often confounded by simultaneous comorbidities and the possibility of undiagnosed disease which may require the measurement of other gonodatropins and hormones.

There is no general agreement on what the lower limit of normal is but the following are all consistent across studies

- TT > 12 nmol/L (350ng/dL) –does not require testosterone substitution
- TT< 8 nmol/L (230ng/dL) – severe hypogonadism and the patient will benefit from testosterone treatment.
- TT 8-12 nmo/L - requires repeat testing of TT with SHBG to calculate the free testosterone or obtain the free testosterone (FT) by equilibrium dialysis
- TT <5.2 nmol/L (150 ng/dL) – check prolactin levels at this level or when secondary hypogonadism is suspected. Elevated prolactin levels may require MRI

Several methods of measuring serum total testosterone are available. Currently methods based on mass spectrometry are the most accurate and precise however immunometric methods are also capable of distinguishing between normal and hypogonadal males[8, 48].

7.4. Free Testosterone

Measurement of *free* (FT) or *bio-available testosterone* should be considered when TT is not diagnostic or is borderline especially in obese males as it can help to confirm the diagnosis of TD. FT levels below 225 pmol/L (655 pg/ml) are indicative to commence testosterone replacement treatment. Methods of measuring free testosterone can vary and equilibrium dialysis is the current gold standard, however it is often difficult, time consuming and is not widely used clinically [35]. FT assays based on analogue displacement assays are widely available but they should not be used as they can give unreliable results [49].

7.5. Calculated Free Testosterone

On the other hand, *calculated free testosterone* which correlates well with FT measured by equilibrium dialysis, can be calculated using Vermuelen's formula [50]. A free testosterone calculator can be found at http://www.issam.ch/freetesto.htm . This calculation uses serum SHBG (nmol/L) levels and TT (ng/dL) and serum albumin (g/dL). Total testosterone should be from a reliable method, unfortunately methodology for assessment of SHBG is not standardised around the world and so calculated free testosterone may vary among different areas of the world. Using this method studies showed an increase in the diagnosis of hypogonadism from 4.8% using TT to over 17% using calculated free testosterone of men with erectile dysfunction [51].

7.6. Other Endocrine Markers

Patients with a *TT* <8 nmol/L (350ng/dL) may benefit from the measurement of serum Luetenizing Hormone (LH), Follicle Stimulating Hormone (FSH) and prolactin. These measurements help ascertain if the hypogonadism is hypogonadotropic or hypergonadotropic. For example, in hypogonadotropic hypogonadism serum LH and FSH levels are low or normal with a low serum testosterone, further evaluation including iron studies, thyroid function tests and serum prolactin (to check for pituitary adenoma) need to be conducted to identify a cause for secondary hypogonadism.

The significance of changes in other endocrine systems with aging is not well understood and current guidelines do not recommend routine determinations of estradiol, thyroid hormones, cortisol, DHES, DHEA-S, melatonin, growth hormone (GH) and insulin-like growth factor (IGF)-I unless other endocrine disorders are suspected [12].

8. Testosterone and Testosterone Replacement Therapy (TRT)

Testosterone (T) is vital for normal functioning throughout a man's life [52]. Signs and symptoms of deficiency in the hypogonadal male include diminished levels of energy, sense of vitality, or sense of well-being, or increased fatigue, as well as depression, reduced muscle mass and strength, reduced bone density, anaemia, frailty, and sexual symptoms such as diminished libido, erectile dysfunction (ED), difficulty achieving orgasm, diminished intensity of the experience of orgasm, and diminished penile sensation.[53, 54] In addition to its effects in restoring physiological levels of T potential benefits of TRT include increases in overall health and survival, strength, sexual desire, energy, emotional well-being, cognition, bone mineral density, glycaemic control, cardiovascular health and erectile function, improvement in some metabolic syndrome symptoms, and reduction in body fat.[52]

ED and hypogonadism are two disease processes often co-exist together. ED prevalence increases as men age, and is strongly associated with common co-morbid conditions of ageing men, such as diabetes and depression [55]. These co-morbidities can increase the prevalence and severity of ED. Over the last few years, there has been a renewed interest in the place of T in ED [56]. Recent studies have provided clear evidence that there are powerful effects of T on the anatomical and physiological substrate of penile erection [35, 57]. T therapy is appropriate only when clinical symptoms and biochemical evidence of hypogonadism co-exist. Hypogonadal men with specific sexual dysfunctions such as ED, diminished libido, or both are candidates for T replacement therapy. T monotherapy is an effective treatment for hypogonadal ED and may correct sexual dysfunction caused by hypogonadism. TRT should be considered the first-line treatment in patients with ED who have hypogonadism and T replacement therapy (TRT) should be started after a definitive diagnosis has been established. It is required and should continue lifetime with appropriate monitoring. The absence, however, of an adequate response after appropriate therapy may require further evaluation to exclude associated co-morbidities, such as those causing vasculogenic or neurogenic ED.[58]

Data derived from studies evaluating the effect of TRT on patients with ED have yielded positive, albeit, mixed results due to the limitations and pitfalls evidences in each study [58, 59]. In a meta-analysis of 17 randomized, placebo controlled clinical trials carried out over the past 30 years, Isidori et al.[60] have shown that, in comparison to placebo, TRT resulted in a significant, but moderate, improvement in all aspects of sexual function (improved erection, ejaculation, orgasm and penile sensation) in men with low or low–normal mean concentrations of T at baseline (<12 nmol/l [<346 ng/dl]) [59]. The magnitude of the effect on erectile function was inversely related to the baseline concentration of T. Hence, the more severe the hypogonadism pre-treatment, the more benefit the patient will gain with TRT [59]. More severe hypogonadism is associated not only with functional defects in the penis (such as impaired regulation of NO and PDE5), but also with structural changes, which could explain why TRT induces such positive effect, in terms of improvement in erectile function, in more severe instances of hypogonadism [18–20, 59].

An evaluation of eight observational studies, which included 258 hypogonadal (total T <10.4 nmol/l [<300 ng/dl]) men who received consultation for ED, reported that TRT improved erectile function in only 36% of cases [80]. Interestingly, these patients were older and their hypogonadism was generally mild (fewer than 2% had total T <7 nmol/l [<200 ng/dl]) when compared to the pool data derived from the meta-analysis described above.

In another meta-analysis focused on the effects of TRT in hypogonadal (total T <10.4 nmol/l [300 ng/dl]) men with sexual dysfunction at baseline, Bolona et al [81] reported similar conclusions. There is generally a lack of data on whether androgen therapy is effective or safe in older men without proven androgen deficiency who do not meet the clinical definition of hypogonadism [59]. Although hypogonadism can be the main cause of ED in younger patients, ED in the elderly comprises only one of many elements in the older patients, in whom ED is often multi-factorial [61-64]. For this reason, combined therapy using T and PDE5 inhibitors is often required for ED in older men (see below).[59].

8.1. Safety, Tolerability and Specific Considerations

Age is not a contraindication to initiate T treatment. Individual assessment of co-morbidities (as possible causes of symptoms) and potential risks versus benefits of T treatment is particularly important in elderly men.[65] When choosing which TRT to prescribe, the primary care physician should consider the availability, safety, tolerability, efficacy and preference of the product. Typically a patient taking TRT properly will perceive a benefit after 3 months [52].

A high prevalence of ED in patients with diabetes has been attributed mainly to vascular and neurologic conditions, but also to hypogonadism, which has been observed to occur commonly with type II diabetes and in middle-aged men with metabolic syndrome [58].

Regular patient follow up is paramount after initiation and continued TRT. Therapy is generally well-tolerated if tailored to the individual patient but can be associated with a wide variety of side-effects including activation side-effects (irritation, redness and rash), acne, changes in mood, depression, sleep disturbance, breast enlargement, hair loss, baldness headache, prolonged or painful erections, aggression or aggressive behaviour, breast pain, weight gain and dizziness. Other potential side-effects include fluid retention, worsening of sleep apnoea and polycythemia.

Many of the components of the metabolic syndrome (obesity, hypertension, dyslipidemia, impaired glucose regulation, and insulin resistance) are also present in hypogonadal men and associated with low plasma T levels. The effects of testosterone administration on glycemic control of men with diabetes mellitus are much less certain [66-68]. It is premature to recommend testosterone treatment for the metabolic syndrome or diabetes mellitus in the absence of laboratory and other clinical evidence of hypogonadism. In men with hypogonadism and diabetes and or the metabolic syndrome, testosterone treatment for traditional hypogonadal symptoms may have other unproven benefits for their metabolic status [65].

There is no conclusive evidence that T therapy increases the risk of prostate cancer or benign prostatic hyperplasia (BPH) [69, 70]. There is also no strong evidence that T treatment will convert subclinical prostate cancer to clinically detectable prostate cancer.[65] There is, however, unequivocal evidence that T can stimulate growth and aggravate symptoms in men with locally advanced and metastatic prostate cancer [71, 72].

Prior to therapy with T, and as part of their medical workup, the risk of prostate cancer must be assessed using by a digital rectal examination (DRE) and determination of serum prostate-specific antigen (PSA) [65]. Pre-treatment prostate ultrasound examinations or biopsies are not recommended as routine requirements unless their cancer risk is deemed sufficiently high. After initiation of T treatment, patients should be monitored for prostate disease at 3–6 months, 12 months, and at least annually thereafter [65]. Men successfully treated for prostate cancer and suffering from confirmed symptomatic hypogonadism are potential candidates for T substitution after a prudent interval if there is no clinical or laboratory evidence of residual cancer. Long-term outcome data are not available. Clinicians must exercise good clinical judgment together with adequate knowledge of advantages and drawbacks of T therapy in this patient group.

Severe lower urinary tract symptoms (LUTS) due to BPH represent a relative contraindication (although there are no compelling data to suggest that T treatment exacerbates LUTS or promotes acute urinary retention). After successful treatment of lower urinary tract obstruction, this contraindication is no longer applicable. Other contraindications to commencing TRT such as breast cancer need to be excluded. Underlying pituitary or testicular disorders should also be treated appropriately [73]. Men with significant erythrocytosis (hematocrit >52%), untreated obstructive sleep apnea, or untreated severe congestive heart failure should not be started on treatment with T without prior resolution of the co-morbid condition [74]. Periodic hematological assessment is indicated to keep hematocrit below 52–55% [74].

8.2. Testosterone and PDE5 Inhibitors (PDE5-I) in ED

The introduction of PDE5 inhibitors in the late 1990s has revolutionized the treatment of ED and the field of sexual medicine. Several clinical trials have demonstrated the efficacy and tolerability of these drugs and they are currently first-line monotherapy for ED [56]. They work by inhibiting PDE5 enzyme, which maintains intracavernosal levels of cGMP, subsequently producing vasodilatation and penile erection. Due to the multi-factorial and complex nature of ED, a significant proportion of patients with complex ED will be therapeutic non-responders to PDE5-I monotherapy [59].

Up to 30% of erectile dysfunction (ED) patients treated with PDE5 inhibitors continue to have poor sexual function, which may be due in part to low serum T [82]. T is not only important for allowing cGMP formation, through a positive modulation of NOS, but also for increasing PDE5 expression in human cavernous tissue [83]. This androgen-induced two step regulation of NO activity and cGMP formation in penile tissue as well as its actions on hypothalamus-spinal dopaminergic and oxitocinergic pathways may be relevant to timely synchronization of penile erections in sexual acts, which are clearly androgen dependent [58,

83]. Therefore, it's easy to speculate that PDE5 inhibitors require that normal levels of androgens are present in order to be effective [59, 84-86].

Results from several studies have suggested that hypogonadism is likely to contribute to a reduced efficacy of PDE5 inhibitors [58, 87-100]. These studies have shown that in men with ED who do not respond to treatment with PDE5 inhibitors alone, the addition of T to restore plasma T to the normal range may be efficacious in treating ED [58, 87-100].

Many studies showed that the combination of TRT and PDE5 inhibitors can lead to an overall efficacy of therapy in the range 34–100% [58, 59, 87-100]. They provide evidence that the combination of PDE-5I with T is more efficacious than either alone. These observations confirm the concept that hypogonadism must be ruled out or, if present, adequately treated, before PDE5 inhibitors are prescribed for ED [59]. Therefore TRT should be considered as the first line treatment in patients with ED who have hypogonadism. Overall, these studies did not identify any additional adverse effects to discourage combined treatment. The safety and efficacy of the three currently available PDE5 inhibitors have been evaluated extensively in patients with ED and concomitant cardiovascular disease, hypertension, hyperlipidemia, or diabetes, with or without additional risk factors. Overall, these studies have shown that there was no evidence of cardiovascular risk from using these agents. However, because ED is known to be a harbinger of cardiovascular events in some men, the presence of ED should prompt investigation and intervention for cardiovascular risk factors. Caution should be taken regarding potential infertility caused by long-term exogenous androgen replacement in young men [55].

The quality of these trials however is questionable due to several limitations such as study design, short follow-up periods and small patient numbers (see Table 2) [55]. Currently there is no sufficient evidence to support routine use of androgen supplementation with PDE5 inhibitor for patients with low to low–normal T levels and ED [55]. Current recommendations suggest therapeutic synergism between TRT and PDE5 inhibitors in hypogondal or borderline eugonadal men [65]. The combination treatment should be considered in hypogonadal patients with ED failing to respond to either treatment alone [65]. In addition, other populations at risk for hypogonadism, such as those with diabetes type II, metabolic syndrome, chronic renal failure and other chronic diseases, are potential candidates who could benefit from combination therapies if response to monotherapy is not sufficient [49, 58].

In the recently published TRiUS trial, which set out to determine if 12-month treatment with a T gel improves sexual function in hypogonadal men, as measured by the Brief Male Sexual Function Inventory (BMSFI), and in subgroups defined by low T, PDE5 inhibitor use, and prior TRT, the results appear promising. In hypogonadal patients, 12-month administration of topical T gel resulted in increased TT and FT levels and significantly improved sexual function. All subgroups studied, including men taking PDE5 inhibitors for ED and those previously on TRT, demonstrated significant improvement in sexual function from baseline scores [82].

Its still unclear whether men with hypogonadism and ED should be treated initially with PDE5 inhibitors, T or the combination of the two [65]. Multi-centre, double-blinded, placebo controlled, longitudinal studies of adequate duration are needed to better clarify whether or not hypogonadism hinders the effects of PDE5 inhibitors [59].

Table 2. Trials of Combined TRT & PDE5-Is in men with ED

Study	Number of patients	Baseline T (nmol/l)	Treatment	Overall efficacy rate
*Aversa et a.[100]l**	20	<13.9 (400ng/dl)	T-patch, sildenafil 100mg	80%
Kalichenki et al.[93]$	120	<11.8 (340ng/dl)	Oral TU, sildenafil 100mg	70%
Shabsigh et al.[90]$*	75	<13.9 (<400ng/dl)	T-Gel, sildenafil 100mg	70%
Chatterjee et a.[97]l	12	9 patients <13.9 (<400ng/dl)	T-im, sildenafil 50-100mg	100%
Foresta et al.[96]	15	<6.94 (<200ng/dl)	T-patch, sildenafil 50mg	Normalised NPT
Shamloul et a.[89]l$	40	<11.8 (<340ng/dl)	Oral TU, Sildenafil 50-100mg	Improvement
Greenstein et al.[95]	31	<13.9 (<400ng/dl)	T-Gel, sildenafil 100mg	63%
Tas et al.[88]	23	<13.9 (<400ng/dl)	T-im, sildenafil 50-100mg	34%
*Rochira et al.[92]**	24	<6.94 (<200ng/dl)	T-im, sildenafil 50mg	Improvement in NPT
Hwang et al.[94]$	32	<10.4 (<300ng/dl)	Oral TU, sildenafil 100mg	57%
Rosenthal et al.[91]$	90	<12.1 (<350ng/dl)	T-Gel, Sildenafil 100mg	92%
Yassin et al.[87]$	69	<11.8 (<340ng/dl)	T-Gel, Tadalafil 20mg	65%
*Buvat et al.[98]**	73	<10.4 (<300ng/dl)	T-Gel, Tadalafil 10mg/day	51%

* Indicates placebo-controlled studies. $Indicates PDE5-I treatment had previously failed. Abbreviations: NPT, nocturnal penile test; T-Gel, testosterone gel; T-im, intramuscular testosterone; T-Patch, testosterone patch; TRT, testosterone replacement therapy; TU, testosterone undecanoate.

Adapted from [59].

References

[1] Lue TF, Giuliano F, Montorsi F, Rosen RC, Andersson KE, Althof S, et al. Summary of the recommendations on sexual dysfunctions in men. *Journal of Sexual Medicine*. 2004;1(1):6-23.

[2] *Slag MF; Morley JE; Elson MK; Trence DL; Nelson CJ; Nelson AE; Kinlaw WB; Beyer HS; Nuttall FQ; Shafer RB*. Impotence in medical clinic outpatients. *Journal of American Medical Association*. 1983;249(13): 1736-40.

[3] Konstantinos Hatzimouratidis a, Edouard Amar b, Ian Eardley c, Francois Giuliano d,, Dimitrios Hatzichristou a FMe, Yoram Vardi f, Eric Wespes. Guidelines on Male Sexual Dysfunction: Erectile Dysfunction and Premature Ejaculation. *European Urology*. 2010;57:804-14. Epub 20 Feb 2010.

[4] Laumann EO, West S, Glasser D, Carson C, Rosen R, Kang JH. Prevalence and correlates of erectile dysfunction by race and ethnicity among men aged 40 or older in the United States: From the Male Attitudes Regarding Sexual Health survey. *Journal of Sexual Medicine*. 2007;4(1):57-65.

[5] Corona G, Lee DM, Forti G, O'Connor DB, Maggi M, O'Neill TW, et al. Age-Related Changes in General and Sexual Health in Middle-Aged and Older Men: Results from the European Male Ageing Study (EMAS). *Journal of Sexual Medicine*. 2010;7(4):1362-80.

[6] Rosen RC, Friedman M, Kostis JB. Lifestyle management of erectile dysfunction: The role of cardiovascular and concomitant risk factors. *American Journal of Cardiology*. 2005;96(12B):76M-9M.

[7] Derby CA, Mohr BA, Goldstein I, Feldman HA, Johannes CB, McKinlay JB. Modifiable risk factors and erectile dysfunction: Can lifestyle changes modify risk? *Urology*. 2000;56(2):302-6.

[8] Measurement of Total Serum Testosterone in Adult Men: Comparison of Current Laboratory Methods Versus Liquid Chromatography-Tandem Mass Spectrometry. *The Journal of Clinical Endocrinology & Metabolism*. 2004;89(2):534-43.

[9] Rajfer J. Relationship Between Testosterone and Erectile Dysfunction. *Reviews in Urology*. 2000;2(2):122-8.

[10] Lunenfeld AMaB. Investigation, treatment and monitoring of late-onset hypogonadism in males. Official Recommendations of ISSAM. *The aging male* 2002;5:74-86.

[11] Alvaro Morales Jpwh, Culley C. Carson Iii. Andropause: A Misnomer For A True Clinical Entity. *The Journal of Urology*. march 2000;163(3):705-12.

[12] Wang C, Nieschlag E, Swerdloff R, Behre HM, Hellstrom WJ, Gooren LJ, et al. Investigation, Treatment, and Monitoring of Late-Onset Hypogonadism in Males: ISA, ISSAM, EAU, EAA, and ASA Recommendations. *European Urology*. 2009;55(1):121-30.

[13] Travison TG, Morley JE, Araujo AB, O'Donnell AB, McKinlay JB. The relationship between libido and testosterone levels in aging men. *Journal of Clinical Endocrinology & Metabolism*. 2006;91(7):2509-13.

[14] Harman SM, Tsitouras PD. Reproductive Hormones In Aging Men .1. Measurement Of Sex Steroids, Basal Luteinizing-Hormone, And Leydig-Cell Response To Human

Chorionic-Gonadotropin. *Journal of Clinical Endocrinology & Metabolism.* 1980;51(1):35-40.

[15] Wu AH, Whittemore AS, Kolonel LN, John EM, Gallagher RP, West DW, et al. Serum Androgens And Sex Hormone-Binding Globulins In Relation To Life-Style Factors In Older African-American, White, And Asian Men In The United-States And Canada. *Cancer Epidemiology Biomarkers & Prevention.* 1995;4(7):735-41.

[16] Kai Sun1 G-QL, Xiang-Feng Chen1, Ping Ping1, Wen-Liang Yao1, Shi-Jun Zhang1, Bo Wang3,, Li Y-HSaZ. Survey for late-onset hypogonadism among old and middle-aged males in Shanghai communities. Asian Journal of Andrology. 2011;10(advance online publication, 28 November 2011; doi:10.1038/aja.2011.171):171.

[17] Stephen R. Plymate JST, and William J. Bremner. Circadian Variation in Testosterone, Sex Hormone- Binding Globulin, and Calculated Non-Sex Hormone-Binding Globulin Bound Testosterone in Healthy Young and Elderly Men. *Journal of Andrology.* 1989;10(5):366-71.

[18] Vermeulen A, Kaufman JM, Deslypere JP, Thomas G. Attenuated Luteinizing-Hormone (LH) Pulse Amplitude But Normal Lh Pulse Frequency, And Its Relation To Plasma Androgens In Hypogonadism Of Obese Men. *Journal of Clinical Endocrinology & Metabolism.* 1993;76(5):1140-6.

[19] Vermeulen A, Kaufman JM. Aging of the Hypothalamopituitary Testicular Axis in Men. *Hormone Research.* 1995;43(1-3):25-8.

[20] Morley JE, Charlton E, Patrick P, Kaiser FE, Cadeau P, McCready D, et al. Validation of a screening questionnaire for androgen deficiency in aging males. *Metabolism-Clinical and Experimental.* 2000;49(9):1239-42.

[21] Abdulmaged M. Traish MMM, Abraham Morgentaler, Michael Zitzmann. Testosterone Deficiency. *The American Journal of Medicine.* 2011;124:578-87.

[22] Bettocchi C, Palumbo F, Cormio L, Ditonno P, Battaglia M, Selvaggi FP. The effects of androgen depletion on human erectile function: a prospective study in male-to-female transsexuals. *International Journal of Impotence Research.* 2004;16(6):544-6.

[23] Rosen Rc. Pharmacological Effects on Nocturnal Penile Tumescence (NPT). In: Bancroft J, editor. *Pharmacology of Sexual Function and Dysfunction*, 1995. p. 277-93.

[24] Maggi GCaM. The role of testosterone in erectile dysfunction. Nature Reviews Urology. 2010;7:46-57.

[25] Melis MR, Argiolas A. Central control of penile erection: A re-visitation of the role of oxytocin and its interaction with dopamine and glutamic acid in male rats. *Neuroscience and Biobehavioral Reviews.* 2011;35(3):939-55.

[26] Andersson KE. Mechanisms of Penile Erection and Basis for Pharmacological Treatment of Erectile Dysfunction. *Pharmacological Reviews.* 2011;63(4):811-59.

[27] Swaab DF. Sexual differentiation of the brain and behavior. *Best Practice & Research Clinical Endocrinology & Metabolism.* 2007;21(3):431-44.

[28] Allard J, Edmunds NJ. Reflex penile erection in anesthetized mice: an exploratory study. *Neuroscience.* 2008;155(1):283-90.

[29] Burnett AL. Role of nitric oxide in the physiology of erection. *Biology of reproduction.* 1995;52(3):485-9.

[30] Francis SH, Busch JL, Corbin JD. cGMP-Dependent Protein Kinases and cGMP Phosphodiesterases in Nitric Oxide and cGMP Action. Pharmacological Reviews. 2010;62(3):525-63.

[31] Shabsigh R. Testosterone therapy in erectile dysfunction. The aging male : *the official journal of the International Society for the Study of the Aging Male*. 2004;7(4):312-8.

[32] Gonzalez-Cadavid NF, Burnett AL, Magee TR, Zeller CB, Vernet D, Smith N, et al. Expression of penile neuronal nitric oxide synthase variants in the rat and mouse penile nerves. *Biology of reproduction*. 2000;63(3):704-14.

[33] Hurt KJ, Sezen SF, Champion HC, Crone JK, Palese MA, Huang PL, et al. Alternatively spliced neuronal nitric oxide synthase mediates penile erection. *Proceedings of the National Academy of Sciences of the United States of America*. 2006;103(9):3440-3.

[34] Mancina R, Filippi S, Morelli A, Vignozzi L, Salonia A, Montorsi F, et al. Expression and functional activity of phosphodiesterase type 5 in human and rabbit vas deferens. *Molecular Human Reproduction*. 2005;11(2):107-15.

[35] Traish AM, Goldstein I, Kim NN. Testosterone and erectile function: From basic research to a new clinical paradigm for managing men with androgen insufficiency and erectile dysfunction. *European Urology*. 2007;52(1):54-70.

[36] Zhang XH, Filippi S, Morelli A, Vignozzi L, Luconi M, Donati S, et al. Testosterone restores diabetes-induced erectile dysfunction and sildenafil responsiveness in two distinct animal models of chemical diabetes. *Journal of Sexual Medicine*. 2006;3(2):253-64.

[37] Filippi S, Vignozzi L, Morelli A, Chavalmane AK, Sarchielli E, Fibbi B, et al. Testosterone Partially Ameliorates Metabolic Profile and Erectile Responsiveness to PDE5 Inhibitors in an Animal Model of Male Metabolic Syndrome. *Journal of Sexual Medicine*. 2009;6(12):3274-88.

[38] Morelli A, Filippi S, Mancina R, Luconi M, Vignozzi L, Marini M, et al. Androgens regulate phosphodiesterase type 5 expression and functional activity in corpora cavernosa. *Endocrinology*. 2004;145(5):2253-63.

[39] Singh R, Artaza JN, Taylor WE, Gonzalez-Cadavid NF, Bhasin S. Androgens stimulate myogenic differentiation and inhibit adipogenesis in C3H 10T1/2 pluripotent cells through an androgen receptor-mediated pathway. *Endocrinology*. 2003;144(11):5081-8.

[40] Heinemann LAJ, Saad F, Heinemann K, Thai DM. Can results of the Aging Males' Symptoms (AMS) scale predict those of screening scales for androgen deficiency? The aging male : *the official journal of the International Society for the Study of the Aging Male*. 2004;7(3):211-8.

[41] Emmelot-Vonk MH, Verhaar HJJ, Nakhai-Pour HR, Grobbee DE, van der Schouw YT. Low testosterone concentrations and the symptoms of testosterone deficiency according to the Androgen Deficiency in Ageing Males (ADAM) and Ageing Males' Symptoms rating scale (AMS) questionnaires. *Clinical endocrinology*. 2011;74(4):488-94.

[42] Mohamed O, Freundlich RE, Dakik HK, Grober ED, Najari B, Lipshultz LI, et al. The quantitative ADAM questionnaire: a new tool in quantifying the severity of hypogonadism. *International Journal of Impotence Research*. 2010;22(1):20-4.

[43] Corona G, Mannucci E, Petrone L, Balercia G, Fisher AD, Chiarini V, et al. ANDROTEST (c): A structured interview for the screening of hypogonadism in patients with sexual dysfunction. *Journal of Sexual Medicine*. 2006;3(4):706-15.

[44] Rosen RC, Riley A, Wagner G, Osterloh IH, Kirkpatrick J, Mishra A. The international index of erectile function (IIEF): A multidimensional scale for assessment of erectile dysfunction. *Urology*. 1997;49(6):822-30.

[45] Shaw JW, Reardon G, Sandor DW, Rosen RC, Ferguson DM. Validation of Stopwatch Measurements of Erection Duration against Responses to the Sexual Encounter Profile and International Index of Erectile Function in Patients Treated with a Phosphodiesterase Type 5 Inhibitor. *Journal of Sexual Medicine*. 2010;7(3):1147-59.

[46] Armagan A, Kim NN, Goldstein I, Traish AM. Dose-response relationship between testosterone and erectile function: Evidence for the existence of a critical threshold. *Journal of Andrology*. 2006;27(4):517-26.

[47] Zitzmann M, Faber S, Nieschlag E. Association of specific symptoms and metabolic risks with serum testosterone in older men. *Journal of Clinical Endocrinology & Metabolism*. 2006;91(11):4335-43.

[48] Sikaris K, McLachlan RI, Kazlauskas R, de Kretser D, Holden CA, Handelsman DJ. Reproductive hormone reference intervals for healthy fertile young men: Evaluation of automated platform assays. *Journal of Clinical Endocrinology & Metabolism*. 2005;90(11):5928-36.

[49] Swerdloff RS, Wang C. Free testosterone measurement by the analog displacement direct assay: Old concerns and new evidence. *Clinical Chemistry*. 2008;54(3):458-60.

[50] Vermeulen A, Verdonck L, Kaufman JM. A critical evaluation of simple methods for the estimation of free testosterone in serum. *Journal of Clinical Endocrinology & Metabolism*. 1999;84(10):3666-72.

[51] Martinez-Jabaloyas JM, Queipo-Zaragoza A, Pastor-Hernandez F, Gil-Salom M, Chuan-Nuez P. Testosterone levels in men with erectile dysfunction. *BJU international*. 2006;97(6):1278-83.

[52] Barkin J. Erectile dysfunction and hypogonadism (low testosterone). *Canadian Journal of Urology*. 2011;18:2-7.

[53] Morales A, Lunenfeld B, International Society for the Study of the Aging M. Investigation, treatment and monitoring of late-onset hypogonadism in males. Official recommendations of ISSAM. International Society for the Study of the Aging Male. The aging male : *the official journal of the International Society for the Study of the Aging Male*. 2002;5(2):74-86.

[54] Lunenfeld B, Saad F, Hoesl CE. ISA, ISSAM and EAU recommendations for the investigation, treatment and monitoring of late-onset hypogonadism in males: scientific background and rationale. The aging male : *the official journal of the International Society for the Study of the Aging Male*. 2005;8(2):59-74.

[55] Dhir RR, Lin HC, Canfield SE, Wang R. Combination therapy for erectile dysfunction: an update review. *Asian Journal of Andrology*. 2011;13(3):382-90.

[56] Garcia JA, Sanchez PE, Fraile C, Escovar P. Testosterone undecanoate improves erectile dysfunction in hypogonadal men with the metabolic syndrome refractory to treatment with phosphodiesterase type 5 inhibitors alone. *Andrologia*. 2011;43(5):293-6.

[57] Traish AM, Guay AT. Are androgens critical for penile erections in humans? Examining the clinical and preclinical evidence. *Journal of Sexual Medicine*. 2006;3(3):382-404.

[58] Greco EA, Spera G, Aversa A. Combining Testosterone and PDE5 Inhibitors in Erectile Dysfunction: Basic Rationale and Clinical Evidences. *European Urology*. 2006;50(5):940-7.

[59] Corona G, Maggi M. The role of testosterone in erectile dysfunction. Nature reviews: *Urology*. 2010;7(1):46-56.

[60] Isidori AM, Giannetta E, Gianfrilli D, Greco EA, Bonifacio V, Aversa A, et al. Effects of testosterone on sexual function in men: results of a meta-analysis. *Clinical endocrinology*. 2005;63(4):381-94.

[61] Traish AM, Saad F, Feeley RJ, Guay AT. The dark side of testosterone deficiency: III. cardiovascular disease. *Journal of Andrology*. 2009;30:477-94.

[62] Corona G, Mannucci E, Forti G, Maggi M. Hypogonadism, ED, metabolic syndrome and obesity: a pathological link supporting cardiovascular diseases. *Int J Androl*. 2009;32:587-98.

[63] Diaz-Arjonilla M, Schwarcz M, Swerdloff RS, Wang C. Obesity, low testosterone levels and erectile dysfunction. *International Journal of Impotence Research*. 2009;21:89-98.

[64] Yassin AA, Saad F, Gooren LJ. Metabolic syndrome, testosterone deficiency and erectile dysfunction never come alone. *Andrologia*. 2008;40:259-64.

[65] Wang C, Nieschlag E, Swerdloff R, Behre HM, Hellstrom WJ, Gooren LJ, et al. Investigation, Treatment, and Monitoring of Late-Onset Hypogonadism in Males: ISA, ISSAM, EAU, EAA, and ASA Recommendations. *European Urology*. 2009;55(1):121-30.

[66] Kapoor D, Goodwin E, Channer KS, Jones TH. Testosterone replacement therapy improves insulin resistance, glycaemic control, visceral adiposity and hypercholesterolaemia in hypogonadal men with type 2 diabetes. *European Journal of Endocrinology*. 2006;154(6):899-906.

[67] Basu R, Man CD, Campioni M, Basu A, Nair KS, Jensen MD, et al. Effect of 2 years of testosterone replacement on insulin secretion, insulin action, glucose effectiveness, hepatic insulin clearance, and postprandial glucose turnover in elderly men. *Diabetes Care*. 2007;30(8):1972-8.

[68] Corrales JJ, Burgo RM, Garcia-Berrocal B, Almeida M, Alberca I, Gonzalez-Buitrago JM, et al. Partial androgen deficiency in aging type 2 diabetic men and its relationship to glycemic control. *Metabolism-Clinical and Experimental*. 2004;53(5):666-72.

[69] Carpenter WR, Robinson WR, Godley PA. Getting over testosterone: Postulating a fresh start for etiologic studies of prostate cancer. *Journal of the National Cancer Institute*. 2008;100(3):158-U5.

[70] Roddam AW, Allen NE, Appleby P, Key TJ, Endogenous Hormones Prostate C. Endogenous sex hormones and prostate cancer: A collaborative analysis of 18 prospective studies. *Journal of the National Cancer Institute*. 2008;100(3):170-83.

[71] Fowler JE, Whitmore WF. Considerations for the use of testosterone with systemic chemotherapy in prostatic-cancer. *Cancer*. 1982;49(7):1373-7.

[72] McConnell JD. Prostatic growth - New insights into hormonal-regulation. British Journal of Urology. 1995;76:5-10.

[73] Handelsman DJ, Zajac JD. 11: Androgen deficiency and replacement therapy in men. *The Medical journal of Australia*. 2004;180(10):529-35.

[74] Calof OM, Singh AB, Lee ML, Kenny AM, Urban RJ, Tenover JL, et al. Adverse events associated with testosterone replacement in middle-aged and older men: A meta-analysis of randomized, placebo-controlled trials. *Journals of Gerontology Series a-Biological Sciences and Medical Sciences*. 2005;60(11):1451-7.

[75] Qoubaitary A, Swerdloff RS, Wang C. Advances in male hormone substitution therapy. *Expert Opin Pharmacother.* 2005;6:1493-506.

[76] Seftel A. Testosterone replacement therapy for male hypogonadism: part III. Pharmacologic and clinical profiles, monitoring, safety issues, and potential future agents. *International Journal of Impotence Research.* 2007;19:2-24.

[77] Edelstein D, Sivanandy M, Shahani S, Basaria S. The latest options and future agents for treating male hypogonadism. *Expert Opin Pharmacother.* 2007;8:2991-3008.

[78] Pfeil E, Dobs AS. Current and future testosterone delivery systems for treatment of the hypogonadal male. *Expert Opinion on Drug Delivery.* 2008;5(4):471-81.

[79] Behre HM, Wang C, Handelsman DI, Nieshlag E. Testosterone: Action, Deficiency, Substitution. Nature Publishing Group; 2004. p. 405-44.

[80] Saad F, Kamischke A, Yassin A, Zitzmann M, Schubert M, Jockenhovel F, et al. More than eight years' hands-on experience with the novel long-acting parenteral testosterone undecanoate. *Asian Journal of Andrology.* 2007;9(3):291-7.

[81] Giagulli VA, Triggiani V, Corona G, Carbone D, Licchelli B, Tafaro E, et al. Evidence-based Medicine Update on Testosterone Replacement Therapy (TRT) in Male Hypogonadism: Focus on New Formulations. *Current Pharmaceutical Design.* 2011;17(15):1500-11.

[82] Khera M, Bhattacharya RK, Blick G, Kushner H, Nguyen D, Miner MM. Improved sexual function with testosterone replacement therapy in hypogonadal men: real-world data from the Testim Registry in the United States (TRiUS). *The journal of sexual medicine.* 2011;8(11):3204-13.

[83] Mills TM, Lewis RW. The role of androgens in the erectile response: A 1999 perspective. *Molecular Urology.* 1999;3(2):75-80.

[84] Morelli A. Androgens regulate phosphodiesterase type 5 expression and functional activity in corpora cavernosa. *Endocrinology.* 2004;145:2253-63.

[85] Zhang XH. Testosterone regulates PDE5 expression and in vivo responsiveness to tadalafil in rat corpus cavernosum. *Eur Urol.* 2005;47:409-16.

[86] Traish AM. Effects of castration and androgen replacement on erectile function in a rabbit model. *Endocrinology.* 1999;140:1861-8.

[87] Yassin AA, Saad F, Diede HE. Testosterone and erectile function in hypogonadal men unresponsive to tadalafil: results from an open-label uncontrolled study. *Andrologia.* 2006;38:61-8.

[88] Tas A, Ersoy A, Ersoy C, Gullulu M, Yurtkuran M. Efficacy of sildenafil in male dialysis patients with erectile dysfunction unresponsive to erythropoietin and/or testosterone treatments. *International Journal of Impotence Research.* 2006;18:61-8.

[89] Shamloul R, Ghanem H, Fahmy I, El-Meleigy A, Ashoor S, Elnashaar A, et al. Testosterone therapy can enhance erectile function response to sildenafil in patients with PADAM: A pilot study. *Journal of Sexual Medicine.* 2005;2(4):559-64.

[90] Shabsigh R, Kaufman JM, Steidle C, Padma-Nathan H. Randomized study of testosterone gel as adjunctive therapy to sildenafil in hypogonadal men with erectile dysfunction who do not respond to sildenafil alone. *Journal of Urology.* 2004;172(2):658-63.

[91] Rosenthal BD, May NR, Metro MJ, Harkaway RC, Ginsberg PC. Adjunctive use of AndroGel (testosterone gel) with sildenafil to treat erectile dysfunction in men with

acquired androgen deficiency syndrome after failure using sildenafil alone. *Urology*. 2006;67(3):571-4.

[92] Rochira V, Balestrieri A, Madeo B, Granata ARM, Carani C. Sildenafil improves sleep-related erections in hypogonadal men: Evidence from a randomized, placebo-controlled, crossover study of a synergic role for both testosterone and sildenafil on penile erections. *Journal of Andrology*. 2006;27(2):165-75.

[93] Kalinchenko SY, Kozlov GI, Gontcharov NP, Katsiya GV. Oral testosterone undecanoate reverses erectile dysfunction associated with diabetes mellitus in patients failing on sildenafil citrate therapy alone. *Aging Male*. 2003;6:94-9.

[94] Hwang TIS, Chen HE, Tsai TF, Lin YC. Combined use of androgen and sildenafil for hypogonadal patients unresponsive to sildenafil alone. *International Journal of Impotence Research*. 2006;18(4):400-4.

[95] Greenstein A, Mabjeesh NJ, Sofer M, Kaver I, Matzkin H, Chen JZ. Does sildenafil combined with testosterone gel improve erectile dysfunction in hypogonadal men in whom testosterone supplement therapy alone failed? *Journal of Urology*. 2005;173(2):530-2.

[96] Foresta C, Caretta N, Rossato M, Garolla A, Ferlin A. Role of androgens in erectile function. *Journal of Urology*. 2004;171:2358-62.

[97] Chatterjee R. A novel therapy with testosterone and sildenafil for erectile dysfunction in patients on renal dialysis or after renal transplantation. *J Fam Plann Reprod Health Care*. 2004;30:88-90.

[98] Buvat J. What is the threshold level of the PDE5 inhibitors' testosterone-dependence in men? A double blind placebo-controlled trial of testosterone gel addition in men with ED non responding to tadalafil 10 mg once a day alone [abstract #376]. *Eur Urol Suppl*. 2009;8:214.

[99] Berridge MJ. Smooth muscle cell calcium activation mechanisms. *J Physiol*. 2008;586:5047-61.

[100] Aversa A, Isidori AM, Spera G, Lenzi A, Fabbri A. Androgens improve cavernous vasodilation and response to sildenafil in patients with erectile dysfunction. *Clinical endocrinology*. 2003;58(5):632-8.

In: Androgens: Production, Functions and Disorders
Editors: B. F. Thompson and D. J. Robinson
ISBN: 978-1-62081-145-0

Chapter IV

Roles of Androgen Receptor in Normal Prostate Development and Prostate Cancer Progression

Yuanjie Niu, Juan Wang and Zhiqun Shang
Chawnshang Chang Sex Hormone Research Center,
Tianjin Institute of Urology, Tianjin Medical University,
Tianjin, China

1. Introduction

Prostate cancer is the most commonly occurring malignancy and the second highest cause of cancer-related mortality in men in the United States [1]. The dependence of prostate cancer on androgens has been recognized for more than 7 decades. Huggins and Hodges [2] provided the first solid in vivo evidence that androgen–androgen receptor (AR) signals function as a proliferation stimulator leading to prostate cancer progression. This became a commonly accepted view, supported by subsequent early success of androgen deprivation therapy (ADT) in treating patients with prostate cancer [3]. However, although the majority of patients with advanced prostate cancer respond initially to ADT, relapse due to progress to castration resistant prostate cancer (CRPC).

Significant efforts have been focused on understanding the mechanisms involved in the development and progression of CRPC. The prostate is a glandular organ comprised of three distinct cell types (basal, luminal and neuroendocrine cells) embedded in a fibro-muscular stroma. Prostatic development, growth and secretory function requires androgens action mediated by AR [5,6]. Testosterone and its more active metabolite, 5α-dihydrotestosterone (DHT) exert their biological effects through binding to the AR and inducing AR transactivation [4-8]. The AR is expressed and remains functional throughout prostate cancer progression and its expression persists in the majority of patients with CRPC [9-15]. Therefore, the eventual failure of ADT cannot be simply explained by loss of the AR. Recent studies have demonstrated that the AR could function as both a proliferation stimulator and a tumor suppressor depending on its expression in individual prostatic cells. This chapter discusses findings that AR plays dual yet opposite roles in individual prostatic cells to exert its diverse and differential functions for prostate development and prostate cancer progression.

2. Dual Roles of the AR in Normal Prostate Development

2.1. Stromal AR: Proliferation Stimulator and Survival Factor

The ability of androgen to promote epithelial cell growth in both normal and neoplastic prostate suggests that the AR might function as a proliferation stimulator. As reviewed by Cunha *et* al [16], in all species the prostate develops from the endodermal unrogenital sinus (UGS). The UGS is surrounded by an embryonic connective tissue called the urogenital sinus mesenchyme (UGM). In response to testosterone secreted from fetal testis, epithelial buds emerge from the wall of the UGS and grow into the surrounding UGM. During prenatal development, AR is initially detected solely in the UGM prior to and during prostate bud formation, and is undetectable in the epithelial compartment [17], which suggests that mesenchymal AR, but not epithelial AR, is critically involved in the early phases of prostate development. In rodents, the solid prostate buds elongate and undergo branching morphogenesis in the perinatal period, and the process continues until maturation at the end of puberty [16-18]. Beginning at the neonatal period, the epithelial cords undergo ductal canalization during which the epithelial cells differentiate into the luminal and basal cells [16]. Prostatic epithelial differentiation is accompanied by differentiation of the mesenchyme into smooth muscle cells and fibroblasts [19, 20].

In 1978 Cunha and Lung [21] concluded from the development of kidney capsule transplants of various combination of UGM and embryonic bladder epithelium (BLE) of wild type (Wt) mice and testicular feminized mice (Tfm, a mouse model lacking functional AR) that the AR in the UGM, not in the epithelium, controls the initiation of prostatic ductal morphogenesis and epithelial cytodifferentiation. Since then, it has been generally believed that androgens-AR signaling regulates prostatic epithelial cell proliferation via a paracrine mechanism independent of the epithelial AR.

Subsequent tissue recombination studies using UGM and UGS epithelium (UGE) indicated that the epithelial AR is required for expression of some prostatic secretory proteins [22,23], and prostatic epithelium induces UGM to differentiate into smooth muscle [24-26]. Postnatal development and maturation of the prostate is believed to be dependent on a mesenchymal-epithelial or stromal-epithelial interactions [27], which continues in adult prostate to maintain cellular homeostasis of the organ [27,28] and remains dependent on androgens-AR signaling, as castration of the adults resulted in prostate involution that is attributed to epithelial cell apoptosis [29]. Androgen supplementation in castrated animals resulted in restoration of prostatic size and ductal morphology [30], indicating that the growth of adult prostate epithelial cells remains dependent on AR signaling. A study of apoptosis of prostate tissues recombined from rat UGM + wt epithelium vs. + Tfm epithelium suggested that survival of prostate epithelial cells is dependent on the stromal but not epithelial AR signaling [31].

A recent study of stromal smooth muscle cell- or fibroblast-specific AR knockout (FSP-ARKO) mice indicated that reducing AR signaling in the prostatic stroma resulted in altered epithelial morphology with more cuboidal and flattened shapes. Further assays showed the decreased proliferation, increased apoptosis, and defective expression of luminal epithelial genes including PSP94, NKX3.1, and probasin in FSP-ARKO mouse prostate [32]. These observations strongly suggest the requirement of the stromal fibroblasts AR for the normal development and differentiation of prostate epithelium. The ability of stromal AR to regulate epithelial cell proliferation and survival supports the consensus of view that the AR can function as a proliferation stimulator and a survival factor for the normal prostate development and tissue homeostasis.

2.2. Epithelial AR: Survival Factor for Epithelial Luminal Cells

The study of prostate epithelium-specific ARKO (pes-ARKO) mice has revealed that prostatic epithelial AR also plays a role as a survival factor for epithelial luminal cells. By crossing floxed AR mice [33] with a probasin-Cre mice [34], Wu *et* al [35] have generated pes-ARKO mice, in which the epithelial probasin promoter transgene is expressed increasingly from 2 to 7 weeks with expression sustained within the epithelial cells throughout life [34]. Consistent with increasing probasin-Cre expression, the epithelial AR expression gradually decreased in the prostate of pes-ARKO mice beginning at 6-weeks old, and was below detection at 24-weeks of age. At 24-weeks or older the ventral prostates of pes-ARKO mice were enlarged compared to their Wt littermates. Histological analysis of the ventral prostate in pes-ARKO mice indicated that there was progressive decrease in epithelial height, loss of glandular infolding, and an increase in epithelial luminal cell apoptosis through 6- to 32-weeks of age.

Prostatic epithelial cells develop from stem cells through proliferation stepwise into basal (cytokeratins (CK) 5-positive, CK8-negative) and intermediate (CK5/CK8-double positive) cells that finally differentiate into epithelial luminal [36] (CK8-positive, CK5-negative) cells. It was found that mice lacking the prostate epithelial AR lost its normal function of expressing and secreting probasin, an androgen-regulated protein specifically expressed in the differentiated prostate epithelial cells. Knockout of epithelial AR led to increased apoptosis in the CK8+ luminal cells, as well as decreased proliferation in the CK8+ luminal cells. The expression of homeobox protein NKX3.1, that plays important roles in the maintenance of normal morphogenesis of prostate, was also found to be decreased in the epithelium of pes-ARKO prostate. These observations strongly suggest that the epithelial AR is an important survival factor for epithelial luminal cells.

2.3. Epithelial AR: Suppressor for Epithelial Basal Intermediate Cell Proliferation

The increased basal intermediate cells population observed in pes-ARKO mice and enlargement of the ventral prostate [35] might be a result of increased proliferation in epithelial basal cells albeit the increasing apoptotic loss of epithelial luminal cells. The prostatic glands of pes-ARKO mice at 24-weeks of age exhibited one layer of undifferentiated epithelial cells with increased incorporation of 5-bromodeoxyuridine (BrdU) indicating an increase in epithelial cell proliferation. BrdU-positive cells were primarily co-localized with CK5-positive basal cells and to a much lesser extent with CK8-positive cells. The increased apoptotic CK8-positive luminal cells and increased proliferative CK5-positive basal cells led to the expansion of CK5/CK8-double positive intermediate cells in the epithelium. Interestingly, because these expanded cells are also both p63-positive and CK5-positive, they may therefore be also defined as expanded stem/progenitor cells in the epithelium. The normal prostatic morphology and glandular histology were reverted back after restoring a functional mutant murine AR [37] in the AR(T857A)-pes-ARKO double transgenic mice [35].

Several studies have reported that stable transfection of the AR into AR-negative prostatic epithelial cells also resulted in the suppression of cell growth [38, 39], thereby supporting the notion that the epithelial AR is a suppressor of epithelial basal intermediate cell proliferation.

These observations from both knockout and knock-in mouse models not only support the view that the AR in prostatic epithelial cells is required for epithelial cell differentiation [22], but also suggest that the epithelial AR functions as a suppressor for epithelial basal intermediate cell proliferation and a survival regulator for differentiated epithelial luminal cells.

2.4. Epithelial AR: Survival Factor for Stromal Smooth Muscle Cell Differentiation

Tissue recombination studies using UGM and UGS epithelium indicated that prostatic epithelium induces UGM to differentiate into smooth muscle[24-26]. Our recent findings showed that the epithelial AR might be involved in this process. Thinner stromal smooth muscle layer of ventral prostates was found in 24-weeks of age pes-ARKO mice. The mature stromal smooth muscle markers, smooth muscle α-actin and calponin were significantly decreased in the ventral prostates of 24-weeks of age pes-ARKO mice compared with Wt mice. TGFβ1 signals may induce the differentiation of prostate stroma that leads to stimulation of the mature smooth muscle cells. The expression of TGFβ1 and its downstream signals were lower in the epithelium of pes-ARKO mice than those from Wt mice. By double staining TGFβ1 and calponin, we demonstrated that the loss of TGFβ1 expression in the epithelium of pes-ARKO mice might coincide with the thinning of the surrounding stromal smooth muscle layer.

After treating the primary cultured human stromal cells (ps-1) with different concentrations of TGFβ1, both smooth muscle α-actin and calponin staining showed that the number of mature stromal smooth muscle cells was significantly dependent on the higher concentration and longer duration of TGFβ1 treatment. MyoD and myogenin expression, which were deeply involved in stromal smooth muscle maturation and differentiation, was also modulated by TGFβ1 stimulation. Altogether, impaired epithelial differentiation by knockout of epithelial AR led to poor stromal differentiation that might involve the modulation of TGFβ1 signals [40].

3. AR Dual Roles in Prostate Cancer Progression and Metastasis

3.1. Stromal AR: Stimulator for Tumor Progression and Metastasis

The observation of Huggins and Hodges [2] was the first evidence that androgens-AR function as a proliferation stimulator for prostate cancer progression. The subsequent ADT for prostate cancer patients with early success appears to support the common view, that androgens-AR function as a proliferator to stimulate the prostate tumor progression. However, in which prostate cells the AR my function as a proliferation stimulator remains unclear. Recent findings suggest that the stromal AR may function as proliferation stimulator to promote the prostate cancer progression and metastasis.

Stromal-epithelial interaction remains important for tumor initiation, progression, and metastasis [41]. During tumor progression, the stromal cells, including fibroblasts, myofibroblasts, endothelial, and immune cells, form a microenvironment supporting the progression, survival, and metastasis of the tumor. In normal human prostate, the stroma is constituted mainly of smooth muscle cells that also express the AR. In prostatic carcinoma, the tumor stroma is constituted mainly of fibroblastic and myofibroblastic cells [42], suggesting that changes have occurred in both the stroma and the epithelium during tumorigenesis. Previous tissue recombination studies by Thompson *et* al [43] with oncogene-carrying virus infected UGM and UGE cells have indicated that prostatic tumorigenesis requires changes in both the stroma and the epithelium, as supported by studies of Cunha and co-workers [42,44-46]. Thus, prostate cancer associated fibroblasts (CAF) are capable of supporting tumor formation of SV40 Tag-immortalized human prostatic epithelial cells, BPH-1 [42,44]. Similarly, UGMs from rat and mouse were capable of eliciting tumor formation from BPH-1 cells stimulated by testosterone + 17β-estradiol (E2) [45,46]. However, CAF in castrated host and UGM from Tfm mice were unable to promote tumor formation from BPH-1 cells (Ricke W. A., personal communication), indicating that the stromal AR signaling is required for prostate tumor initiation and progression.

Human prostatic stromal WPMY1 cells [47] were stably transfected with AR-siRNA (WPMY1-ARsiRNA) to knockdown the endogenous AR expression [48]. The results from co-culture of PC3 with WPMY1 cells in a Boyden chamber invasion assay [49] indicated that PC3 cells were less invasive when co-cultured with WPMY1-ARsiRNA cells than with WPMY1-v cells [48], indicating that androgens-AR signaling in prostatic stromal cells is a stimulator of cancer cell invasion.

PC3 cells co-cultured with WPMY1-v or WPMY-ARsi cells were then inoculated orthotopically into the prostate of nude mice and the results showed that PC3 cells produced smaller primary and pelvic lymph node (PLN) metastatic tumors when in combination with WPMY1-ARsi than with WPMY1-v cells [48]. These *in vivo* observations suggest that stromal AR signaling can provide growth factor(s) for AR-negative prostate cancer PC3 cells and influence their progression, invasion, and metastasis. Such androgen-AR dependent growth *in vivo,* has also been observed in other androgen-independent prostate cancer cell lines [50,51].

3.2. Epithelial AR: Survival Factor for Tumor Progression and Metastasis in Luminal Cells

The transgenic adenocarcinoma mouse prostate (TRAMP) model is a prostate cancer model that develops prostate tumor spontaneously at 10- to 12-weeks of age [52]. pes-ARKO-TRAMP mice [48,53] were generated from TRAMP mice in a similar fashion as pes-ARKO mice. Prostatic epithelial cells develop from stem cells through proliferation stepwise into basal and intermediate cells that finally differentiate into epithelial luminal cells [54]. The AR is expressed in about 50% of basal cells and all of the luminal cells [55]. The AR knockdown from these cells in pes-ARKO-TRAMP mice [48] proceeded gradually with age in a fashion very similar to that of pes-ARKO mice [35].

The prostates of pes-ARKO-TRAMP mice exhibited elevated epithelial cell apoptosis (18%) over the prostates of Wt TRAMP mice (2%) at 16-weeks of age. The apoptotic cells were CK8-positive epithelial luminal cells[48]. These *in vivo* data suggest that the AR in epithelial luminal cells may function as a survival factor to protect prostate cancer cells from apoptosis.

3.3. Epithelial AR: Suppressor for Tumor Progression and Metastasis in Basal Intermediate Cells

In pes-ARKO-TRAMP mice compared to Wt TRAMP mice [53], there was an increasing number of proliferating cells in the basal compartment as indicated by the expression of Ki67 (a proliferation marker) immunoreactivity and BrdU incorporation. The increased BrdU-incorporating cells [48,53] were CK5-positive basal cells [36]. The increased epithelial luminal cells apoptosis and epithelial basal cell proliferation [48,53] resulted in the cell population changes with an increase in the CK5/CK8-double positive basal intermediate cells [56].

Furthermore, pes-ARKO-TRAMP mice developed larger and less differentiated primary tumors in the ventral prostate than Wt TRAMP mice at 16-weeks of age [53]. The primary tumors of pes-ARKO-TRAMP exhibited a higher population of CK5/CK8- and CD44-positive (a basal [57] and early progenitor cells marker [58]) intermediate-like cells than tumors of Wt TRAMP mice. These results suggest knockout of the epithelial AR resulted in cell population changes with expansion of intermediate-like tumor cells and decreased secretory luminal cells in the prostate of pes-ARKO-TRAMP mice [53]. Moreover, these results confirm the observations in pes-ARKO mice that the epithelial AR functions as a proliferation suppressor for epithelial basal intermediate cells and a survival factor for epithelial luminal cells. As prostate carcinoma arises from epithelial cells, the opposite functions of the epithelial AR in various cells at different stages might also influence prostate cancer progression in TRAMP mice by favoring survival of differentiated tumor epithelium while suppressing proliferation of epithelial basal intermediate cells thereby retarding tumor growth.

3.4. Epithelial AR: Suppressor of Prostate Cancer Metastasis

The androgens-AR signaling in prostate cancer epithelial cells also influence metastasis of the tumor [48]. Thus, the size of metastatic tumors in the PLN of pes-ARKO-TRAMP mice were found increased compared to their Wt TRAMP littermates at 24-weeks of age [48]. In addition, more prostate cancer metastatic foci were observed within the livers of pes-ARKO-TRAMP than Wt TRAMP mice [48]. Primary cultured PLN tumor cells isolated from pes-ARKO-TRAMP mice were more invasive than those from Wt TRAMP mice [48] in the Boyden chamber invasion assay. Importantly, restoring a functional AR via virus infection reduced the invasiveness of PLN tumor cells from pes-ARKO-TRAMP mice [48]. These observations appeared to conform to the observation that castrated TRAMP mice developed PLN metastatic tumors that were poorly differentiated and more aggressive than tumors of intact mice [52].

4. AR Dual Roles in Human Prostate Cancer Cell Lines

4.1. AR in PC3 Cells: Suppressor for Proliferation and Metastasis

The PC3 cells, originally isolated from a bone marrow metastasis [59], express both CK5 and CK8/18, but not the AR, and thus are basal intermediate-like tumor cells [60] that are highly tumorigenic. Ectopic expression of the AR in PC3 cells resulted in androgen-dependent suppression of cell proliferation [61-65] or mild promotion of growth [66], depending on the expression vector being used. These contrasting results suggest that *in vitro* cell growth assay might not adequately represent the behavior of prostate cancer cells *in vivo*.

Further comparison of PC3-v vs PC3-AR9 cells found that PC3-AR9 cells were less invasive in the Boyden chamber invasion assay and produced less osteolytic lesions in a bone-wafer resorption assay [67] than PC3-v cells [48]. After injection into the tibia of athymic nude mice, PC3-v tumors grew more aggressively and more invasively than PC3-AR9 tumors [48]. Collectively, these gain-of-function data from knock-in of the functional human AR into PC3 cells suggest that in bone, lack of AR signaling in prostate cancer cells promotes invasion, both *in vitro* and *in vivo*.

PC3-v or PC3-AR9 cells were further orthotopically inoculated into the anterior prostate of nude mice. Mice inoculated with PC3-v cells developed bigger prostate primary tumors that were less differentiated [53] and larger PLN metastatic tumors [48] than mice inoculated with PC3-AR9 cells. Moreover, when PC3-v or PC3-AR9 cells were co-cultured with WMPY1 stromal cells and orthotopically transplanted in nude mice, PC3-AR9 cells still produced smaller primary and metastatic tumors than PC3-v cells, suggesting that in the presence of stromal AR stimulation, the suppressor function of the epithelial AR remains effective.

Using transfectants of PC3 cells with an inducible AR-expressing transgene, Nelius *et* al [68] also observed that induction of AR expression resulted in an androgen-dependent decrease in invasion *in vitro* and decreased tumorigenecity due to decreased microvescular density and increased tumor cell apoptosis after subcutaneous inoculation of the transfectant in nude mice.

Together, these results showed that loss of human prostatic epithelial AR leads to the development of more invasive and metastatic prostate cancers and gain of AR function reverses these characteristics.

4.2. AR in LNCaP Cells: Proliferation Stimulator and Suppressor

The LNCaP cells, isolated from a lymph node metastatic tumor [69] and express a mutated AR(T877A) [70], are CK5-negative and CK8/18-positive and thus are luminal-like tumor cells [60]. These cells may respond to androgens-AR signaling differentially depending on different environments. For example, the AR-expressing LNCaP-FGC and LNCaP-LNO cell lines isolated from the same lymph node metastatic tumor [63] responded differentially toward androgen treatment [65]. Proliferation of LNCaP-FGC cells was stimulated by low concentrations of androgen and suppressed by high concentrations of androgen, whereas proliferation of LNCaP-LNO cells was suppressed by the physiological concentration of androgen, which caused G_0/G_1 cell cycle arrest [65]. These different cell phenotypes might be the result of prostate cancer cells adapting to their different androgenic environment.

Furthermore, androgen-dependent LNCaP cells cultured in long-term absence of androgen [72] or merely after prolonged passages (>80) [73] resulted in the development of sublines that acquired androgen-independent proliferation. Addition of androgen to these androgen-independent variants of LNCaP cells suppressed proliferation [74] and promoted apoptosis [75]. Such adaptive changes of AR function from proliferation stimulator to suppressor have also been observed in LNCaP xenografts *in vivo* after castration of the host [76-78].

Down-regulation of the AR with anti-sense oligonucleotides [79] or siRNA [80,81] may also result in the suppression of cell proliferation or promotion of apoptosis in both androgen-dependent and androgen-independent lines of LNCaP cells. *In vivo* studies also found that an AR-specific anti-sense oligonucleotide could suppress the growth of LNCaP xenografts [82].

Interestingly, recurrent androgen-independent tumors, developed from orthotopic primary tumors of LNCaP cells after castration of SCID mice hosts, exhibited increased proliferation and decreased apoptosis compared to the androgen-dependent primary tumors of LNCaP cells, a change that is associated with decreased AR protein expression [77]. Thus, it appeared that once LNCaP cells progressed to the androgen-independent state *in vivo*, they might be relieved from epithelial AR suppression due to down-regulation of the AR. Nevertheless, these observations suggest that, dependent on cell environment, the AR may function as a proliferation stimulator, or suppressor, or cell survival factor for LNCaP cell growth and invasion.

4.3. AR in CWR22Rv1 Cells: Proliferation Stimulator and Suppressor

The CWR22Rv1 cells, derived as a recurrent tumor [83] after ADT of CWR22 xenograft originally established from a human prostate primary tumor [84], is an AR and CK8/18-expressing tumor cell line [60].

CWR22rv1-$AR^{+/-}$ cells, which were created using a homologous gene recombination strategy [85] to knockdown the AR in CWR22rv1 cells, expressed much less AR with negligible AR transactivation and displayed suppressed growth rate compared to the parental CWR22rv1-$AR^{+/+}$ cells. On the other hand, CWR22rv1-$AR^{+/-}$ cells produced bigger primary and PLN metastatic tumors than CWR22rv1-$AR^{+/+}$ cells upon orthotopic transplantation (Niu *et al*, manuscript in preparation). This contrasting observation indicates that the *in vitro* growth pattern of prostate cancer cells might not accurately represent the role of the AR *in vivo*.

CWR22rv1-$AR^{+/-}$ cells were more invasive than the parental cells in the Boyden chamber invasion assay. Using an AR-specific siRNA to knockdown the AR in CWR22rv1-$AR^{+/+}$ cells also rendered the resultant CWR22rv1-$AR^{+/+}$-ARsi cells more invasive than the parental cells transfected with scrambled RNA [48].

Expression of a functional human AR (hAR) in CWR22rv1-$AR^{+/-}$ cells through a retroviral vector rendered the CWR22rv1-$AR^{+/-}$-hAR cells less invasive than the parental cells transfected with the empty vector (CWR22rv1-$AR^{+/-}$-v) [48]. Together, these observations support the notion that the AR in prostatic cancer epithelial cells functions as a suppressor for the prostate cancer progression and metastasis.

4.4. AR in PC346C Prostate Cancer Cells: Proliferation Stimulator and Suppressor

The PC346C cells, established from a human primary tumor via xenograft [86, 87], is an androgen-dependent prostate cancer cell line expressing Wt AR and CK8/18 [60]. After long-term androgen ablation, another androgen-independent subline, PC346DCC, with a 95% decrease in AR expression was generated from PC346C cells.

Upon orthotopic transplantation, PC346DCC tumors grew more rapidly in intact than in castrated hosts, suggesting stromal influence in addition to androgen-independent growth [59].

Treatment of PC346C cells with the anti-androgen, 5-hydroxyflutamide, resulted in two new sublines: one (PC346Flu1) with overexpression of the AR and its proliferation suppressed by androgen *in vitro,* was more tumorigenic in castrated than in intact nude mice, whereas the other (PC346Flu2) with the AR(T877A) mutation behaved similarly to LNCaP cells [50]. These contrasting results again demonstrated that androgen-AR signals could be either proliferation stimulator or suppressor in the similar human prostate cancer cells.

However, it should be noted that data obtained solely from *in vitro* studies of human prostate cancer cell lines might not reliably predict the role of the AR in prostate cancer progression and metastasis.

Moreover, several current available human prostate cancer cells were generated from long term culture in the absence of androgen [74], which may not represent the *in vivo* human prostate condition based on the reports by Titus et al [88] showing that even after ADT treatment, DHT concentration left in the human prostate tissue is about 10% of the normal level and is still sufficient to elicit AR transactivation. Together, it is important to conduct *in vivo* animal studies of prostate cancer cells to delineate the pathophysiologic roles of the AR in prostate cancer progression and metastasis.

5. AR Dual Roles in Prostate Cancer Stem Cell Progression

Issacs and Coffey [89] have proposed that the prostate gland in a castrated animal can be regenerated upon androgen supplementation from a population of prostate stem cells (PSC) in the remnant basal compartment that gives rise to proliferating progenitors of transit amplifying intermediate cells, which in turn proliferate and differentiate into basal and luminal cells. It has been proposed that cancer arises from neoplastic transformation of stem cells [90]. Thus, prostate cancer may be considered to derive from neoplastic transformation of PSC to form cancer stem cells (CSC) that generate progenitor cells, which in turn progress into an epithelial tumor [54].

Recent studies have made considerable progress toward identifying PSC [91,92] and CSC [91]. The putative human PSC are a subpopulation (~1%) of basal cells, which can be enriched by selecting integrin $\alpha_2\beta_1^{hi}$ cells [93]. The $\alpha_2\beta_1^{hi}$ cells can be further separated into the $\alpha_2\beta_1^{hi}/CD133^+$ putative stem cells and $\alpha_2\beta_1^{hi}/CD133^-$ transient amplifying intermediate cell populations [94], with the latter expressing the AR protein at a low level detectable only in the presence of a proteasome inhibitor or after stimulation with an androgen [95]. Furthermore, $\alpha_2\beta_1^{hi}/CD133^+$ cells isolated from human prostate tumors express various cell surface markers and displayed different proliferative potentials and tumorigenisity [96,97].

These results suggest that prostate cancer progression from prostatic CSC involve different stem/progenitor cells with different AR expression levels and varying proliferation, differentiation, tumorigenic, and metastatic potentials. In addition, the distribution of tumor stem/progenitor cells in prostate cancer might vary with patient and tumor stage. The different AR expression in these various cells at different stages should render them able to respond differentially to the dual functions of the AR. For example, the lack of AR in tumor cells would prevent them from suppression effects of epithelial AR for proliferation and metastasis, while allow them to be responsive to the stimulatory functions of the stromal AR.

6. AR Dual Roles and Human Prostate Cancer Clinical Data

The proliferation-stimulating role of the AR is at the center of the premise for ADT in prostate cancer treatment [2]. ADT with either surgical or medical castration usually results in a response rate of 70-80% with a 12- to 33-month duration of progression-free survival [98]. However, after an average of 24 months, the tumor almost always recurs and no longer responds to ADT [99], even though the AR remains to be expressed in the prostate cancer [100,101]. Interestingly, cell sorting of ADT-refractory tumors found that after ADT, the prostatic epithelial basal cell marker, CK5 increased from 29% to 75% [36,102], an observation consistent with the expansion of basal intermediate-like tumor cells observed in pes-ARKO-TRAMP and ind-ARKO-TRAMP mice [53].

Recent clinical findings from 254 prostate cancer patients indicated that treatment with ADT resulted in promotion of prostate metastatic tumors [103]. We have evaluated AR expression in primary prostate tumors (97 cases) and prostate metastases (28 cases). The results showed a significant difference between AR expression in primary tumors (91.75%) and metastatic tumors (67.86%) ($P<0.01$). These *in vivo* clinical data were consistent with a recent study [104] that utilized tissue arrays from prostate cancer patients treated with radical prostatectomy, where it was concluded that AR expression was significantly decreased in metastatic prostate tumors as compared to primary prostate cancers (mean 1.30 vs. 3.49, $p<0.01$).

Together, these clinical findings all support the suppressor roles of the AR in epithelial basal intermediate cell proliferation and metastasis as demonstrated in the above human prostate cancer cells and various mouse models.

7. The Impact of AR Dual Roles on Current Clinical ADT

7.1. Challenge to Current ADT

The fact that the stromal and epithelial ARs have opposite functions to modulate prostate cancer cell proliferation and metastasis may challenge the current treatment of prostate cancer with ADT. As ADT suppresses the functions of both stromal and epithelial AR, its treatment effect would depend on which AR plays a more significant role in a given stage of prostate cancer progression. At early stages of prostate cancer when the carcinoma cells are dependent predominantly on stromal AR-derived signals for growth and metastasis, ADT would result in regression of the tumor due to apoptosis of the tumor epithelial cells, leaving CSC and some transit amplifying progenitors intact. Subsequently, CSC and its progenitors may adapt to the low androgenic condition due to ADT and develop, with increasing neuroendocrine differentiation, AR somatic mutations, and/or changes in AR coregulator expression, into more aggressive tumors capable of androgen-independent growth. Transition of prostate cancer into a more aggressive phenotype under low androgen condition has been reported during tumorigenesis in Nkx3-Pten mutant mice [105].

After prostate cancers have developed to a later stage (with or without ADT), androgen ablation would accelerate tumor metastasis due to ablation of the metastasis suppressor function of the epithelial AR and thus results in failure of ADT. Consistent with this view is the observation of Kleeberger *et* al [103] who found that the expression of nestin, which is linked to cancer cell migration and metastasis, in prostate cancer cells is associated with ADT. Moreover, nestin expression was detected in tumors of 75% of prostate cancer patients with deadly androgen-independent disease. On the other hand, nestin was undetectable in tumors from patients without ADT.

7.2. Target the AR at an Early Stage

It is apparent that ADT will have some beneficial effects on early prostate cancer, but will eventually fail in view of the dual roles of androgens-AR signaling in prostate cancer progression. Therefore, it may be better to target AR at an early stage of prostate cancer when the progression and metastasis of the tumor is predominantly dependent on stromal AR function. This approach is best exemplified by the timing of induction of knockdown AR in ind-ARKO-TRAMP mice, as early induction at 4- to 12-weeks of age resulted in inhibition of prostate tumor growth and metastasis, while late induction at 20-weeks old failed to inhibit tumor growth and metastasis [53]. Thus, due to the dual functions of the AR, treatment of prostate cancer patients with ADT may have good response at the early stage of cancer but worsen the prognosis at later stages of cancer progression when the tumors are less dependent on the stromal AR derived signals for growth, a scenario apparently observed in castrated TRAMP mice [106]. For example, combined ADT of advanced prostate cancer provided at best a 5% increase in 5 year survival rate over ADT alone [107], but did not cure the disease. On the other hand, therapies aiming at maximal eradication of cancer cells including the CSC and its progenitors such as radical prostatectomy [108-110] and radiation therapy [111-114], in conjunction with ADT, should provide a better clinical outcome than ADT alone, particularly in localized prostate cancer.

Strategies targeting the tumor stroma [115,116], also should be beneficiary to prostate cancer patients as exemplified by the better treatment effect of combining therapy of ADT with anti-angiogenic agents targeting stromal derived vascular factors [117] in the Dunning rat prostate cancer model. Similarly, combining ADT with Trk tyrosine kinase inhibitors targeting the receptor for stroma-derived AR-independent nerve growth factor was reported to prolong tumor regression in rat prostate cancer model [118]. However, if the disease recurs after these treatments, the adjuvant ADT may still fail because the recurrent tumor may become independent of the stromal AR for growth and more invasive due to abrogation of the metastasis suppressor function of the epithelial AR.

7.3. Concomitant Treatment of AR Antagonism with Anti-Metastasis Agents

As AR antagonism and adjunct treatments cannot prevent development of metastatic prostate tumors, it will be better to combine ADT with anti-metastatic agents to improve treatment effects. In metastatic tumors of pes-ARKO-TRAMP and PC3-v cells, the expression of various tumor metastasis-related genes were significantly changed in favor of invasion and metastasis as compared to the tumors of wt TRAMP mice and PC3-AR9 cells, respectively. Thus, several pro-metastasis genes such as cyclooxgenase-2 (Cox-2), matrix metaloproteinase-9 (MMP-9), interleukin-6 (IL-6), and tumor necrosis factor-α (TNF-α) were elevated, while anti-metastasis genes such as neutral endopeptidase (NEP) and P27(kip1) were decreased in the tumors of pes-ARKO-TRAMP mice and PC3-v orthotopic grafts compared to the tumors of their AR-expressing counterparts [48]. Part of these effects of the AR appeared to be mediated through the suppression of TGFβ1, as expression of the AR in prostate cancer cells resulted in suppression of TGFβ1 expression, and overexpression of TGFβ1 resulted in increased invasiveness and similarly altered expressions of the above metastasis-related genes [48].

As patients with recurrent prostate cancer usually die of metastatic disease, their survival can be extended if tumor metastasis can be delayed or prevented through additional treatment regimens. Treatments targeting metastasis-related genes down stream to the epithelial AR signaling should improve clinical results of adjunct ADT. For example, treatments targeting

TGFβ1 and/or its receptor have been currently under clinical trial [119]. Current available agents targeting Akt, COX-2, and MMP-9, or Vitamin E and other relevant anti-metastasis agents as previously proposed [120,121], in conjunction with ADT might have clinical benefits for patients with recurrent prostate cancer [48]. In addition, agents targeting bone metastases such as endothelin receptor antagonists [122] appears to be an encouraging treatment to combat metastatic prostate cancer.

Therefore, it may be concluded that the discovery of the dual roles of the AR on prostate cancer progression and metastasis undoubtedly will provoke consideration of new treatment strategies to combat this deadly disease.

Future Direction

It is apparent that successful treatment of prostate cancer will rely on complete eradication of prostate cancer cells. This treatment objective can be achieved only by prostatectomy when the disease is localized. Unfortunately detection of prostate cancer at such an early stage is less frequent than more advanced diseases. Since the AR functions as an important regulator of proliferation and metastasis during prostate cancer progression, it remains necessary to target the AR for prostate cancer treatment. In view of the dual yet opposite roles of the stromal and epithelial AR as discussed earlier, it becomes apparent that targeting the stromal AR specifically in an early stage would benefit the patient better than ADT that targets the whole body's androgens-AR signal. Thus, future development of stromal AR-specific antagonists, such as stromal cell-specific delivery of AR anti-sense oligonucleotides, siRNA, or antagonists that target specifically the stromal AR-AR coregulator complex, would provide a better treatment regimen for delaying development of metastatic prostate cancer and prolong the survival of patients.

Conclusion

ADT aiming at reducing androgens-AR signaling has been employed for treatment of prostate cancer patients during the past 7 decades. Although ADT is always effective initially, it invariably fails after long-term treatment albeit the AR remains functionally active in the recurrent tumor. Recent studies with prostate stromal- and epithelial-specific ARKO TRAMP models as wells as *in vitro* and *in vivo* studies of various human prostate cancer cells with knock-in and knock-out of the AR have indicated that prostatic stromal and epithelial AR play dual yet opposite roles to modulate prostate development, tissue homeostasis, prostate cancer progression and metastasis. In normal prostate, the stromal AR functions as a stimulator of prostate epithelial cell proliferation and survival factor for luminal epithelial cells, whereas the epithelial AR function as a suppressor of proliferation and a survival factor for luminal cells. In prostate cancer, the stromal AR functions as a stimulator of prostate cancer cell proliferation and metastasis, whereas the epithelial AR functions as a suppressor of prostate cancer cell proliferation and metastasis. The relative influences of the opposite functions of the stromal and epithelial AR may depend on the stages of prostate cancer progression. The failure of ADT appears due to inhibition of the suppressor roles of the epithelial AR, which becomes the dominant factor at a late stage of cancer progression. Therefore, future treatment of prostate cancer should consider targeting the stromal AR specifically at an early stage, while applying ADT in combination with appropriate anti-metastasis agents.

References

[1] Jemal, A. *et al.* Cancer statistics 2005, *CA Cancer J. Clin.* 55,10-30 (2005).

[2] Huggins, C. and Hodges, C. V. Studies on prostatic cancer. I. The effect of castration, of estrogen and of androgen injection on serum phosphatases in metastatic carcinoma of the prostate. *Cancer Res.* 1, 293-297 (1941).

[3] Denis, L. J. and Griffiths, K. Endocrine treatment in prostate cancer. *Sem. Surg. Oncol.* 18, 52-74 (2000).

[4] Heinlein, C. A. and Chang, C. Androgen receptor in prostate cancer. *Endocr. Rev.* 25, 276-308 (2004).

[5] Roy, A. K. *et al.* Regulation of androgen action. *Vit. Horm.* 55, 309-352 (1999).

[6] Heinlein, C. A. and Chang, C. Androgen receptor (AR) coregulators: an overview. *Endocr. Rev.* 23, 175-200 (2002).

[7] Rahman, M., Miyamoto, H. and Chang, C. Androgen receptor coregulators in prostate cancer: mechanism and clinical implication. *Clin. Cancer Res.* 10, 2208-2219 (2004).

[8] Wang, L, Hsu, C-L and Chang, C. Androgen receptor corepressors: an overview. *Prostate* 63, 117-130 (2005).

[9] Buchanan, G., Irvine, R. A., Coetzee, G. A., and Tilley, W. D. Contribution of the androgen receptor to prostate cancer predisposition and progression. *Cancer Metastasis Rev.* 20, 207-223 (2001).

[10] Cunha, G. R. *et al.* The endocrinology and developmental biology of the prostate. *Endocr. Rev.* 8, 338-362 (1987).

[11] Mohler, J. L. *et al* Androgen and glucocorticoid receptors in the stroma and epithelium of prostatic hyperplasia and carcinoma. *Clin Cancer Res.* 2, 889-895 (1996).

[12] van der Kwast, T. H. *et al.* Androgen receptors in endocrine-therapy-resistant human prostate cancer. *Int. J. Cancer* 48, 189-193 (1991).

[13] Sadi, M. V., Walsh, P. C., and Barrack, E. R. Immunohistochemical study of androgen receptors in metastatic prostate cancer. *Cancer* 67, 3057-3064 (1991).

[14] Chodak, G. W., *et al.* Nuclear localization of androgen receptor in heterogeneous samples of normal, hypeplastic and neoplastic human prostate. *J. Urol.* 147, 798-803 (1992).

[15] Hobisch, A. *et al.* Androgen receptor status of lymph node metastases from prostate cancer. *Prostate* 28, 129-135 (1996).

[16] Cunha, G. R. et al. Hormonal, cellular and molecular regulation of normal and neoplastic prostatic development. *J. Steroid Biochem. Mol. Biol.* 92, 221-236 (2004).

[17] Cooke PS, Young P, Cunha GR. Androgen receptor expression in developing male reproductive organs. *Endocrinology* 128(6), 2867–2873 (1991).

[18] Kinbara, H. and Cunha, G. R. Ductal heterogeneity in rat dorsal-lateral prostate. *Prostate* 28, 58-64 (1995).

[19] Hayward, S. W. *et al.* Stromal development in the ventral prostate, anterior prostate and seminal vesicle of the rat. *Acta Anatom.* 155, 94-103 (1996).

[20] Hayward, S. W., Rosen MA, Cunha GR. Stromal-epithelial interactions in normal and neoplastic prostate. *Br. J. Urol.* 79 (Suppl. 2), 18-26 (1997).

[21] Cunha, G. R. and Lung, B. The possible influences of temporal factors in androgenic responsiveness of urogenital tissue recombinants from wild-type and androgen-insensitive (Tfm) mice. *J. Exp. Zool.* 205, 181-194 (1978).

[22] Donjacour, A. A. and Cunha, G. R. Assessment of prostatic protein secretion in tissue recombinants made of urogenital sinus mesenchyme and urothelium from normal or androgen-insensitive mice. *Endocrinology* 131, 2342–2350 (1993).

[23] Cunha, G.R. and Young, P. Inability of Tfm (testicular feminization) epithelial cells to express androgen-dependent seminal vesicle secretory proteins in chimeric tissue recombinants. *Endocrinology* 128, 3293–3298 (1991).

[24] Donjacour, A. A. and Cunha, G. R. The effect of androgen deprivation on branching morphogenesis in the mouse prostate. *Dev. Biol.* 128, 1-14 (1988).

[25] Hayward, S. W., Cunha, G. R. and Dahiya, R. Normal development and carcinogenesis of the prostate. A unifying hypothesis. *An. N. Y. Acad. Sci.* 784, 50–62 (1996).

[26] Cunha, G. R., *et al.* Role of epithelial–mesenchymal interactions in the differentiation and spatial organization of visceral smooth muscle. *Epithel Cell Biol.* 1, 76–83 (1992).

[27] Hayward, S. W., *et al.* Interactions between adult human prostatic epithelium and rat urogenital sinus mesenchyme in a tissue recombination model. *Differentiation* 63, 131–40 (1998).

[28] Hayward, S. W., Del Buono, R., Deshpande, N. and Hall, P. A. A functional model of adult human prostate epithelium: the role of androgen and stroma in architectural organization and the maintenance of differentiated secretory function. *J. Cell Sci.* 102, 361-372 (1992).

[29] Issacs, J. T. Antagonistic effect of androgen on prostatic cell death. *Prostate* 5, 545-557 (1984).

[30] Sugimura, Y., Cunha, G. R. and Donjacour, A. A. Morphological and histological study of castration-induced degeneration and androgen-induced regeneration in the mouse prostate. *Biol. Reprod.* 34, 973-983 (1986).

[31] Kyprianou, N., and Isaacs J. T. Activation of programmed cell death in the rat ventral prostate after castration. *Endocrinology* 122, 552–562 (1988).

[32] Yu S, Yeh CR, Niu Y, Chang HC, Tsai YC, Moses HL, Shyr CR, Chang C, Yeh S. Altered prostate epithelial development in mice lacking the androgen receptor in stromal fibroblasts. *Prostate.* Jul 7. doi: 10.1002/pros.21445 (2011).

[33] Yeh, S. *et al.* Generation and characterization of androgen receptor knockout (ARKO) mice: an *in vivo* model for the study of androgen functions in selective tissues. *Proc. Natl. Acad. Sci. USA* 99, 13498-13503 (2002).

[34] Wu X. *et al*, Generation of a prostate epithelial cell-specific Cre transgenic mouse model for tissue-specific gene ablation. *Mech. Dev.* 101, 61–69 (2001).

[35] Wu, C-T., *et al.* Increased prostate cell proliferation and loss of cell differentiation in mice lacking epithelial androgen receptor. *Pro. Natl. Acad. Sci. USA.* 104, 12679-12684 (2007).

[36] van Leenders, G.J.L.H., Aalders, T. W., Hulsbergen-van de Kaa, C. A., Ruiter, D. J. and Schalken, J. A. Expression of basal cell keratins in human prostate cancer metastases and cell lines. *J. Pathol.* 195, 563-570 (2001).

[37] Han, G. *et al.* Mutation of the androgen receptor causes oncogenic transformation of the prostate. *Proc. Natl. Acad. Sci. USA*. 102, 1151-1156 (2005).

[38] Ling M. T., Chan K. W., and Choo, CK. Androgen induces differentiation of a human papillomavirus 16 E6/E7 immortalized prostate epithelial cell line. *J. Endocrinol.* 70, 287-296 (2001).

[39] Whitacre, D. C., *et al.* Androgen induction of in vitro prostate cell differentiation. *Cell Growth Differ.* 13, 1-11 (2002).

[40] Niu Y, *et al.* Increased CK5/CK8-positive intermediate cells with stromal smooth muscle cell atrophy in the mice lacking prostate epithelial androgen receptor. *PLoS One.* 6(7):e20202 (2011).

[41] Condon, M. S. The role of the stromal microenviroment in prostate cancer. *Sem. Cancer Biol.* 15, 132-137 (2005).

[42] Cunha, G. R., Hayward, S. W., Wang, Y. Z. and Ricke, W. A. Role of the stromal microenvironment in carcinogenesis of the prostate. *Int. J. Cancer* 107, 1-10 (2003).

[43] Thompson, T. C. *et al.* Genetic predisposition and mesenchymal-epithelial interactions in ras+myc-induced carcinogenesis in reconstituted mouse prostate. *Mol. Carcinog.* 7, 165-179 (1993).

[44] Olumi, A. F. *et al.* Carcinoma-associated fibroblasts direct tumor progression in initiated human prostatic epithelium. *Cancer Res.* 59, 5002-5011 (1999).

[45] Wang, Y. *et al.* A human prostatic epithelial model of hormonal cacinogenesis. Cancer Res. 61, 6064-6072 (2001).

[46] Ricke, W. A. *et al.* Steroid hormones stimulate human prostate cancer progression and metastasis. *Int. J. Cancer* 118, 2123-2131 (2006).

[47] Webber, M. M. *et al.* A human prostatic stromal myofibroblast cell line WPMY-1: a model for stromal-epithelial interactions in prostatic neoplasia. *Carcinogenesis* 20, 1185-1192 (1999).

[48] Niu, Y., *et al.* Androgen receptor is a tumor suppressor and stimulator in prostate cancer metastasis. *Proc. Natl. Acad. Sci. USA*. 105(34), 12182-7 (2008).

[49] Pilatus, U. *et al.* Imaging prostate cancer invasion with multi-nuclear magnetic resonance methods: the Metabolic Boyden Chamber. *Neoplasia* 2, 273-279 (2000).

[50] Marques, R. B., *et al.* Androgen receptor modifications in prostate cancer cells upon long-term androgen ablation and antiandrogen treatment. *Int. J. Cancer* 117, 221–229 (2005).

[51] Halin, S., Hammarsten, P., Wikström, P. and Bergh, A. Androgen-independent prostate cancer cells transiently respond to castration treatment when growing in an androgen-dependent prostate environment. *Prostate* 67, 370-377 (2007).

[52] Gingrich, J. R., *et al.* Androgen-independent prostate cancer progression in the TRAMP model. *Cancer Res.* 57, 4687-4691 (1997).

[53] Niu Y., et al. Targeting the stromal androgen receptor in primary prostate tumors at earlier stages. *Proc. Natl. Acad. Sci. USA*. 105(34), 12188-93 (2008).

[54] Litvinov, I. V., De Marzo, A. M., and Isaacs, J. T. Is the Achilles' heel for prostate cancer therapy: a gain of function in androgen receptor signaling? *J Clin Endocrinol Metab.* 88, 2972-2982 (2003).

[55] Mirosevich, J., *et al.* Androgen receptor expression in proliferating basal and luminal cells in adult murine ventral prostate. *J. Endocrinol.* 162, 341-350 (1999).

[56] van Leenders, G. J. L. H., and Schalken J. A. Epithelial cell differentiation in the human prostate epithelium: Implications for the pathogenesis and therapy of prostate cancer. *Oncol. Hemat.* 46, S3–S10 (2003).

[57] Liu, A. Y., Roudier, M. P. and True L. D. Heterogeneity in primary and metastatic prostate cancer as defined by cell surface CD profile. *Am. J. Pathol.* 165, 1543-1556 (2004).

[58] Bhatia, B., *et al.* Cell-autonomous induction of functional tumor suppressor 15-lipoxygenase 2 (15-LOX2) contributes to replicative senescence of human prostate progenitor cells. *Oncogene* 24, 3583–3595 (2005).

[59] Kaighn, M. E., Narayan, K. S., Ohnuki, Y., Lechner, J. F. and Jones, L. W. Establishment and characterization of a human prostatic carcinoma cell line (PC3). *Invest. Urol.* 17, 16-23 (1979).

[60] van Bokhoven, A., et al. Molecular characterization of human prostate carcinoma cell lines. Prostate 57, 205-225 (2003).

[61] Yuan, S., *et al.* Androgen induced inhibition of cell proliferation in an androgen-insensitive prostate cancer cell line (PC3) transfected with a human androgen receptor complementary DNA. *Cancer Res.* 53, 1304-1311 (1993).

[62] Garcia-Arenas, R., *et al.* The expression of prostatic acid phosphotase is transcriptionally regulated in human prostate carcinoma cells. *Mol. Cell. Endocrinol.* 111, 29-37 (1995).

[63] Yeh, S., *et al.* Increase of androgen-induced cell death and androgen receptor transactivation by BRCA1 in prostate cancer cells. *Proc. Natl. Acad. Sci. USA*. 97, 11256-11261 (2000).

[64] Litvinov, I. V., Antony, L, and Isaacs, J. T. Molecular characterization of an improved vector for evaluation of the tumor suppressor versus oncogene abilities of the androgen receptor. *Prostate 61*, 299-304 (2004).

[65] Litvinov, I. V., *et al.* PC3, but not DU145, human prostate cancer cells retain the coregulators required for tumor suppressor ability of androgen receptor. *Prostate* 66, 1329-1338 (2006).

[66] Altuwaijri, S., *et al.* Expression of human AR cDNA driven by its own promoter results in mild promotion, but ot suppression, of growth in human prostate cancer PC3 cells. *Asian J. Androl.* 9, 181-188 (2007).

[67] Goater, J. J., O'Keefe, R. J., Rosier, R. N., Puzas, J. E. and Schwarz, E. M. Efficacy of *ex vivo* OPG gene therapy in preventing wear debris induced osteolysis. *J. Orthop. Res.* 20, 169-173 (2002).

[68] Nelius, T., *et al.* Androgen receptor target NFkB and TSP1 to suppress prostate tumor growth *in vivo*. *Int. J. Cancer* 121, 999-1008 (2007).

[69] Horoszewicz, J. S., *et al.* LNCaP model of human prostatic carcinoma. *Cancer Res.* 43, 1809-1818 (1983).

[70] Veldscholte, J., *et al.* A mutation in the ligand binding domain of the androgen receptor of human LNCaP cells affects steroid binding characteristics and response to anti-androtens. *Biochem. Biophys. Res. Commun.* 17, 534-540 (1990).

[71] Soto, A. M., et al. Variants of the human prostate LNCaP cell line as tools to study discrete components of the androgen-mediated proliferative response. *Oncol. Res.* 7, 545-558 (1995).

[72] Kokontis, J. M., Takakura, K., Hay, N. and Liao, S. Increased androgen receptor activity and altered c-myc expression in prostate cancer cells after long-term androgen deprivation. *Cancer Res.* 54, 1566-1573 (1994).

[73] Igawa, T. *et al.* Establishment and characterization of androgen-independent human prostate cancer LNCaP cell model. *Prostate* 50, 222-235 (2002).

[74] Kokontis, J. M., Hay, S., and Liao, S. Progression of LNCaP prostate tumor cells during androgen deprivation: hormone-independent growth, repression of proliferation by androgen, and role for $p27^{Kip1}$ in androgen-induced cell cycle arrest. *Mol. Endocrinol.* 12, 941-953 (1998).

[75] Joly-Pharaboz, M. O., *et al.* Inhibition of growth and induction of apoptosis by androgens of a variant of LNCaP cell line. *J. Steroid Biochem. Mol. Biol.* 73, 237-249 (2000).

[76] Chuu, C-P., *et al.* Inhibition of tumor growth and progression of LNCaP prostate cancer cells in athymic mice by androgen and liver X receptor agonist. *Cancer Res.* 66, 6482-6486 (2006).

[77] Zhou, JR., Yu, L., Zerbini, L. F., Libermann, T. A. and Blackburn, G. L. Progression to androgen-independent LNCaP human prostate tumors: cellular and molecular alterations. *Int. J. Cancer* 110, 800-806 (2004).

[78] Thalmann, G. N. *et al.* LNCaP progression model of human prostate cancer: androgen-independence and osseous metastasis. *Prostate* 44, 91-103 (2000).

[79] Eder, I. E., et al. Inhibition of LNCaP prostate cancer cells by means of androgen receptor antisense oligonucleotides. *Cancer Gene Ther.* 7, 997-1007 (2000).

[80] Liao, X., Tang, S., Thrasher, J. B., Griebling, T. L. and Li, B. Small interfering RNA-induced androgen receptor silencing leads to apoptotic cell death in prostate cancer. *Mol. Cancer Ther.* 4, 505-515 (2005).

[81] Hååg, P., Bektic, J., Bartsch, G., Klocker, H. and Eder, I. E. Androgen receptor down regulation by small interference RNA induces cell growth inhibition in androgen sensitive as well as androgen independent prostate cancer cells. *J. Steroid Biochem. Mol. Biol.* 96, 251-258 (2005).

[82] Eder, I. E., *et al.* Inhibition of LNCaP prostate tumor growth in vivo by an antisense oligonucleotide directed against the human androgen receptor. *Cancer Gene Ther.* 9, 117-125 (2002).

[83] Nagabhushan, M., *et al.* CWR22: the first human prostate cancer xenograft with strongly androgen-dependent and relapsed strains both in vivo and in soft agar. *Cancer Res.* 56, 3042–3046 (1996).

[84] Wainstein M. A. *et al.* A., CWR22: androgen-dependent xenograft model derived from a primary human prostatic carcinoma. *Cancer Res.* 54, 6049-6052 (1994).

[85] Yeh, S. *et al.* Abnormal mammary gland development and growth retardation in female mice and MCF7 breast cancer cells lacking androgen receptor. *J. Exp. Med.* 198, 1899-1908 (2003).

[86] van Weerden, W. M., *et al.* Development of seven new human prostate tumor xenograft models and their histopathological characterization. *Am. J. Pathol.* 149, 1055-1062 (1996).

[87] van Weerden, W. M., and Romijn, J. C. Use of nude mouse xenograft models in prostate cancer research. *Prostate* 43, 263-271 (2000).

[88] Titus, M. A., Schell, M. J., Lih, F. B., Tomer, K. B. and Mohler, J. L. Testosterone and dihydrotestosterone tissue levels in recurrent prostate cancer. *Clin. Cancer Res.* 11, 4653-4657 (2005).

[89] Issacs, J. T. and Coffey D. S. Etiology and disease process of benign prostate hyperplasia. *Prostate suppl.* 2, 33-50 (1989).

[90] Reya, T., Morrison, S. J., Clarke, M. F. and Weissman, I. L. Stem cells, cancer, and cancer stem cells. *Nature* 414, 105-111 (2001).

[91] Lawson, D. A. and Witte O. N. Stem cells in prostate cancer initiation and progression. *J. Clin. Invest.* 117, 2044-2050 (2007).

[92] Collins, A. T. and Maitland N. J. Prostate cancer stem cells. *Eur. J. Cancer* 42, 1213-1218 (2006).

[93] Collins, A. T., Habib, F. K., Maitland, N. J. and Neal, D. E. Identification and isolation of human prostate epithelial stem cells based on alpha(2)beta(1)-integrin expression. *J. Cell. Sci.* 114, 3865-3872 (2001).

[94] Richardson, G. D., *et al.* CD133, a novel marker for human prostatic epithelial stem cells. *J. Cell Sci.* 117, 3539-3545 (2004).

[95] Heer, R., Robson, C. N., Shenton, B. K. and Leung, H. Y. Role of androgen in determining differentiation and regulation of androgen receptor expression in the human prostatic epithelial transient amplifying population. *J. Cell. Physiol.* 212, 572-578 (2007).

[96] Patrawala L., *et al.* Highly purified $CD44^+$ prostate cancer cells from xenograft human tumors are enriched in tumorigenic and metastatic progenitor cells. *Oncogene* 25, 1696-1708 (2006).

[97] Collins, A. T., Berry, P. A., Hyde, C., Stower, M. J. and Maitland, N. J. Prospective identification of tumorigenic prostate cancer stem cells. *Cancer Res.* 65, 10946-10951 (2005).

[98] Bochoysky N. (1993) in Cancer Medicine. 3rd ed. Holland, J. F., et al. (eds), Lea and Febiger, Philadelphia, pp.884-896.

[99] Eisenberger, M. A., *et al.* Bilateral orchiectomy with or without flutamide for metastatic prostate cancer. *N. Engl. J. Med.* 339, 1036-1042 (1998).

[100] Mostaghel, E. A., *et al.* Intraprostatic androgens and androgen-regulated gene expression persist after testosterone suppression: therapeutic implications for castrationresistant prostate cancer. *Cancer Res* 67, 5033-5041 (2007).

[101] Visakorpi, T., *et al. In vivo* amplification of the androgen receptor gene and progression of human prostate cancer. *Nat Genet* 9:401-406 (1995).

[102] Bruchovsky, N., *et al.* (1990) Effects of androgen withdrawal on the stem cell composition of the Shionogi carcinoma. *Cancer Res* 50, 2275-2282 (1990).

[103] Kleeberger, W., *et al.* Roles of the stem cell-associated intermediate filament nestin in prostate cancer migration and metastasis. *Cancer Res.* 67, 9199-9206 (2007).

[104] Li, R., *et al.* High level of androgen receptor is associated with aggressive clinicopathologic features and decreased biochemical recurrence-free survival in prostate: cancer patients treated with radical prostatectomy. *Am. J. Surg. Pathol.* 28, 928-934 (2004).

[105] Banach-Petrosky, W., *et al.* Prolonged exposure to reduced levels of androgen accelerates prostate cancer progression in Nkx3.1; Pten mutant mice. *Cancer Res.* 67, 9089-9096 (2007).

[106] Johnson, M. A., *et al.* Castration triggers growth of previously static androgen-independent lesions in the transgenic adenocarcinoma of the mouse prostate (TRAMP) model. *Prostate* 62, 322-338 (2005).

[107] Schmitt, B., *et al.* Combined androgen blockade with non-steroidal anti-androgens for advanced prostate cancer. A systemic review. *Urology* 57, 727-732 (2001).

[108] DeKernion, J. B., *et al.* Prognosis of patients with stage D1 prostate carcinoma following radical prostatectomy with and without early endocrine therapy. *J. Urol.* 144, 700-703 (1990).

[109] Zincke, H., *et al.* Stage D1 prostate cancer treated by radical prostatectomy and adjuvant hormonal treatment: evidence for favorable survival in patients with DNA diploid tumors. *Cancer* 70(Suppl.1), 311-323 (1992).

[110] Messing, E. M., *et al.* Immediate hormonal therapy compared with observation after radical prostatectomy and pelvic lymphadenectomy in men with node-positive prostate cancer. *N. Eng. J. Med.* 341, 1781-1788 (1999).

[111] Bolla, M., *et al.* Long-term results with immediate androgen suppression and external irradiation in patients with locally advanced prostate cancer (an EORTC study): a phase II randomized trial. *Lancet* 360, 103-108 (2002).

[112] Pilepich, M. V., *et al.* Androgen suppression adjuvant to definitive radiotherapy in prostate carcinoma—long-term results of phase III RTOG 85–31. *Int. J. Radiat. Oncol. Biol. Phys.* 61, 1285-1290 (2005).

[113] Horwitz, E. M., *et al.* Subset analysis of RTOG 85-31 and 86-10 indicates an advantage for long-term vs.shor-term adjuvant hormones for patients with locally advanced nonmetastatic prostate cancer treated with radiation therapy. *Int. J. Radiat. Oncol. Biol. Phys.* 49, 947-956 (2001).

[114] Hanks, G. E., *et al.* Phase III trial of long-term adjuvant androgen deprivation after neoadjuvant hormonal cytoreduction and radiotherapy in locally advanced carcinoma of the prostate: the Radiation Therapy Oncology Group Protocol 92-02. *J. Clin. Oncol.* 21, 3972-3978 (2003).

[115] Hofmeister, V., Schrama, D. and Becker, J. C. Anti-cancer therapies targeting the tumor stroma. *Cancer Immunol. Immunother.* 57, 1-17 (2008).

[116] Bouzin, C. and Feron, O. Targeting tumor stroma and exploiting mature tumor vasculature to improve anti-cancer drug delivery. *Drug Resistance Updates* 10, 109–120 (2007).

[117] Johansson, A., *et al.* A stroma targeted therapy enhances castration effects in a transplantable rat prostate cancer model. *Prostate* 67, 1664–1676 (2007).

[118] George, D. J., *et al.* Sustained *in vivo* regression of Dunning H rat prostate cancers treated with combination of androgen ablation and Trk tyrosine kinase inhibitors, CEP-751 (KT-6587) or CEP-701 (KT-5555). *Cancer Res.* 59, 2395-2401 (1999).

[119] Pinkas, J. and Teicher, B. A. TGF-□ in cancer and as a therapeutic target. *Biochem. Pharmacol.* 72, 523-529 (2006).

[120] Miyamoto, H., Altuwaijri, S., Cai, Y., Messing, E. M. and Chang, C. Inhibition of the Akt, cyclooxygenase-2, and matrix metalloproteinase-9 pathways in combination with

androgen deprivation therapy: potential therapeutic approaches for prostate cancer. *Mol. Carcinog*. 44, 1-10 (2005).

[121] Fujimoto, N., Chang, C., Nomura, M. and Matsumoto, T. Can we prevent prostate cancer? Rationale and current status of prostate cancer chemoprevention. *Urol. Int*. 74, 289-297 (2005).

[122] Carducci, M. A. and Jimeno, A. Targeting bone metastasis in prostate cancer with endothelin receptor antagonists. *Clin. Cancer Res*. 12(20 Suppl.) 6296s-6300s (2006).

In: Androgens: Production, Functions and Disorders
Editors: B. F. Thompson and D. J. Robinson
ISBN: 978-1-62081-145-0

Chapter V

Endogenous Pathways for Steroidogenesis and Androgen Signalling in Prostate Cancer Cells

***Nigel C. Bennett*[1], *John D. Hooper*[2] *and Glenda C. Gobe*[*1]**
[1] School of Medicine, University of Queensland,
Princess Alexandra Hospital, Brisbane, Australia
[2] Mater Medical Research Institute, South Brisbane, Australia

Summary

In normal prostatic epithelial cells, androgen receptor (AR) expression and activity are regulated by androgens like dihydrotestosterone (DHT) and testosterone. These androgens are mainly produced in the testis. Prostate cancer (PCa) retains the ability to respond to androgens in the early stage of development, in contrast with advanced stages that become independent of exogenous androgens while maintaining expression of functional AR. Targeting of androgens and AR has become the foremost approach for successful PCa palliation, especially for metastatic disease, with androgen ablation inhibiting PCa growth. Chemical castration, surgery (orchidectomy/orchiectomy), or androgen deprivation therapies are effective in most cases. This outcome, however, is temporary because an important effect of androgen deprivation is that recurrent PCa evolves to a castrate-resistant phenotype. It has recently become apparent that PCa cells in advanced stages of the disease synthesize androgens from cholesterol, and adrenal precursors, *de novo* and hence survive independently of testicular androgens. A corollary is that the steroidogenic enzymes required for androgen synthesis from cholesterol need to be present in PCa for this to happen. Our recent data indicate these are present but to varying extents in different patient samples. This book chapter will discuss the current knowledge about presence and extent of steroidogenic

[*] Corresponding author: Assoc Prof Glenda C Gobe, Centre for Kidney Disease Research School of Medicine, University of Queensland, Princess Alexandra Hospital, Woolloongabba, Brisbane, AUSTRALIA, 4102, Phone: 61-7-31765655; Fax: 61-7-31762970, Email: g.gobe@uq.edu.au.

enzymes in PCa compared with normal prostate cells and present a theory about the importance of individualising patient therapies according to the steroidogenic potential of their PCa as it develops.

Introduction

Recent reports have revealed that prostate cells, tissues and malignancies, under certain therapeutic conditions, possess the potential to produce steroids, more specifically androgens, endogenously. This ability removes the inherent reliance of prostatic tissues upon testicular androgens for survival and has major implications for treatment of prostate cancer (PCa).

In a genetically-normal human male, the androgen testosterone is synthesised during steroidogenesis and secreted by the Leydig cells in the testes [1]. Side chain cleavage from cholesterol by the CYP11A1 gene product, P450scc, is the rate-limiting step in steroidogenesis. This process is stimulated by luteinizing hormone (LH) and regulated by a feedback mechanism involving the action of testosterone on the pituitary and hypothalamus. Testosterone, the main circulating androgen, is carried by blood in one of three ways: as free androgen (approximately 2%); through low affinity binding to serum albumin (≥60%); or through high affinity binding to sex hormone-binding globulin (SHBG) (≥40%) [2]. When delivered to the cellular bilayer of the prostatic glandular epithelium, testosterone is converted to 5α-dihydrotestosterone (DHT) by the enzyme 5α-reductase type 2 [3]. The androgen DHT is 2-10 times more biologically-potent than testosterone and binds AR with greater affinity.[4] Both of these androgens are bound by the ligand binding domain (LBD) of the androgen receptor (AR) and predominantly activate AR genomic functions. AR function and regulation by androgen is essential not only for development of male sexual characteristics and the prostate but for the development of many organs in both men and women [5]. In addition, in adult males, androgen signalling is necessary to maintain muscle mass, muscle strength, libido, bone density, erythropoiesis and spermatogenesis [6]. AR activity, via androgen signalling, mediates cellular proliferation, differentiation, apoptosis and metabolism thus functioning in tissue maintenance and homeostasis [5].

Misregulation of the androgen signalling axis can induce detrimental effects. Consequently, development of diseases such as hypogonadism [7], benign prostatic hyperplasia (BPH) [8, 9], male pattern baldness [10, 11], androgen insensitivity syndrome (AIS) [12], and prostate cancer (PCa) [13] are common. Additionally, aberrant production of androgens in PCa tissues can aid disease progression and hinder treatment regimens. This chapter will endeavour to explain the integral role that androgens and their receptor play in PCa development and progression. PCa is the main prostatic disease discussed in this chapter, with special consideration for *de novo* steroidogenesis.

The Prostate

The prostate is located at the base of the bladder, surrounds the urethra, and is anterior to the ampulla of the rectum. It is the largest accessory gland in the male reproductive system. Two thirds of the prostate is predominantly glandular while the other is fibromuscular. The glandular ducts produce prostatic fluid which equates to approximately 20% of seminal volume. The luminal epithelial cells of prostatic glands are believed to be the cells of origin for prostate cancer (PCa). However recent data using a xenograft mouse model show human primary basal cells mimic the histological and molecular features of human PCa, with loss of glandular basal cells, over-proliferation of luminal cells and expression of prostate specific antigen (PSA). These results illustrate that characterization of PCa using histology alone is not indicative of cellular origins, and significant phenotypic changes may occur even in very early PCa stages [14].

Androgen Receptor

To appreciate the relationship between AR, androgen signalling and steroidogenesis in cells throughout the human body, the complexities of AR function and regulation need to be considered. Androgens entering the cytoplasm of cells in the prostatic epithelium are bound by AR and elicit effects that are both genomic and non-genomic (i.e. cytoplasmic) in nature. And the variations of cellular regulation imposed upon AR activity are vast in number. AR is a nuclear transcription factor integral in mediating androgen signalling and is expressed in androgen target tissues such as reproductive tissues (e.g. seminal vesicle, testes and prostate), liver, central nervous system and skeletal muscle. The prostate is one of the highest expressing AR tissues as determined by real-time polymerase chain reaction (PCR) assays [15]. Androgen binding to AR affects AR conformation, AR-protein interactions, induces AR nuclear shuttling, induces chromatin remodelling and encourages AR-DNA interactions.

Briefly, inactive AR does not contain bound androgen. It is located in the cytoplasm associated with heat shock proteins (HSPs) [16-18], the cytoskeleton [19], and chaperones [18, 20]. Androgen binding to AR induces a conformational change in AR, resulting in the release of HSPs. Other proteins such as importin-α and the 70 kDa AR-associated (ARA70) protein are then recruited to induce AR stabilisation and nuclear translocation. In the nucleus, AR dimerises and binds numerous possible co-activators, which along with the AR-DNA-binding domain (AR-DBD) proteins facilitate DNA binding in gene promoter regions. DNA binding in turn promotes recruitment of SRC/p160 co-activator family members and the histone acetyltransferases (CBP/p300, P/CAF) resulting in chromatin remodelling. Chromatin remodelling is essentially an unwinding or loosening of chromatin structure which allows binding of TATA binding protein then consequent recruitment of general transcription factors and RNApolII for transcription initiation.

The human AR gene is found on the X chromosome (q11-12) and consists of eight exons, generating a cDNA approximately 2760 nucleotides in length and a protein consisting of approximately 920 amino acids. Like the other nuclear receptor superfamily members, AR is a modular protein consisting of four structurally and functionally distinct domains. These are the poorly conserved amino-terminal domain (NTD), a highly conserved DBD, a small hinge region and a reasonably conserved carboxyl-terminal ligand-binding domain (LBD) [21]. There are several excellent reviews that describe the activation process propagated by AR-androgen interaction [16-18, 22], and the structure and function of the AR domains [23-29].

The AR-NTD (αα 1-558) is constitutively active with an inherent ability to activate transcription without androgen in LBD negative AR mutants [30, 31]. It contains a polymorphic tri-nucleotide (CAG) repeat region encoding for poly-glutamine (poly-Q) [32]. The CAG repeat varies in length between 11 and 31 tri-nucleotide repeats [33]. It has been proposed that this variation in repeat length is associated with higher risk of BPH (≤19 repeats), defective spermatogenesis (≥28 repeats) and PCa (≤18 repeats) [7, 34] with supportive and contradicting data available for PCa development [35, 36].

The AR-DBD (αα 559–624) facilitates binding to androgen response elements (ARE) in gene enhancer and promoter regions. There is a high degree of amino acid sequence similarity between AR-DBD and other steroid hormone receptor family members, for example, glucocorticoid receptors (GR), mineralocorticoid receptors (MR) and progesterone receptors (PR). This similarity contributes to recognition and binding of other hormone response elements (HREs) [37]. This family of receptors dimerises in head-to-head and head-to-tail arrangements to form homodimers that bind palindromic nucleotide sequences (e.g. 5'-TGTTCT-3') assembled as reversed repeats separated by three nucleotides (IR3) [26]. Highly specific AREs also exist, consisting of two hexameric sequences assembled as a direct repeat with a three nucleotide spacer (ADR3) [38]. Variations in dimerisation arrangements and recognition of multiple but different HREs and AREs most likely have significant influence upon transcriptional activity [39, 40].

A bipartite nuclear localisation signal (NLS) located at the junction of the AR-DBD and a small hinge region (located between the AR-DBD and AR-LBD), facilitates nuclear translocation following androgen binding and consequent conformational changes of AR [40-42]. Treatment of PCa with the anti-androgens hydroxyflutamide or bicalutamide will induce nuclear localisation of AR, but disrupt normal transcription initiation by androgen signalling [43, 44]. However, in some patients these compounds agonise AR and subsequently induce a clinical anti-androgen withdrawal response in PCa patients whereby serum PSA levels decline after treatment has ceased [45]. A promising new second generation AR antagonist, a diarylthiohydantoin called MDV3100, is currently being tested in a phase III clinical trial for the treatment of CRPC. MDV3100 has a five fold greater binding affinity for AR compared to *bicalutamide* and does not promote AR nuclear *translocation* like bicalutamide thus preventing binding to DNA and nuclear *co-activator*s [45, 46]. This highlights how the targeted cellular localisation of an AR-ligand complex is specifically regulated by the chemical structure of bound ligand, whether an agonist or antagonist. Overall, AR conformation and stability induced by a specific ligand ultimately dictate whether transcription is initiated, and the amount of transcription that is initiated.

The AR-LBD (αα 677–919) is responsible for ligand binding. Binding of the androgens testosterone and DHT increases the half-life of AR by more than three times in the LNCaP PCa cell line [21]. Androgen bound to the AR-LBD promotes AR protein stability and encourages AR nuclear translocation [47]. The crystal structure of the AR-LBD bound with DHT or R1881 (metribolone, a powerful synthetic anabolic steroid) illustrates α-helix number twelve juxtaposed to the ligand-binding pocket [48, 49]. The position of helix twelve most likely prevents ligand dissociation, thus increasing the period and impact of ligand-activated transcriptional regulation by AR. Of great interest when considering the use of AR inhibitors for treatment of PCa, ARA70 in DU145 PCa cells expands the group of AR activating ligands to include estrogens and the non-steroidal anti-androgen hydroxyflutamide [21, 50, 51]. Apparent AR ligand binding promiscuity illustrates one possible mechanism by which PCa cells circumvent therapies designed to disrupt normal androgen signalling. Located within the

AR-LBD is the nuclear export signal which, following normal dissociation of androgen, facilitates shuttling of AR from the nucleus to the cytoplasm [52].

AR functions primarily as a homodimer but additional transcriptional regulation occurs if AR forms heterodimers with orphan nuclear receptor testicular receptor 4 (TR4) and estrogen receptor alpha (ERα) isoform [41, 53]. Interestingly, AR is also a downstream target of G-protein coupled receptor (GPCR) signalling and is phosphorylated by receptor tyrosine kinases, such as HER-2/neu, which can activate AR without androgen [54-57]. Phosphorylation at serine residues (Ser80, Ser93 and Ser641) promotes androgen binding, mediates AR activity and protects AR from proteolytic degradation [58, 59]. Additional phosphorylation (Ser213, Ser506 and Ser650) by mitogen activated protein kinases (MAPK), especially extracellular signal-regulated protein kinase (ERK) and p38 and c-Jun N-terminal kinase (JNK), and Akt (protein kinase B/PKB) enhances AR response to low levels of androgens and recruitment of chromatin remodelling by nuclear AR co-activators [60].

Prostate Cancer and Treatment

PCa is the most common diagnosed registerable cancer, and is the second leading cause of cancer deaths in Australian males, behind lung cancer (Australian Bureau of Statistics). Before identification of serum PSA as a PCa marker, acid phosphatase levels were monitored with a discernible rise being associated with PCa dissemination and skeletal metastases [61-63]. Huggins and Hodges (1941) found that acid phosphatase levels were significantly reduced in castrated patients with primary or secondary disease [64]. This observed androgen response in PCa patients still dominates selection of therapies for malignant PCa today.

Many treatments for localised and advanced PCa have been developed, including surgery to remove the prostate (prostatectomy), local radiation (external beam and internal seeds (brachytherapy)) and chemotherapies. The most commonly used therapy for advanced or malignant PCa targets inhibition of the androgen signalling axis (Figure 1). More specifically, testicular androgen synthesis is inhibited by chemical castration, using treatment with gonadotropin-releasing hormone (GRH) super-agonists or luteinizing hormone-releasing hormone (LHRH). In addition, AR activity can be competitively inhibited by antagonists such as the non-steroidal anti-androgens bicalutamide and flutamide. These androgen-targeted therapies can be used individually, or in conjunction, to provide complete androgen blockade [65-67]. These therapies, however, do not prevent development of castrate-resistant PCa (CRPC) which is accompanied by a five year survival rate of approximately 20% [65, 67].

Development of malignant PCa relies upon androgen and a functional AR [68]. AR is expressed in the majority (>80%) of recurrent CRPC and it is thought that CRPC depends upon activation of anti-apoptotic and proliferative genes by AR [7, 13, 69]. Disruption of the androgen-AR relationship is targeted during all stages of PCa progression, including neo-adjuvant androgen deprivation therapy (ADT) with external radiation therapy for intermediate and high-risk PCa through to palliative treatment in CRPC. Associated side effects from ADT include decreased libido, impotence, fatigue, hot flushes and morbidities such as anemia, cognitive impairment, altered lipid profiles and osteoporosis [70].

The 5α reductase type 2 inhibitors, finasteride and dutasteride, although effective at preventing the reduction of testosterone to DHT, have proven controversial for PCa prevention [71]. Finasteride and dutasteride were studied in the Prostate Cancer Prevention Trial and Reduction by Dutasteride of Prostate Cancer Events (REDUCE) trial, respectively. Both agents were reported to reduce PCa incidence; finasteride by 25% and dutasteride by 23% [72, 73]. Although these drugs are approved to treat conditions such as BPH, and likely can preferentially prevent low-grade PCa, the United States Food and Drug Administration (FDA) issued a warning to healthcare professionals (June 2011) that finasteride and dutasteride have a potential risk for developing high-grade PCa [74].

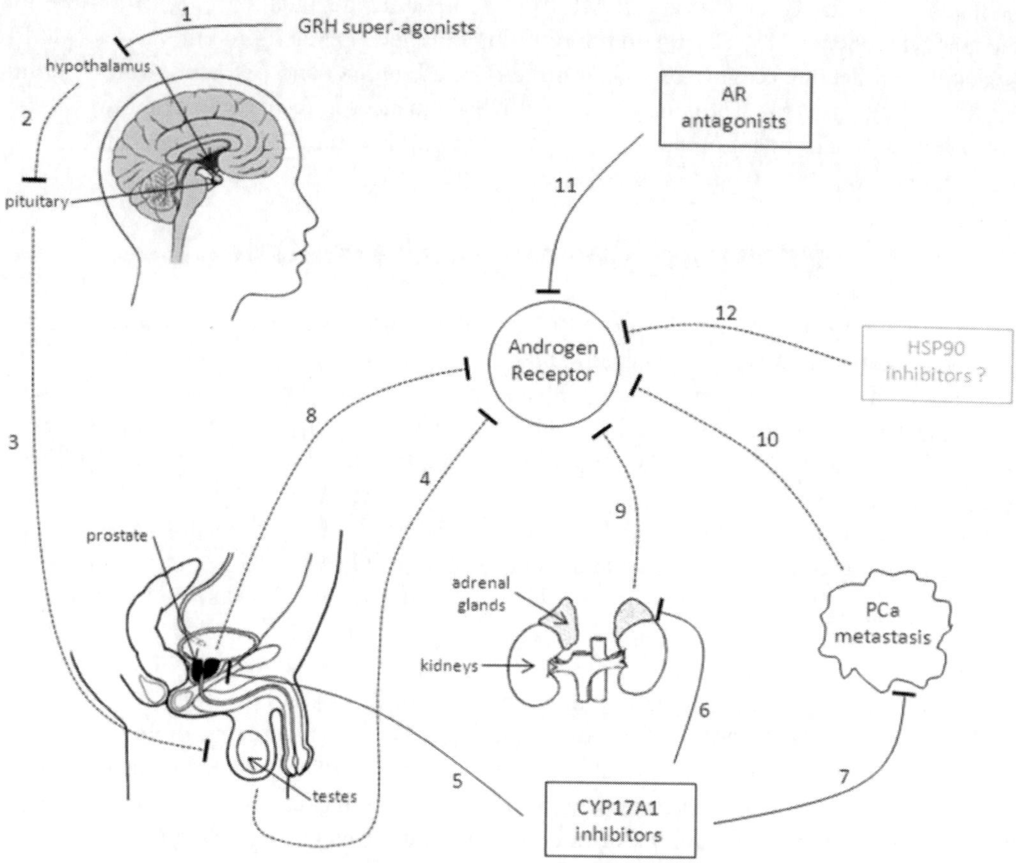

Figure 1. Prostate cancer therapies designed to disrupt androgen synthesis and signaling. Gonadotropin-releasing hormone (GRH) super-agonists inhibit release of luteinizing hormone releasing hormone from the hypothalamus (1), which in turn inhibits release of luteinizing hormone from the pituitary (2). This, in turn, inhibits testosterone synthesis in the testes (3), which inhibits androgen receptor (AR) activity (4). CYP17A1 inhibitors (e.g. Abiraterone acetate) block conversion of pregnenolone and progesterone to androgens in tissues containing this enzyme (5-7), which in turn inhibits AR activity (8-10). AR antagonists (e.g. MDV3100, flutamide, bicalutamide) act directly upon AR by competitively binding to its ligand binding pocket, which in turn inhibits AR activity (11). Heat shock protein 90 (HSP90) inhibitors (e.g. 17-AAG and IPI-504) potentially destabilise AR structure, leading to protein degradation and loss of AR numbers (12). PCa = prostate cancer.

Links between Androgen Levels and Prostate Cancer Risk

Serum testosterone levels decline in human males from approximately 30 years of age from $\geq$ 40 nM to $\leq$ 10 nM by age 90. In addition, SHBG levels rise, thus further compounding the decrease in free testosterone [75]. A recent report from Mearini et al (2011) suggests that serum testosterone levels are predictive of PCa and that PCa is frequently associated with low testosterone serum levels [76]. Irrespective of falling androgen concentrations in serum, levels of testosterone and DHT in prostate tissue are substantially higher than in serum [77, 78]. The most potent prostatic androgen, DHT, may have a substantial role in PCa development. A 2003 study found that men treated with a selective 5α-reductase inhibitor in the Prostate Cancer Prevention Trial had a 24% decreased risk of developing PCa. This result suggests that reduced DHT levels in prostate tissues may slow or inhibit cancerous transformations [72]. Contradictory to this finding is the knowledge that ADT, although initially inducing disease remission in most patients, promotes progression to the typically-fatal CRPC phenotype.

Steroidogenesis and Prostate Cancer

Steroidogenesis in prostate, PCa tissues and adrenals may supply CRPC with androgens, and thus provide a survival advantage. Phenotypic transition to CRPC is expected during ADT although ADT promotes tumour atrophy and leads to remission in the majority of cases. These inconsistent effects emphasize ADT as a double-edged sword. If considered an evolutionary stress factor, ADT selects for tumour cells that have castrate-resistant capacity prior to commencement of ADT or induces genes required for steroidogenesis within PCa tumours.

In patients undergoing ADT for recurrent PCa, AR activity is generally inhibited. Until recently, prostate tumours that circumvented ADT were thought to be androgen-independent or androgen refractory. Similar to these terms, when discussing PCa today, the term 'castrate-resistant' suggests a proliferative phenotype independent of testicular androgen. However, the vast majority of CRPC maintain AR expression and activity, as evidenced by expression of androgen-regulated proteins such as PSA [79]. In an attempt to explain how CRPC tumours survive without androgens, various hypotheses were developed [65]. In ADT patients with castrate levels of circulating androgens, a surprising finding was that testosterone and DHT remained at sufficient levels for AR activation [80, 81]. Furthermore, it was confirmed that human PCa cells are capable of *de novo* androgen synthesis, as shown using an LNCaP xenograft mouse model of CRPC [82]. Interestingly, ADT is also associated with elevated levels of circulating insulin and development of metabolic syndrome. For example, treatment of PCa cell lines (LNCaP, 22Rv1, VCaP) with insulin increased mRNA and protein expression of steroidogenic enzymes, and significantly increased intracellular and secreted levels of androgens. AR activity could be observed indirectly following insulin treatment by increased expression of PSA mRNA [83].

CRPC may utilise circulating adrenal androgens or possibly perform *de novo* steroidogenesis to produce testosterone. For cells or tissues to be classified as steroidogenic they must express cholesterol monooxygenase (P450scc) from the CYP11A1 gene. P450scc cleaves a hydrocarbon side chain from cholesterol thus converting it to pregnenolone, a reaction necessary for all steroids hormones to be produced. Androgen-dependent and independent cell lines (LNCaP, PC3, DU145) express P450scc [84]. They also express the gene products from CYP17A1, 17β3HSD and 3β2HSD which are required for converting pregnenolone to testosterone [84-87]. Human BPH tissue and cancerous prostatic tissue express mRNA and protein of steroidogenic enzymes, indicating that they harbour steroidogenic potential [88]. One contradictory report exists, whereby the majority of samples of normal human prostate, locally advanced PCa, local CRPC and lymph node metastases revealed low or absent expression of mRNA for steroidogenic enzymes required for *de novo* steroidogenesis [89]. mRNA expression of enzymes from AKR1C3 and SRD5A1 genes, responsible for converting androstenedione to testosterone and testosterone to dihydrotestosterone respectively, were highly expressed. These data suggest intra-tumoral *de novo* steroidogenesis is minimal compared to late pathway androgenesis via utilisation of circulating adrenal androgens [89]. If so, blocking late pathway androgenesis, by inhibiting upstream enzymes such as CYP17A1, may prove favourable for CRPC patients. This approach has gained a great deal of momentum via clinical trials and recent (April 2011) FDA approval of abiraterone acetate, a CYP17A1 inhibitor, for the treatment of CRPC.

The role of LH, which is responsible for stimulating testosterone synthesis in Leydig cell, and the LH receptor (LHR), may contribute to endogenous steroidogenesis in PCa cells. The PCa cell lines LNCaP, C4-2B and 22RV1 express mRNA and protein for LH and LHR. When exposed to additional LH, these cell lines significantly up-regulated expression of several steroidogenic genes including StAR, CYB5B, CYP11A, and 3βHSD. LH also stimulated testosterone and progesterone synthesis in LNCaP cells and enhanced viability [90]. If these data are indicative of pathways that may be active in clinical tissue, PCa may indeed be androgen self-sufficient, with autocrine and paracrine LH signalling maintaining intra-tumoral steroidogenesis.

Benign primary human prostate stromal cells may also possess steroidogenic potential. A recent study that stimulated Hedgehog signalling in prostate stromal cells found steroidogenic activity significantly increased [91]. Hedgehog stimulation induced expression of genes involved in cholesterol and steroid synthesis (CYP17A1 and CYP19A1), metabolism, and transport. It also significantly increased testosterone synthesis in prostate stromal cells supplemented with dihydroepiandrosterone, as determined by enzyme-linked immuno-sorbent assay. The authors surmise that stromal cells in the PCa tumour microenvironment may promote or influence CRPC development. [91]

Knowledge of intratumoral steroidogenesis in CRPC has opened a new avenue for development of drugs designed specifically for targeting these mechanisms. Abiraterone acetate specifically targets the CYP17A1 gene product (17alpha-Hydroxyprogesterone aldolase). Abiraterone acetate is a pregnenolone analogue that binds irreversibly to 17α-hydroxyprogesterone aldolase and potently blocks the C17,20-lyase function of this enzyme [92]. Consequently, conversion of pregnenolone or progesterone to androgens and estrogens is inhibited. Abiraterone acetate holds some promise for improved treatment regimens for managing CRPC [93, 94]. Unfortunately, abiraterone acetate does not cure CRPC, and the continued growth of CRPC in PCa patients treated with abiraterone acetate suggests abiraterone resistance establishes quickly or already exists. Human CRPC mice xenografts treated with abiraterone had suppressed serum PSA and tumour levels of testosterone and

DHT. However, abiraterone treatment increased expression of full-length and truncated AR variants, CYP17A1, other enzymes involved in androgen synthesis, for example, HSD3B2, HSD17B3 SDR5A2, and CYP17A1 in human biopsies [95, 96]. These data suggest CRPC that is resistant to CYP17A1 inhibition is still androgen dependent. Additionally, tumours harbouring ligand-promiscuous AR, due to LBD binding pocket mutations, may not be CYP17A1-dependent or subsequently androgen-dependent. [96] Mechanisms conducive to abiraterone resistance may be supported by continued and up-regulated expression of steroidogenic enzymes, AR, ligand-independent AR splice variants and mutant AR. Elevated serum insulin levels, as a result of ADT-induced metabolic syndrome, may add to abiraterone resistance by enhancing steroidogenesis [83].

Despite the use of GRH super-agonists or LHRH agonists (testicular testosterone synthesis inhibitors), AR antagonists (MDV3100, flutamide, bicalutamide), and now adrenal and intratumoral CYP17A1 inhibitors, sustained inactivation of AR in CRPC continues to be an elusive ambition. Combinations of GRH agonists and AR antagonists are used to treat CRPC, which inhibit AR activity via two unrelated mechanisms, but with limited and temporary success. The addition of further AR inhibitory therapies that function via unrelated mechanisms, such as CYP17A1 inhibitors, in combination therapies may enhance CRPC treatment regimen efficacy. However the likelihood of increasing the number and severity of treatment-associated morbidities is a major consideration. Additionally, another unrelated mechanism to be considered is disruption of AR structure and maturation by HSP90 inhibitors. Unfortunately, current HSP90 inhibitors such as 17-allylamino-17-demethoxygeldanamycin (17-AAG) and retaspimycin hydrochloride (IPI-504), both of which have been through Phase II clinical trials, have had limited effect on reducing PSA levels and cancer burden, and both induce significant morbidities [97, 98]. Although, development of HSP90 inhibitors as CRPC therapeutics has been ineffective to date, future compounds, perhaps in combination with currently-used AR inhibitory strategies, may prove advantageous (Figure 1).

Conclusion

There has been continuing discussion in the literature regarding the progression of PCa from an androgen-dependent to an androgen-independent phenotype. More recently, the question has arisen regarding whether a true androgen-independent phenotype exists. The majority (>80%) of androgen-independent PCa or CRPC continue to express a functional AR. Referring to advanced PCa as "androgen-independent" has become misleading when considering data published recently about *de novo* steroidogenesis in PCa. 'Castrate-resistant' PCa is a more accurate diagnosis. This term indicates that growth of a specific type of prostate tumour is independent of testicular androgen. If still reliant upon androgens, castrate-resistant tumours are most likely surviving by *de novo* androgenesis from adrenal or endogenous precursors and could well be described as 'androgen self-sufficient'. The recent discovery of endogenous androgenesis in prostate tissues and cancers is significant. Future steroidogenesis investigations should also examine endogenous synthesis of estrogens, glucocorticoids, mineralocorticoids and progestins, and their respective receptors, and the impact they may be having upon PCa development and survival.

References

[1] Chemes, H.E., Infancy is not a quiescent period of testicular development. *Int J Androl,* 2001. 24(1): p. 2-7.

[2] Dalton, J.T. and W. Gao, Androgen Receptor, in *Nuclear Receptors, Proteins and Cell Regulation*, C.M. Bunce and M.J. Campbell, Editors. 2010, Springer Science+Business Media B.V. .

[3] Radmayr, C., et al., 5-alpha-reductase and the development of the human prostate. *Indian J Urol*, 2008. 24(3): p. 309-12.

[4] Oettel, M., Testosterone metabolism, dose-response relationships and receptor polymorphisms: selected pharmacological/toxicological considerations on benefits versus risks of testosterone therapy in men. *Aging Male*, 2003. 6(4): p. 230-56.

[5] Nelson, P.S., et al., The program of androgen-responsive genes in neoplastic prostate epithelium. *Proc Natl Acad Sci U S A,* 2002. 99(18): p. 11890-5.

[6] Mooradian, A.D., J.E. Morley, and S.G. Korenman, Biological actions of androgens. *Endocr Rev,* 1987. 8(1): p. 1-28.

[7] Montgomery, J.S., D.K. Price, and W.D. Figg, The androgen receptor gene and its influence on the development and progression of prostate cancer. *J Pathol,* 2001. 195(2): p. 138-46.

[8] Koivisto, P.A., et al., Androgen receptor gene alterations and chromosomal gains and losses in prostate carcinomas appearing during finasteride treatment for benign prostatic hyperplasia. *Clin Cancer Res,* 1999. 5(11): p. 3578-82.

[9] Zitzmann, M., et al., Prostate volume and growth in testosterone-substituted hypogonadal men are dependent on the CAG repeat polymorphism of the androgen receptor gene: a longitudinal pharmacogenetic study. *J Clin Endocrinol Metab,* 2003. 88(5): p. 2049-54.

[10] Loprinzi, C.L., et al., Megestrol acetate for the prevention of hot flashes. *N Engl J Med,* 1994. 331(6): p. 347-52.

[11] Strum, S.B., et al., Anaemia associated with androgen deprivation in patients with prostate cancer receiving combined hormone blockade. *Br J Urol,* 1997. 79(6): p. 933-41.

[12] Brinkmann, A.O., Molecular basis of androgen insensitivity. *Mol Cell Endocrinol,* 2001. 179(1-2): p. 105-9.

[13] Han, G., et al., Mutation of the androgen receptor causes oncogenic transformation of the prostate. *Proc Natl Acad Sci U S A,* 2005. 102(4): p. 1151-6.

[14] Goldstein, A.S., et al., Identification of a cell of origin for human prostate cancer. *Science*, 2010. 329(5991): p. 568-71.

[15] Keller, E.T., W.B. Ershler, and C. Chang, The androgen receptor: a mediator of diverse responses. *Front Biosci,* 1996. 1: p. d59-71.

[16] He, B., et al., Activation function 2 in the human androgen receptor ligand binding domain mediates interdomain communication with the NH(2)-terminal domain. *J Biol Chem*, 1999. 274(52): p. 37219-25.

[17] He, B., J.A. Kemppainen, and E.M. Wilson, FXXLF and WXXLF sequences mediate the NH2-terminal interaction with the ligand binding domain of the androgen receptor. *J Biol Chem,* 2000. 275(30): p. 22986-94.

[18] Loy, C.J., K.S. Sim, and E.L. Yong, Filamin-A fragment localizes to the nucleus to regulate androgen receptor and coactivator functions. *Proc Natl Acad Sci U S A*, 2003. 100(8): p. 4562-7.

[19] Veldscholte, J., et al., Hormone-induced dissociation of the androgen receptor-heat-shock protein complex: use of a new monoclonal antibody to distinguish transformed from nontransformed receptors. *Biochemistry*, 1992. 31(32): p. 7422-30.

[20] Ozanne, D.M., et al., Androgen receptor nuclear translocation is facilitated by the f-actin cross-linking protein filamin. *Mol Endocrinol,* 2000. 14(10): p. 1618-26.

[21] Rahman, M., H. Miyamoto, and C. Chang, Androgen receptor coregulators in prostate cancer: mechanisms and clinical implications. *Clin Cancer Res,* 2004. 10(7): p. 2208-19.

[22] Saporita, A.J., et al., Identification and characterization of a ligand-regulated nuclear export signal in androgen receptor. *J Biol Chem,* 2003. 278(43): p. 41998-2005.

[23] Bennett, N.C., et al., Molecular cell biology of androgen receptor signalling. *Int J Biochem Cell Biol,* 2010. 42(6): p. 813-27.

[24] Gelmann, E.P., Molecular biology of the androgen receptor. *J Clin Oncol,* 2002. 20(13): p. 3001-15.

[25] Reid, J., et al., The androgen receptor transactivation domain: the interplay between protein conformation and protein-protein interactions. *Biochem Soc Trans,* 2003. 31(Pt 5): p. 1042-6.

[26] Verrijdt, G., A. Haelens, and F. Claessens, Selective DNA recognition by the androgen receptor as a mechanism for hormone-specific regulation of gene expression. *Mol Genet Metab,* 2003. 78(3): p. 175-85.

[27] Shen, H.C. and G.A. Coetzee, The androgen receptor: unlocking the secrets of its unique transactivation domain. *Vitam Horm*, 2005. 71: p. 301-19.

[28] Centenera, M.M., et al., The contribution of different androgen receptor domains to receptor dimerization and signaling. *Mol Endocrinol,* 2008. 22(11): p. 2373-82.

[29] Claessens, F., et al., Diverse roles of androgen receptor (AR) domains in AR-mediated signaling. *Nucl Recept Signal,* 2008. 6: p. e008.

[30] Jenster, G., et al., Domains of the human androgen receptor involved in steroid binding, transcriptional activation, and subcellular localization. *Mol Endocrinol*, 1991. 5(10): p. 1396-404.

[31] Simental, J.A., et al., Transcriptional activation and nuclear targeting signals of the human androgen receptor. *J Biol Chem,* 1991. 266(1): p. 510-8.

[32] Choong, C.S. and E.M. Wilson, Trinucleotide repeats in the human androgen receptor: a molecular basis for disease. *J Mol Endocrinol,* 1998. 21(3): p. 235-57.

[33] Edwards, A., et al., Genetic variation at five trimeric and tetrameric tandem repeat loci in four human population groups. *Genomics*, 1992. 12(2): p. 241-53.

[34] Mononen, N., et al., Androgen receptor CAG polymorphism and prostate cancer risk. *Hum Genet,* 2002. 111(2): p. 166-71.

[35] Platz, E.A., et al., Racial variation in prostate cancer incidence and in hormonal system markers among male health professionals. *J Natl Cancer Inst,* 2000. 92(24): p. 2009-17.

[36] Lange, E.M., et al., The androgen receptor CAG and GGN repeat polymorphisms and prostate cancer susceptibility in African-American men: results from the Flint Men's Health Study. *J Hum Genet,* 2008. 53(3): p. 220-6.

[37] Jenster, G., et al., Identification of two transcription activation units in the N-terminal domain of the human androgen receptor. *J Biol Chem,* 1995. 270(13): p. 7341-6.

[38] Shaffer, P.L., et al., Structural basis of androgen receptor binding to selective androgen response elements. *Proc Natl Acad Sci U S A,* 2004. 101(14): p. 4758-63.

[39] Luisi, B.F., et al., Crystallographic analysis of the interaction of the glucocorticoid receptor with DNA. *Nature*, 1991. 352(6335): p. 497-505.

[40] Schoenmakers, E., et al., Differential DNA binding by the androgen and glucocorticoid receptors involves the second Zn-finger and a C-terminal extension of the DNA-binding domains. *Biochem J*, 1999. 341 (Pt 3): p. 515-21.

[41] Zhou, Z.X., et al., A ligand-dependent bipartite nuclear targeting signal in the human androgen receptor. Requirement for the DNA-binding domain and modulation by NH2-terminal and carboxyl-terminal sequences. *J Biol Chem*, 1994. 269(18): p. 13115-23.

[42] Zilliacus, J., et al., Determinants for DNA-binding site recognition by the glucocorticoid receptor. *J Biol Chem,* 1992. 267(35): p. 24941-7.

[43] Farla, P., et al., Antiandrogens prevent stable DNA-binding of the androgen receptor. *J Cell Sci,* 2005. 118(Pt 18): p. 4187-98.

[44] Klokk, T.I., et al., Ligand-specific dynamics of the androgen receptor at its response element in living cells. *Mol Cell Biol*, 2007. 27(5): p. 1823-43.

[45] Tran, C., et al., Development of a second-generation antiandrogen for treatment of advanced prostate cancer. *Science*, 2009. 324(5928): p. 787-90.

[46] Scher, H.I., et al., Antitumour activity of MDV3100 in castration-resistant prostate cancer: a phase 1-2 study. *Lancet*, 2010. 375(9724): p. 1437-46.

[47] Lee, D.K. and C. Chang, Endocrine mechanisms of disease: Expression and degradation of androgen receptor: mechanism and clinical implication. *J Clin Endocrinol Metab,* 2003. 88(9): p. 4043-54.

[48] Matias, P.M., et al., Structural evidence for ligand specificity in the binding domain of the human androgen receptor. Implications for pathogenic gene mutations. *J Biol Chem,* 2000. 275(34): p. 26164-71.

[49] Sack, J.S., et al., Crystallographic structures of the ligand-binding domains of the androgen receptor and its T877A mutant complexed with the natural agonist dihydrotestosterone. *Proc Natl Acad Sci U S A,* 2001. 98(9): p. 4904-9.

[50] Miyamoto, H., et al., Promotion of agonist activity of antiandrogens by the androgen receptor coactivator, ARA70, in human prostate cancer DU145 cells. *Proc Natl Acad Sci U S A,* 1998. 95(13): p. 7379-84.

[51] Miyamoto, H. and C. Chang, Antiandrogens fail to block androstenedione-mediated mutated androgen receptor transactivation in human prostate cancer cells. *Int J Urol,* 2000. 7(1): p. 32-4.

[52] Heinlein, C.A. and C. Chang, The roles of androgen receptors and androgen-binding proteins in nongenomic androgen actions. *Mol Endocrinol,* 2002. 16(10): p. 2181-7.

[53] Lee, Y.F., et al., Convergence of two repressors through heterodimer formation of androgen receptor and testicular orphan receptor-4: a unique signaling pathway in the steroid receptor superfamily. *Proc Natl Acad Sci U S A,* 1999. 96(26): p. 14724-9.

[54] Cao, X., et al., Regulator of G-protein signaling 2 (RGS2) inhibits androgen-independent activation of androgen receptor in prostate cancer cells. *Oncogene*, 2006. 25(26): p. 3719-34.

[55] Yeh, S., et al., From HER2/Neu signal cascade to androgen receptor and its coactivators: a novel pathway by induction of androgen target genes through MAP kinase in prostate cancer cells. *Proc Natl Acad Sci U S A,* 1999. 96(10): p. 5458-63.

[56] Manin, M., et al., Androgen receptor expression is regulated by the phosphoinositide 3-kinase/Akt pathway in normal and tumoral epithelial cells. *Biochem J,* 2002. 366(Pt 3): p. 729-36.

[57] Scheid, M.P. and J.R. Woodgett, PKB/AKT: functional insights from genetic models. *Nat Rev Mol Cell Biol,* 2001. 2(10): p. 760-8.

[58] Gioeli, D., et al., Androgen receptor phosphorylation. Regulation and identification of the phosphorylation sites. *J Biol Chem*, 2002. 277(32): p. 29304-14.

[59] Blok, L.J., P.E. de Ruiter, and A.O. Brinkmann, Forskolin-induced dephosphorylation of the androgen receptor impairs ligand binding. *Biochemistry*, 1998. 37(11): p. 3850-7.

[60] Rochette-Egly, C., Nuclear receptors: integration of multiple signalling pathways through phosphorylation. *Cell Signal*, 2003. 15(4): p. 355-66.

[61] Kutscher, W. and H. Wolbergs, Prostate phosphatase. *Hoppe-Seylers Zeitschrift Fur Physiologische Chemie*, 1935. 236: p. 237-240.

[62] Gutman, A.B. and E.B. Gutman, An "acid" phosphatase occurring in the serum of patients with metastasizing carcinoma of the prostate gland. *Journal of Clinical Investigation,* 1938. 17(4): p. 473-478.

[63] Robinson, J.N., E.B. Gutman, and A.B. Gutman, Clinical significance of increased serum "acid" phosphatase in patients with bone metastases secondary to prostatic carcinoma. *Journal of Urology*, 1939. 42(4): p. 602-618.

[64] Huggins, C. and C.V. Hodges, Studies on prostatic cancer - I The effect of castration, of estrogen and of androgen injection on serum phosphatases in metastatic carcinoma of the prostate. *Cancer Research,* 1941. 1(4): p. 293-297.

[65] Feldman, B.J. and D. Feldman, The development of androgen-independent prostate cancer. *Nat Rev Cancer,* 2001. 1(1): p. 34-45.

[66] Labrie, F., et al., Treatment of prostate cancer with gonadotropin-releasing hormone agonists. *Endocr Rev*, 1986. 7(1): p. 67-74.

[67] Newling, D.W.W., Maximum androgen blockade: a case study report. *Prostate Cancer and Prostatic Diseases,* 2000. 3(3): p. 203-212.

[68] Tomlins, S.A., et al., Integrative molecular concept modeling of prostate cancer progression. *Nat Genet,* 2007. 39(1): p. 41-51.

[69] Russell, P.J., S. Bennett, and P. Stricker, Growth factor involvement in progression of prostate cancer. *Clin Chem,* 1998. 44(4): p. 705-23.

[70] Holzbeierlein, J.M., M.D. McLaughlin, and J.B. Thrasher, Complications of androgen deprivation therapy for prostate cancer. *Curr Opin Uro*l, 2004. 14(3): p. 177-83.

[71] Paba, S., et al., Steroid 5alpha-reductase as a novel therapeutic target for schizophrenia and other neuropsychiatric disorders. *Curr Pharm Des*, 2011. 17(2): p. 151-67.

[72] Thompson, I.M., et al., The influence of finasteride on the development of prostate cancer. *N Engl J Med*, 2003. 349(3): p. 215-24.

[73] Andriole, G.L., et al., Effect of dutasteride on the risk of prostate cancer. *N Engl J Med,* 2010. 362(13): p. 1192-202.

[74] Hamilton, R.J. and S.J. Freedland, 5-alpha reductase inhibitors and prostate cancer prevention: where do we turn now? *BMC Med*, 2011. 9: p. 105.

[75] Schulman, C. and B. Lunenfeld, The ageing male. *World J Urol,* 2002. 20(1): p. 4-10.

[76] Mearini, L., et al., Low serum testosterone levels are predictive of prostate cancer. *World J Urol*, 2011.

[77] Nishiyama, T., Y. Hashimoto, and K. Takahashi, The influence of androgen deprivation therapy on dihydrotestosterone levels in the prostatic tissue of patients with prostate cancer. *Clin Cancer Res*, 2004. 10(21): p. 7121-6.

[78] Page, S.T., et al., Persistent intraprostatic androgen concentrations after medical castration in healthy men. *J Clin Endocrinol Metab*, 2006. 91(10): p. 3850-6.

[79] Yuan, X., et al., Androgen receptor remains critical for cell-cycle progression in androgen-independent CWR22 prostate cancer cells. *Am J Pathol*, 2006. 169(2): p. 682-96.

[80] Mohler, J.L., et al., The androgen axis in recurrent prostate cancer. *Clin Cancer Res,* 2004. 10(2): p. 440-8.

[81] Titus, M.A., et al., Testosterone and dihydrotestosterone tissue levels in recurrent prostate cancer. *Clin Cancer Res,* 2005. 11(13): p. 4653-7.

[82] Locke, J.A., et al., Androgen levels increase by intratumoral de novo steroidogenesis during progression of castration-resistant prostate cancer. *Cancer Res,* 2008. 68(15): p. 6407-15.

[83] Lubik, A.A., et al., Insulin increases de novo steroidogenesis in prostate cancer cells. *Cancer Res*, 2011. 71(17): p. 5754-64.

[84] Dillard, P.R., M.F. Lin, and S.A. Khan, Androgen-independent prostate cancer cells acquire the complete steroidogenic potential of synthesizing testosterone from cholesterol. *Mol Cell Endocrinol,* 2008. 295(1-2): p. 115-20.

[85] Castagnetta, L.A., et al., Expression of different 17beta-hydroxysteroid dehydrogenase types and their activities in human prostate cancer cells. *Endocrinology*, 1997. 138(11): p. 4876-82.

[86] Negri-Cesi, P., et al., 5alpha-reductase isozymes and aromatase are differentially expressed and active in the androgen-independent human prostate cancer cell lines DU145 and PC3. *Prostate*, 1999. 41(4): p. 224-32.

[87] Handratta, V.D., et al., Potent CYP17 inhibitors: improved syntheses, pharmacokinetics and anti-tumor activity in the LNCaP human prostate cancer model. *J Steroid Biochem Mol Biol*, 2004. 92(3): p. 155-65.

[88] Montgomery, R.B., et al., Maintenance of intratumoral androgens in metastatic prostate cancer: a mechanism for castration-resistant tumor growth. *Cancer Res*, 2008. 68(11): p. 4447-54.

[89] Hofland, J., et al., Evidence of Limited Contributions for Intratumoral Steroidogenesis in Prostate Cancer. *Cancer Res*, 2010. 70(3): p. 1256-64.

[90] Pinski, J., et al., Effect of luteinizing hormone on the steroidogenic pathway in prostate cancer. *Prostate*, 2011. 71(8): p. 892-8.

[91] Levina, E., et al., Paracrine hedgehog increases the steroidogenic potential of prostate stromal cells in a Gli-dependent manner. *Prostate*, 2011.

[92] Reid, A.H., et al., CYP17 inhibition as a hormonal strategy for prostate cancer. *Nat Clin Pract Urol,* 2008. 5(11): p. 610-20.

[93] Attard, G., et al., Selective inhibition of CYP17 with abiraterone acetate is highly active in the treatment of castration-resistant prostate cancer. *J Clin Oncol,* 2009. 27(23): p. 3742-8.

[94] Attard, G., et al., Phase I clinical trial of a selective inhibitor of CYP17, abiraterone acetate, confirms that castration-resistant prostate cancer commonly remains hormone driven. *J Clin Oncol*, 2008. 26(28): p. 4563-71.

[95] Mostaghel, E.A., et al., Resistance to CYP17A1 inhibition with abiraterone in castration-resistant prostate cancer: induction of steroidogenesis and androgen receptor splice variants. *Clin Cancer Res,* 2011. 17(18): p. 5913-25.

[96] Cai, C., et al., Intratumoral de novo steroid synthesis activates androgen receptor in castration-resistant prostate cancer and is upregulated by treatment with CYP17A1 inhibitors. *Cancer Res,* 2011. 71(20): p. 6503-13.

[97] Oh, W.K., et al., Multicenter phase II trial of the heat shock protein 90 inhibitor, retaspimycin hydrochloride (IPI-504), in patients with castration-resistant prostate cancer. *Urology*, 2011. 78(3): p. 626-30.

[98] Heath, E.I., et al., A phase II trial of 17-allylamino-17-demethoxygeldanamycin in patients with hormone-refractory metastatic prostate cancer. *Clin Cancer Res*, 2008. 14(23): p. 7940-6.

In: Androgens: Production, Functions and Disorders
Editors: B. F. Thompson and D. J. Robinson

ISBN: 978-1-62081-145-0

Chapter VI

The Role and Effects of Novel Androgen Therapy in the Treatment of Prostate Cancer

Joelle El-Amm and Jeanny B. Aragon-Ching*
Department of Medicine, Division of Hematology/Oncology,
George Washington University Medical Center, Washington, DC, US

Abstract

Androgen deprivation therapy has been the mainstay of treatment for locally advanced and metastatic prostate cancer. While most patients respond, emergence of castration resistance ultimately ensues, a condition that is ultimately fatal. Recent studies have focused on mechanisms involved in the pathways of castration resistance and have brought about new insights and development of new agents in targeting these androgen pathways.

However, the effects of treatment entail adverse effects unique to androgen deprivation. This review focuses on current thinking on novel mechanisms of castration resistance, with particular attention to intratumoral androgen synthesis and resulting therapeutic maneuvers targeting these pathways and its antecedent adverse effects.

I. Introduction

Androgens are the male hormones that are essential for normal male development including normal prostate glandular growth but they also drive the development of prostate cancer by regulating the growth and differentiation of prostate cancer cells [1].

* Corresponding Author: Jeanny B. Aragon-Ching, M.D. Assistant Professor of Medicine Department of Medicine Division of Hematology and Oncology, George Washington University Medical Center 2150 Pennsylvania Avenue, NW Washington, DC 20037, USA, Tel: 202-741-2478; Fax: 202-741-2487, Email: jaragonching @mfa.gwu.edu.

The secretion of androgens is under the regulation of the hypothalamus which contains neurosecretory cells that produce gonadotropin-releasing hormone (GnRH) or luteinizing-hormone-releasing hormone (LHRH) in a pulsatile manner. In response to GnRH, the anterior lobe of the pituitary secretes luteinizing hormone (LH) which enters the bloodstream and acts on the testes.

The interstitial cells of Leydig that are present in the testes contain on their membranes receptors for LH and produce testosterone in response to the binding of LH to its receptor. Testosterone enters the bloodstream and reaches the prostate cells where it is converted to its more potent form Dihydrotestosterone (DHT) by the enzyme 5 alpha reductase. Both testosterone and DHT initiate glandular growth by binding to the cytoplasmic androgen receptor (AR) of prostate cells.

The AR is a steroid receptor transcription factor composed of 919 amino acids and incorporates into its protein structure an N-terminal regulation domain that is involved in transcriptional activation, a C-terminal domain which includes the ligand binding domain (LBD) and a central DNA-binding domain. [2].

AR is present in the cytoplasm but once activated by ligand binding, a conformational change occurs which makes it dissociate from heat shock proteins and makes it then translocate to the nucleus to bind to Androgen Responsive Elements (AREs) in the promoter regions of multiple genes thus regulating their transcription . Androgens are the ligands that bind to AR stabilizing it and increasing its level. [3,4] Understanding of the normal function of the mature AR leads also to elucidation of the potential mechanisms of resistance since AR activation has traditionally been thought to require the C-terminal domain which contains the LBD for its function.

Androgen deprivation therapy (ADT) has become the mainstay of treatment of prostate cancer since prostate cancer is felt to be a disease that is primarily driven by androgens. ADT is currently indicated for locally advanced high risk prostate cancer, for metastatic prostate cancer and for treatment of recurrent prostate cancer. ADT has traditionally been achieved through different means including surgical castration through bilateral orchiectomy, or chemical castration through GnRH agonists or antagonists, estrogens, androgen receptor antagonists such as flutamide, bicalutamide and nilutamide, antiandrogens or adrenal-ablating drugs such as ketoconazole.

While most patients with prostate cancer would initially respond to ADT, castration resistance eventually ensues which ultimately has a poor prognosis and lethality. Hence, the need for further investigations into the mechanisms as well as treatment in combating such progression is essential. The term castration resistant prostate cancer (CRPC) is defined as progressive rise in PSA despite castrate levels of testosterone which typically is defined as less than 50 ng/dl, or more recently has been increasingly re-defined as less than 20 ng/dl. [5,6] While varying time frames have been described, CRPC usually develops 1 – 3 years after the start of ADT. [7].

It is often recognized by an increase in prostate-specific antigen (PSA) levels, and once this occurs survival ranges between 12-46 months. [8] The mechanism of castration resistance has been a matter of debate historically but recent findings brought about new insights and hence, development of new therapeutic maneuvers targeting the pathways of castration resistance.

II. Emergence of Castration Resistance: Androgen Regulation and Signaling

The importance of the AR signaling in the development of castration resistance has been shown in recent studies. [9] However, increase in overall AR expression is probably insufficient to overcome and activate AR transcription.[10] Thus, several mechanisms that are overlapping and likely interconnected contribute to the development of castration-resistant prostate cancer (CRPC) and have been proposed and demonstrated. [11,12] One of the proposed mechanisms is hypersensitivity of AR through AR gene mutations which cause the AR to become extremely sensitive to very low amounts of androgens that may be present even after castration or blockade by androgens. [13] For instance, AR gene mutations such as T877A and W741C/L, among others, are important mechanisms for castration-resistance as evidenced by earlier investigations in anti-androgen withdrawal. [14-18] However, certain varying splicing variants that are being described and potentially implicated in progression as well that do not require the ligand for activation. These alternative AR splice variants are able to constitutively activate the AR in the absence of the LBD and engage different transcriptional targets. [19-22] One example of such mechanism that renders the AR constitutively active is the AR 1/2/2b. [23] It was found that the presence of these AR isoforms enhanced the endogenous expression of AR-dependent genes given the constitutive activation. These AR variants are also more readily detected in CRPC compared to hormone-naïve cancers, and perhaps occur as a result of AR-selected (ADT) therapy. A second mechanism of resistance is the increased level of the AR protein through either its stabilization or its gene amplification or its increased transcription rate or stabilization of the mRNA. [24] Although not entirely under this mechanism, recent studies have shown that dihydrotestosterone appeared to be generated from conversion of Delta-androstanedione by steroid-5 alpha-reductase isoenzyme-1 to 5 alpha-androstanedione, [25] thus bypassing the testosterone pathway entirely. [26] A third mechanism is the promiscuous activation of AR signaling by various non-androgenic ligands following AR mutations that expand the ligand binding specificity of the AR. [12,18,27] A fourth mechanism is the bypass pathway whereby pathways other than the AR pathway are utilized to stimulate prostate cells through the activation of oncogenes or inactivation of tumor suppressor genes. For example, the activation of the *bcl-2* protein in prostatic intraepithelial neoplasia (PIN) lesions can cause cells to avoid utilizing the AR pathway. The role of neuropeptides secreted by neuroendocrine cells has been studied and showed to promote the proliferative rate of prostate cancer cells in the absence of hormones. [11,28,29] Another potential mechanism related to the bypass pathway is the notable epithelial-mesenchymal transition (EMT) which plays a key role in therapeutic resistance and metastasis in prostate cancer. [30,31] Normal prostate tissue and prostate cancer model LuCap35 exhibited EMT and stem-cell like features following surgical castration possibly mediated by the Zeb1 transcription factor. [32] A fifth proposed mechanism is the outlaw pathway**.** In this mechanism, growth factor pathways such as IGF1 and KGF and Receptor tyrosine kinases such as HER-2 /neu in addition to IL-4 and IL-6 and IL-8 activate the AR pathway either by direct activation or cross-talk. [11,29,33,34] Similarly, other factors such as p300, had also previously been found to be upregulated in those undergoing ADT and may also be regulated by IL-6. [35,36] For instance, inhibition of IGF-1 using an antibody mediates a decrease in AR signaling and nuclear localization. [37] IL-6 is one of the most widely studied of these agents. It is a glycoprotein that has been shown to be elevated in the blood of patients with metastatic prostate cancer, and levels of IL-6 seem to correlate with tumor burden, serum PSA, extent of metastasis and CRPC.[38] It activates the AR in a signal Transducers and activators of Transcription (STAT3) -dependent manner. [39,40] Studies have shown that administration of an anti-IL6 antibody inhibits the conversion of androgen-dependent to castrate-resistant phenotype in mice and that IL-6 can regulate the expression of androgen-responsive genes in a castration –resistant manner, and induces castration-resistant growth of androgen-dependent human prostate cancer cells. [41,42] IL-8 is a chemokine secreted by cancer cells such as prostate cancer cells and promotes angiogenesis by exhibiting chemotactic activity for endothelial cells. [43] IL-8 overexpression in the prostate cancer cell line, PC-P3, transformed the cell line to a more metastatic phenotype through development of angiogenic factors such as MMP-9 which cause increased tumorigenicity. [44] Other studies showed that IL-8 mediates transition of androgen dependence to castration resistance through the use of the tyrosine kinases Src and FAK. [45] HER-2/neu seems also to play a role in AR activation, in part by stabilizing the AR protein and by enhancing its DNA binding activity. [46] In addition, there has been mechanistic cooperation between PTEN loss and HER2 such that patients with low level of PTEN and high HER 2/3 levels tend to have a poor prognosis. This has functional significance since MEK inhibition could suppress tumor proliferation by restoring the PTEN loss-induced cell death. [47] Furthermore, dual epidermal growth factor (EGFR) and HER2 inhibition, in combination with anti-androgen withdrawal, sensitizes prostate cancer cell lines to apoptosis. [48] Other growth factors have also been implicated and might be converging on a common signaling intermediary. [49] A sixth mechanism proposed is activation by androgens converted from adrenal androgens or synthesized intratumorally via the *de novo* route. GnRH agonists or antagonists that suppress testicular androgen synthesis do not influence the production of adrenal androgens because they only suppress the secretion of luteinizing hormone (LH) and the production of adrenal androgens is not under the control of LH. Therefore, following medical or surgical castration, adrenal androgens remain and androgens can be synthesized intratumorally [50].

Androgens are synthesized from cholesterol via multiple enzymatic steps in the testes and adrenal glands. Pregnenolone is formed from cholesterol catalyzed by the enzyme CYP11A1. In the zona reticularis of the adrenal cortex, pregnenolone is converted to 17-hydroxypregnenolone and then to dehydroepiandrosterone (DHEA) by the 17-hydroxylase and 17,20-lyase activities, respectively, of CYP17A1. The 17-hydroxylase activity of CYP17A1 and the type II 3β-hydroxysteroid dehydrogenase (3B-HSD) enzyme, catalyze the conversion of pregnenolone to 17-hydroxyprogesterone. The 17,20-lyase activity of CYP17A1 is then involved in the conversion of 17-hydroxyprogesterone to androstenedione. Since castration does not affect androgen synthesis in the adrenal glands, the adrenal androgens DHEA and androstenedione can be produced in the adrenal glands, enter the bloodstream and reach the prostate cancer cells to get converted to testosterone and DHT. [51] DHEA and androstenedione can also be synthesized from cholesterol within the tumor in CRPC cells and converted to testosterone and DHT through the *de novo* route. Similar to the androgens produced by the testes, the de novo hormones can then stimulate leading to activation of AR-regulated genes. [52] This is the basis of the intracrine androgen metabolism pathway. Recently, the expression of enzymes involved in androgen biosynthesis, such as HSD3B1/2 and AKR1 C1-3 has been shown to be upregulated in CRPC. [53] Intraprostatic androgens can be de novo synthesized from cholesterol or other adrenal gland molecular precursors, mediated by genes encoding enzymes, including AKR1C3 (which converts androstenedione to testosterone in the prostate) , HSD3B2 (which converts DHEA to androstenedione), SRD5A1 (which converts testosterone to dihydrotestosterone) ,in addition to CYP17A1 (which converts Pregnenolone to 17 hydroxypregnenolone),CYP19A1 (which converts androstenedione to estrone and testosterone to estradiol) and UBT2B15 (which converts androstanediol to glucuronides) [54,55].

III. Clinical Studies

The clinical management of advanced CRPC in the past has employed the use of alternative secondary hormonal manipulation following antiandrogen withdrawal. Certainly, with the advent of the use of docetaxel, the landscape of treatment has changed. However, progression beyond use of chemotherapy remains an ongoing challenge. The elucidation of the different mechanisms of castration resistance offers potential strategies for the development of novel therapies targeting those pathways involved in CRPC. Growth factors that activate the AR and intracellular signaling downstream of these growth factors are potential targets. On the other hand, the interaction between the AR DNA binding domain and the AREs DNA could theoretically be inhibited or the activity of heat-shock proteins could be interrupted. [56,57] Inhibition of the steroid-metabolizing enzymes represents a potential target. One important enzyme involved in steroidogenes is the 17, 20 lyase. However, due to common features between this enzyme and other CYP enzymes, the challenge is to offer therapeutic advantage with minimal toxicity. Indeed, ketoconazole, which has 17,20-lyase inhibition activity, has been used clinically in patients with CRPC for a long period of time. [58] However, ketoconazole treatment is often discontinued because of toxicities that arise owing to its low specificity for 17, 20-lyase. [59] Furthermore, ketoconazole while showing objective PSA and tumor responses, has never been shown to improve overall survival, [60,61] except perhaps in the setting of presence of certain markers, such as higher baseline androgenic (androstenedione) levels. [62] Abiraterone Acetate is an oral, potent, selective, [63] and irreversible inhibitor of CYP17 that is 10-to 30-fold more potent than the nonselective inhibitor, ketoconazole and has attained approval from the US FDA on April 28, 2011 for metastatic CRPC after docetaxel failure. [64] Abiraterone is highly protein-bound and mainly hepatically metabolized by CYP3A4 and SULT2A1 to inactive metabolites. Excretion is mainly in the feces (~88%) and urine (~5%) with a terminal half- life of 12 ± 5 hours. [65] Given the promising results in the phase I and II trials of abiraterone, [66,67] even in patients who have received prior ketoconazole, [68] a randomized, multicenter, multinational, phase 3 trial (COU-AA-301) was conducted to compare the safety and efficacy of abiraterone acetate (AA) in men with mCRPC previously treated with docetaxel, with the primary endpoint of OS. [69] One thousand ninety-five patients were randomly assigned in a 2:1 ratio to receive 5 mg of prednisone twice daily with either 1000 mg of abiraterone acetate (797 patients) or placebo (398 patients). After a median follow-up of 12.8 months, the primary end point which was overall survival was longer in the abiraterone acetate–prednisone group than in the placebo–prednisone group (14.8 months vs.10.9 months; HR 0.65; $P<0.001$). All secondary end points, including time to PSA progression (10.2 vs. 6.6

months; P<0.001), progression-free survival (5.6 months vs. 3.6 months; P<0.001), and PSA response rate (29% vs. 6%, P<0.001), favored the treatment group. Furthermore, pain responses were also seen. There was a population of patients (a third of those in the phase II study) who were observed to have discordant serologic and radiographic response in the form of bone scan flares (in 12 of 23 patients, deemed to have later improved), that should alert cautionary interpretation of patients' clinical course while on therapy.[70] The promising results of abiraterone acetate translate its value not only in the post-chemotherapy setting, in which it is currently approved, but to the pre-chemotherapy setting in the COU-302 study, the results of which are eagerly awaited. Certainly, additional mechanisms of resistance, such as induction of steroidogenesis and upregulation of AR splice variants as previously mentioned, must be overcome [71].

Another novel, nonsteroidal 17,20 lyase-specific inhibitor has recently been investigated and showed promising results. [72] TAK-700 is a non-steroidal androgen synthesis inhibitor of the enzyme CYP17A1, found to show high selectivity for 17,20-lyase over 17-hydroxylase in preclinical experiments. In a phase I/II trial, 26 patients with metastatic CRPC were enrolled at five different dose levels ranging from 100 mg to 600 mg twice daily dosing. [73] All patients treated with ≥ 300 mg had a PSA decline. Twelve of fifteen patients (80%) who received TAK-700 ≥ 300 mg for ≥ 3 cycles had PSA reductions of ≥ 50% and 4 of 15 (27%) had reductions ≥ 90%. Updated results on 96 patients was presented showing PSA response rates (≥50% decrease) at 12 weeks of 63%, 52%, 41%, and 62% in the 300 mg twice daily, 400 and 600 mg daily with prednisone, and 600 mg daily groups, respectively. [74] Of the 43 patients with RECIST-evaluable disease, 6 had a partial response, 23 had stable disease, and 9 had disease progression. The mean circulating tumor cell numbers were also found to be decreased in all groups. TAK-700 is currently being investigated in combination with docetaxel in men with metastatic CRPC as well as compared to placebo in chemotherapy-naïve and post-docetaxel treatment.

Another 17,20 lyase inhibitor currently being evaluated in clinical investigations is TOK-001 (formerly VN/124-1), a compound highly similar structurally to abiraterone, [75] but potentially more orally bioavailable, stable and potent. [76] In addition, unlike abiraterone acetate which has predominantly CYP17A1 inhibition, it has been reported that TOK-001 has a downregulatory effect on AR expression *in vitro* and *in vivo* and has AR antagonistic activity in contrast to bicalutamide which showed up-regulation of AR expression. A phase I/II evaluation of TOK-001 in CRPC was initiated in 2009.

Another therapeutic approach has been the development of AR antagonists which would theoretically block the interaction between androgens and ARs. However, first-generation AR antagonists such as bicalutamide show minimal efficacy for CRPC because of their partial agonist activity that favor tumor progression. [27,77] Recently, MDV3100, a second-generation anti-androgen has been investigated in patients with CRPC. [78] MDV3100 has about a five-fold higher binding affinity for the AR compared to bicalutamide and does not promote translocation of AR to the nucleus and in addition prevents binding of AR to DNA and co-activator proteins. *In vitro* LNCaP cells treated with MDV3100 showed expression of androgen dependent genes PSA and TMPRSS2 which were down-regulated in contrast to bicalutamide where the expression was up-regulated. MDV3100 also induces tumor cell apoptosis, and has no agonist activity. In a phase I/II multicenter study, 140 patients with progressive metastatic CRPC underwent dose escalations of MDV3100, starting at 30 mg daily to 600 mg daily. [6] The study population was relatively heavily pretreated. The results demonstrated anti-tumor effects at all doses, including decreases in serum prostate-specific antigen of 50% or more in 56% of the patients studied attaining a plateau at between 150 and 240 mg daily. There were also responses in soft tissue in 22% of patients (13 out of 59) and conversion from unfavorable ($\geq$ 5 CTC/7.5 ml of blood) to favorable (<5 CTC/7.5 ml of blood) circulating tumor cell counts in 49% of patients (25 out of 51). The median time for radiological progression was 47 weeks. The maximum tolerated dose of MDV3100 was 240 mg daily. Long-term follow-up efficacy data from this trial was presented at ASCO-GU in February 2011, which showed MDV 3100 continues to demonstrate durable antitumor activity as evaluated by median times to PSA progression and radiographic progression. An international, multi-center, randomized, placebo-controlled, phase III study (AFFIRM) has finished enrollment with a target accrual goal of 1199 patients to determine whether MDV3100 at a dose of 160 mg will detect an overall survival difference in patients with metastatic castration-resistant prostate cancer who were treated with prior docetaxel. Preliminary results of AFFIRM showed promising improvement in overall survival in the treatment arm thus halting the trial early and offering all patients MDV3100 given the favorable risk-benefit and safety profile. As reported by the Independent Data Monitoring Committee (IDMC), results showed that MDV3100 with prednisone offered a 4.8-month improvement in median overall survival compared to the placebo arm with prednisone. The estimated median survival for men treated with MDV3100 was 18.4 months versus 13.6 months for men treated with placebo, with a 37% risk reduction for death in the MDV3100 arm compared to placebo (Hazard Ratio=0.631). [79] A phase III study (PREVAIL) investigating the use of MDV3100 in a chemotherapy-naïve population who have failed androgen deprivation therapy is currently ongoing. In addition, there is an open phase II trial (TERRAIN) comparing MDV31000 with bicalutamide in prostate cancer patients who have progressed while on leuprolide or underwent surgical castration. The primary endpoint of the trial is progression-free survival. Another phase II trial that is examining the effects of MDV3100 monotherapy on prostate cancer patients who have not had any prior hormonal therapies or chemotherapy is open with target accrual of approximately 60 patients in Europe with PSA as the primary endpoint. BMS-641988 is an AR inhibitor that is highly potent, designed specifically based on AR crystal structure. In preclinical models, it showed higher binding affinity to AR than bicalutamide. The phase I trials results have yet to be reported [80].

The aforementioned drugs have targeted the AR and there are several others that have been studied. [81,82] While these treatments predominantly target the androgen regulation and signaling, as previously mentioned in the emergence of castration resistance, several other compounds and drugs have been studied in efforts to combat advanced CRPC. The use of AT-101, a bcl-2 inhibitor, has been shown to potentiate effects of ADT. [83] It has been combined with chemotherapy but results are not uniformly promising, except perhaps in patients with high-risk disease. [84] Another transcription factor, nuclear factor (NF)-kappaB, functions as a downstream substrate of mitogen-activated protein kinase (MAP kinase), phosphatidylinositol 3-kinase (PI3-kinase), AKT, and protein kinase C and has been found in prostate cancer to affect AR expression by decreasing AR expression levels, prostate-specific antigen secretion, and proliferation of prostate cancer cells in vitro. [85,86] A single phase II study combined bortezomib, an NF-kappa B inhibitor, with hormonal therapy in patients manifesting with early PSA recurrence after local therapy and showed 11 of 15 patients achieving complete responses with an average time to progression of 5.5 months. [87] Other pathways targeting EGFR and the ERB family of tyrosine kinases will are also emerging as important therapeutic targets [88].

IV. Adverse Effects of Targeting Androgen Regulation: Recent Findings

Traditional ADT has concomitant side-effects that range from metabolic, hematologic, bone, cardiovascular, and mood effects. [89] The addition of anti-androgens accentuates these side-effects though differences have been seen with ADT versus anti-androgen monotherapy alone.

In the COU-AA-301 trial, The safety profile observed for patients who received AA was fairly similar with the placebo group with most adverse events occurring as grade 1 and 2 toxicity, with fatigue as the most common, followed by back pain, nausea, constipation, bone pain, and arthralgias.[38] Grade 1 and 2 urinary tract infection was more frequent in the abiraterone acetate group (12%, vs. 7% in the placebo group; P = 0.02). Similar rates of treatment discontinuation due to adverse events were seen in both groups, 19% for AA and 23% for placebo. Given the CYP17 inhibition and the resultant stimulation of adrenocorticotropic hormone (ACTH) leading to increase in corticosterone and deoxycorticosterone levels , adverse events associated with elevated mineralocorticoid levels were expectedly more common in the AA arm compared with placebo.

Hence, fluid retention and edema, hypokalemia, and hypertension, cardiac disorders and liver-function test abnormalities were more common in the abiraterone acetate group than in the placebo group (55% vs. 43%, P<0.001). No individual grade 4 adverse events occurred in 2% or more of patients in either treatment group.

A total of 11% of patients in the abiraterone acetate group and 13% of patients in the placebo group died within 30 days after the last dose of study medication, primarily as a result of disease progression. A lower proportion of patients in the abiraterone acetate group than in the placebo group had an adverse event that resulted in death (12% vs. 15%).

The most common adverse events with TAK 700 were fatigue followed by non-dose-related GI events such as nausea, constipation, and vomiting. [39]The most common treatment related adverse events with MDV3100 was fatigue which occurred around 4 weeks after administration. [45] Grade 3/4 adverse events were seen predominantly at a dose of 360 mg or greater, including fatigue in 11% of patients, which generally responded to dose reduction, asthenia in 2% and seizures in 2%. Due to tolerability issues at doses above 360 mg and the potential concern for seizures a maximum tolerated dose of 240 mg was selected for sustained treatment.

Conclusion

Understanding the cellular mechanisms involved in the androgenic signals and function has led to better molecular targets for the treatment of advanced CRPC. The continuous progress and effort in dissecting the alterations in the AR pathway in castration resistant prostate cancer has led to the development of novel therapies targeting those pathways.

Further research in further elucidating the castration-resistant pathway will potentially open the way for new therapeutic maneuvers.

On the other hand, further understanding for the potential resistance to these newer agents is also constantly evolving. New insights in treatment of castrate resistant prostate cancer are expected to be gained in clinical trials using those novel agents.

Table 1. New therapies targeting mechanisms of resistance in mCRPC

Name of drug	Mechanism of action	Phase III trial	Population Tested	Compared to	Primary End-point	Current indication	Most Common Adverse Effects
Abiraterone Acetate + Prednisone	Irreversible inhibition of CYP17	COU-AA-301	CRPC previously treated with Docetaxel	Placebo + Prednisone	Overall Survival Benefit (14.8 versus 10.9 months)	mCRPC after Docetaxel	Mineralo-corticoid excess
MDV-3100 + prednisone	Androgen receptor signaling inhibitor	AFFIRM	CRPC previously treated with Docetaxel	Placebo + Prednisone	Overall Survival Benefit (18.4 versus 13.6 months)	mCRPC after Docetaxel	Fatigue

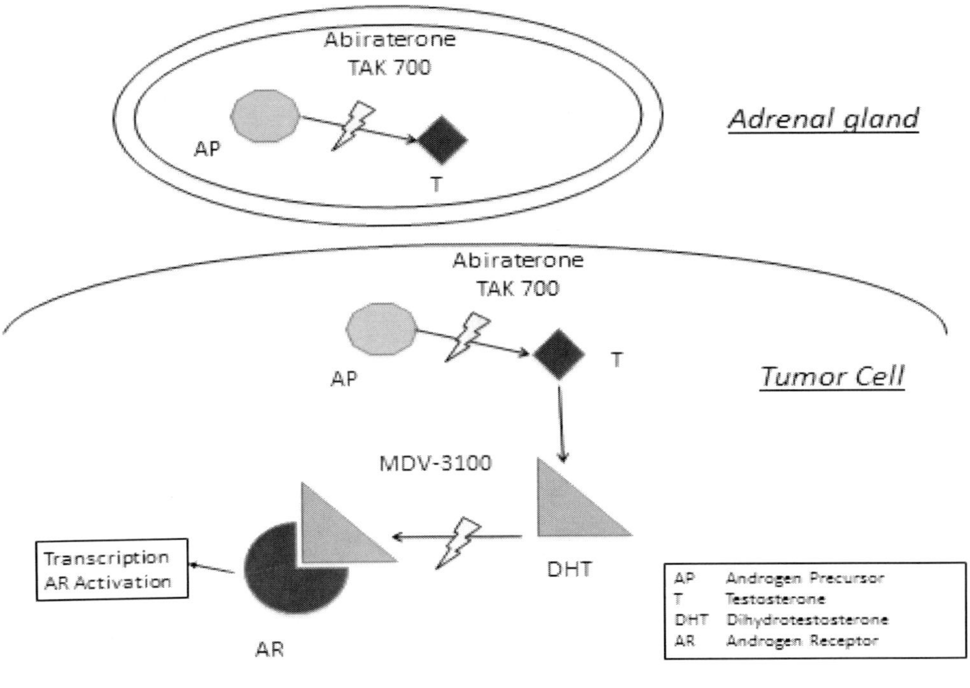

Figure 1. Simplified mechanism of action of novel targeted hormonal therapy agents in prostate cancer.

References

[1] Huggins C, Bergenstal DM: Inhibition of human mammary and prostatic cancers by adrenalectomy. *Cancer Res.* 1952; 12(2):134-41.

[2] Gelmann EP: Molecular biology of the androgen receptor. *J. Clin. Oncol.* 2002; 20(13):3001-15.

[3] Tyagi RK, Lavrovsky Y, Ahn SC, et al: Dynamics of intracellular movement and nucleocytoplasmic recycling of the ligand-activated androgen receptor in living cells. *Mol. Endocrinol.* 2000; 14(8):1162-74.

[4] Zhou ZX, Lane MV, Kemppainen JA, et al: Specificity of ligand-dependent androgen receptor stabilization: receptor domain interactions influence ligand dissociation and receptor stability. *Mol. Endocrinol.* 1995; 9(2):208-18.

[5] Bubley GJ, Carducci M, Dahut W, et al: Eligibility and response guidelines for phase II clinical trials in androgen-independent prostate cancer: recommendations from the Prostate-Specific Antigen Working Group. *J. Clin. Oncol.* 1999; 17(11):3461-7.

[6] Scher HI, Beer TM, Higano CS, et al: Antitumour activity of MDV3100 in castration-resistant prostate cancer: a phase 1-2 study. *Lancet.* 2010; 375(9724):1437-46.

[7] Hellerstedt BA, Pienta KJ: The current state of hormonal therapy for prostate cancer. *CA Cancer J. Clin.*. 2002; 52(3):154-79.

[8] Oefelein MG, Agarwal PK, Resnick MI: Survival of patients with hormone refractory prostate cancer in the prostate specific antigen era. *J. Urol.* 2004; 171(4):1525-8.

[9] Scher HI, Sawyers CL: Biology of progressive, castration-resistant prostate cancer: directed therapies targeting the androgen-receptor signaling axis. *J. Clin. Oncol.* 2005; 23(32):8253-61.

[10] Chen CD, Welsbie DS, Tran C, et al: Molecular determinants of resistance to antiandrogen therapy. *Nat. Med.* 2004; 10(1):33-9.

[11] Dutt SS, Gao AC: Molecular mechanisms of castration-resistant prostate cancer progression. *Future Oncol.* 2009; 5(9):1403-13.

[12] Feldman BJ, Feldman D: The development of androgen-independent prostate cancer. *Nat Rev Cancer.* 2001; 1(1):34-45.

[13] Pienta KJ, Bradley D: Mechanisms underlying the development of androgen-independent prostate cancer. *Clin. Cancer Res.* 2006; 12(6):1665-71.

[14] Suzuki H, Akakura K, Komiya A, et al: Codon 877 mutation in the androgen receptor gene in advanced prostate cancer: relation to antiandrogen withdrawal syndrome. *Prostate.* 1996; 29(3):153-8.

[15] Veldscholte J, Berrevoets CA, Brinkmann AO, et al: Anti-androgens and the mutated androgen receptor of LNCaP cells: differential effects on binding affinity, heat-shock protein interaction, and transcription activation. *Biochemistry.* 1992; 31(8):2393-9.

[16] Fenton MA, Shuster TD, Fertig AM, et al: Functional characterization of mutant androgen receptors from androgen-independent prostate cancer. *Clin. Cancer Res.* 1997; 3(8):1383-8.

[17] Hara T, Kouno J, Nakamura K, et al: Possible role of adaptive mutation in resistance to antiandrogen in prostate cancer cells. *Prostate.* 2005; 65(3):268-75.

[18] Hara T, Miyazaki J, Araki H, et al: Novel mutations of androgen receptor: a possible mechanism of bicalutamide withdrawal syndrome. *Cancer Res.* 2003; 63(1):149-53.

[19] Hu R, Dunn TA, Wei S, et al: Ligand-independent androgen receptor variants derived from splicing of cryptic exons signify hormone-refractory prostate cancer. *Cancer Res.* 2009; 69(1):16-22.

[20] Sun S, Sprenger CC, Vessella RL, et al: Castration resistance in human prostate cancer is conferred by a frequently occurring androgen receptor splice variant. *J. Clin. Invest.* 2010; 120(8):2715-30.

[21] Guo Z, Yang X, Sun F, et al: A novel androgen receptor splice variant is up-regulated during prostate cancer progression and promotes androgen depletion-resistant growth. *Cancer Res.* 2009; 69(6):2305-13.

[22] Hu R, Isaacs WB, Luo J: A snapshot of the expression signature of androgen receptor splicing variants and their distinctive transcriptional activities. *Prostate.* 2011; 71(15):1656-67.

[23] Dehm SM, Schmidt LJ, Heemers HV, et al: Splicing of a novel androgen receptor exon generates a constitutively active androgen receptor that mediates prostate cancer therapy resistance. *Cancer Res.* 2008; 68(13):5469-77.

[24] Edwards J, Krishna NS, Grigor KM, et al: Androgen receptor gene amplification and protein expression in hormone refractory prostate cancer. *Br. J. Cancer.* 2003; 89(3):552-6.

[25] Chang KH, Li R, Papari-Zareei M, et al: Dihydrotestosterone synthesis bypasses testosterone to drive castration-resistant prostate cancer. *Proc. Natl. Acad. Sci. USA.* 2011; 108(33):13728-33.

[26] Sharifi N: The 5alpha-Androstanedione Pathway to Dihydrotestosterone in Castration-Resistant Prostate Cancer. *J. Investig. Med.* 2011.

[27] Yoshida T, Kinoshita H, Segawa T, et al: Antiandrogen bicalutamide promotes tumor growth in a novel androgen-dependent prostate cancer xenograft model derived from a bicalutamide-treated patient. *Cancer Res.* 2005; 65(21):9611-6.

[28] Grossmann ME, Huang H, Tindall DJ: Androgen receptor signaling in androgen-refractory prostate cancer. *J. Natl. Cancer Inst.* 2001; 93(22):1687-97.

[29] Dehm SM, Tindall DJ: Regulation of androgen receptor signaling in prostate cancer. *Expert Rev Anticancer Ther.* 2005; 5(1):63-74.

[30] Xie D, Gore C, Liu J, et al: Role of DAB2IP in modulating epithelial-to-mesenchymal transition and prostate cancer metastasis. *Proc. Natl. Acad. Sci. USA.* 2010; 107(6):2485-90.

[31] Saini S, Majid S, Yamamura S, et al: Regulatory Role of mir-203 in Prostate Cancer Progression and Metastasis. *Clin. Cancer Res.* 2011; 17(16):5287-98.

[32] Sun Y, Wang BE, Leong KG, et al: Androgen Deprivation Causes Epithelial-Mesenchymal Transition in the Prostate: Implications for Androgen-deprivation Therapy. *Cancer Res.* 2011.

[33] Wu JD, Haugk K, Woodke L, et al: Interaction of IGF signaling and the androgen receptor in prostate cancer progression. *J. Cell Biochem.* 2006; 99(2):392-401.

[34] Wu JD, Haugk K, Coleman I, et al: Combined in vivo effect of A12, a type 1 insulin-like growth factor receptor antibody, and docetaxel against prostate cancer tumors. *Clin. Cancer Res.* 2006; 12(20 Pt 1):6153-60.

[35] Debes JD, Schmidt LJ, Huang H, et al: p300 mediates androgen-independent transactivation of the androgen receptor by interleukin 6. *Cancer Res.* 2002; 62(20):5632-6.

[36] Debes JD, Sebo TJ, Lohse CM, et al: p300 in prostate cancer proliferation and progression. *Cancer Res.* 2003; 63(22):7638-40.

[37] Plymate SR, Haugk K, Coleman I, et al: An antibody targeting the type I insulin-like growth factor receptor enhances the castration-induced response in androgen-dependent prostate cancer. *Clin. Cancer Res.* 2007; 13(21):6429-39.

[38] Drachenberg DE, Elgamal AA, Rowbotham R, et al: Circulating levels of interleukin-6 in patients with hormone refractory prostate cancer. *Prostate.* 1999; 41(2):127-33.

[39] Chen T, Wang LH, Farrar WL: Interleukin 6 activates androgen receptor-mediated gene expression through a signal transducer and activator of transcription 3-dependent pathway in LNCaP prostate cancer cells. *Cancer Res.* 2000; 60(8):2132-5.

[40] Aaronson DS, Muller M, Neves SR, et al: An androgen-IL-6-Stat3 autocrine loop re-routes EGF signal in prostate cancer cells. *Mol. Cell Endocrinol.* 2007; 270(1-2):50-6.

[41] Wallner L, Dai J, Escara-Wilke J, et al: Inhibition of interleukin-6 with CNTO328, an anti-interleukin-6 monoclonal antibody, inhibits conversion of androgen-dependent prostate cancer to an androgen-independent phenotype in orchiectomized mice. *Cancer Res.* 2006; 66(6):3087-95.

[42] Lee SO, Lou W, Hou M, et al: Interleukin-6 promotes androgen-independent growth in LNCaP human prostate cancer cells. *Clin. Cancer Res.* 2003; 9(1):370-6.

[43] Araki S, Omori Y, Lyn D, et al: Interleukin-8 is a molecular determinant of androgen independence and progression in prostate cancer. *Cancer Res.* 2007; 67(14):6854-62.

[44] Inoue K, Slaton JW, Eve BY, et al: Interleukin 8 expression regulates tumorigenicity and metastases in androgen-independent prostate cancer. *Clin. Cancer Res.* 2000; 6(5):2104-19.

[45] Lee LF, Louie MC, Desai SJ, et al: Interleukin-8 confers androgen-independent growth and migration of LNCaP: differential effects of tyrosine kinases Src and FAK. *Oncogene.* 2004; 23(12):2197-205.

[46] Mellinghoff IK, Vivanco I, Kwon A, et al: HER2/neu kinase-dependent modulation of androgen receptor function through effects on DNA binding and stability. *Cancer Cell.* 2004; 6(5):517-27.

[47] Ahmad I, Patel R, Singh LB, et al: HER2 overcomes PTEN (loss)-induced senescence to cause aggressive prostate cancer. *Proc. Natl. Acad. Sci. USA.* 2011; 108(39):16392-7.

[48] Chen L, Mooso BA, Jathal MK, et al: Dual EGFR/HER2 inhibition sensitizes prostate cancer cells to androgen withdrawal by suppressing ErbB3. *Clin. Cancer Res.* 2011; 17(19):6218-28.

[49] Weber MJ, Gioeli D: Ras signaling in prostate cancer progression. *J. Cell Biochem.* 2004; 91(1):13-25.

[50] Stanbrough M, Bubley GJ, Ross K, et al: Increased expression of genes converting adrenal androgens to testosterone in androgen-independent prostate cancer. *Cancer Res.* 2006; 66(5):2815-25.

[51] Rainey WE, Nakamura Y: Regulation of the adrenal androgen biosynthesis. *J. Steroid Biochem. Mol. Biol.* 2008; 108(3-5):281-6.

[52] Yamaoka M, Hara T, Kusaka M: Overcoming persistent dependency on androgen signaling after progression to castration-resistant prostate cancer. *Clin. Cancer Res.* 2010; 16(17):4319-24.

[53] Montgomery RB, Mostaghel EA, Vessella R, et al: Maintenance of intratumoral androgens in metastatic prostate cancer: a mechanism for castration-resistant tumor growth. *Cancer Res.* 2008; 68(11):4447-54.

[54] Mostaghel EA, Nelson PS: Intracrine androgen metabolism in prostate cancer progression: mechanisms of castration resistance and therapeutic implications. *Best Pract. Res. Clin. Endocrinol. Metab.* 2008; 22(2):243-58.

[55] Page ST, Lin DW, Mostaghel EA, et al: Persistent intraprostatic androgen concentrations after medical castration in healthy men. *J. Clin. Endocrinol. Metab.* 2006; 91(10):3850-6.

[56] Drysdale MJ, Brough PA, Massey A, et al: Targeting Hsp90 for the treatment of cancer. *Curr. Opin. Drug Discov. Devel.* 2006; 9(4):483-95.

[57] Adachi M, Takayanagi R, Tomura A, et al: Androgen-insensitivity syndrome as a possible coactivator disease. *N. Engl. J. Med.* 2000; 343(12):856-62.

[58] Trachtenberg J, Pont A: Ketoconazole therapy for advanced prostate cancer. *Lancet.* 1984; 2(8400):433-5.

[59] Loose DS, Kan PB, Hirst MA, et al: Ketoconazole blocks adrenal steroidogenesis by inhibiting cytochrome P450-dependent enzymes. *J. Clin. Invest.* 1983; 71(5):1495-9.

[60] Small EJ, Baron A, Bok R: Simultaneous antiandrogen withdrawal and treatment with ketoconazole and hydrocortisone in patients with advanced prostate carcinoma. *Cancer.* 1997; 80(9):1755-9.

[61] Small EJ, Halabi S, Dawson NA, et al: Antiandrogen withdrawal alone or in combination with ketoconazole in androgen-independent prostate cancer patients: a phase III trial (CALGB 9583). *J. Clin. Oncol.* 2004; 22(6):1025-33.

[62] Ryan CJ, Halabi S, Ou SS, et al: Adrenal androgen levels as predictors of outcome in prostate cancer patients treated with ketoconazole plus antiandrogen withdrawal: results from a cancer and leukemia group B study. *Clin. Cancer Res.* 2007; 13(7):2030-7.

[63] Ang JE, Olmos D, de Bono JS: CYP17 blockade by abiraterone: further evidence for frequent continued hormone-dependence in castration-resistant prostate cancer. *Br. J. Cancer.* 2009; 100(5):671-5.

[64] FDA Approval Abiraterone Acetate. 2011.

[65] Barrie SE, Potter GA, Goddard PM, et al: Pharmacology of novel steroidal inhibitors of cytochrome P450(17) alpha (17 alpha-hydroxylase/C17-20 lyase). *J. Steroid. Biochem. Mol. Biol.* 1994; 50(5-6):267-73.

[66] Attard G, Reid AH, Yap TA, et al: Phase I clinical trial of a selective inhibitor of CYP17, abiraterone acetate, confirms that castration-resistant prostate cancer commonly remains hormone driven. *J. Clin. Oncol.* 2008; 26(28):4563-71.

[67] Attard G, Reid AH, A'Hern R, et al: Selective inhibition of CYP17 with abiraterone acetate is highly active in the treatment of castration-resistant prostate cancer. *J. Clin. Oncol.* 2009; 27(23):3742-8.

[68] Ryan CJ, Smith MR, Fong L, et al: Phase I clinical trial of the CYP17 inhibitor abiraterone acetate demonstrating clinical activity in patients with castration-resistant prostate cancer who received prior ketoconazole therapy. *J. Clin. Oncol.* 2010; 28(9):1481-8.

[69] de Bono JS, Logothetis CJ, Molina A, et al: Abiraterone and increased survival in metastatic prostate cancer. *N. Engl. J. Med.* 2011; 364(21):1995-2005.

[70] Ryan CJ, Shah S, Efstathiou E, et al: Phase II study of abiraterone acetate in chemotherapy-naive metastatic castration-resistant prostate cancer displaying bone flare discordant with serologic response. *Clin. Cancer Res.* 2011; 17(14):4854-61.

[71] Mostaghel EA, Marck BT, Plymate SR, et al: Resistance to CYP17A1 inhibition with abiraterone in castration-resistant prostate cancer: induction of steroidogenesis and androgen receptor splice variants. *Clin. Cancer Res.* 2011; 17(18):5913-25.

[72] Kaku T, Hitaka T, Ojida A, et al: Discovery of orteronel (TAK-700), a naphthylmethylimidazole derivative, as a highly selective 17,20-lyase inhibitor with potential utility in the treatment of prostate cancer. *Bioorg. Med. Chem.* 2011; 19(21):6383-99.

[73] Dreicer R, Agus DB, MacVicar GR, et al: Safety, pharmacokinetics, and efficacy of TAK-700 in metastatic castration-resistant prostrate cancer: A phase I/II, open-label study. *J. Clin. Oncol.* 2010; 28(15_suppl):Abstract 3084.

[74] Agus DB, Stadler WM, Shevrin DH: Safety, efficacy, and pharmacodynamics of the investigational agent TAK-700 in metastatic castration-resistant prostate cancer (mCRPC): Updated data from a phase I/II study. *J. Clin. Oncol.* 2011; 29abstr 4531.

[75] Vasaitis T, Belosay A, Schayowitz A, et al: Androgen receptor inactivation contributes to antitumor efficacy of 17{alpha}-hydroxylase/17,20-lyase inhibitor 3beta-hydroxy-17-(1H-benzimidazole-1-yl)androsta-5,16-diene in prostate cancer. *Mol. Cancer Ther.* 2008; 7(8):2348-57.

[76] Bruno RD, Vasaitis TS, Gediya LK, et al: Synthesis and biological evaluations of putative metabolically stable analogs of VN/124-1 (TOK-001): head to head anti-tumor efficacy evaluation of VN/124-1 (TOK-001) and abiraterone in LAPC-4 human prostate cancer xenograft model. *Steroids.* 2011; 76(12):1268-79.

[77] Bales GT, Chodak GW: A controlled trial of bicalutamide versus castration in patients with advanced prostate cancer. *Urology.* 1996; 47(1A Suppl):38-43; discussion 48-53.

[78] Tran C, Ouk S, Clegg NJ, et al: Development of a second-generation antiandrogen for treatment of advanced prostate cancer. *Science.* 2009; 324(5928):787-90.

[79] Medivation and Astellas Announce Positive Survival Data From Interim Analysis of Phase 3 AFFIRM Trial of MDV3100 in Men With Advanced Prostate Cancer. Accessed on November 25, 2011. Available on http://investors.medivation.com /releasedetail.cfm?releaseid=620500.

[80] Attar RM, Jure-Kunkel M, Balog A, et al: Discovery of BMS-641988, a novel and potent inhibitor of androgen receptor signaling for the treatment of prostate cancer. *Cancer Res.* 2009; 69(16):6522-30.

[81] Jones JO, Bolton EC, Huang Y, et al: Non-competitive androgen receptor inhibition in vitro and in vivo. *Proc. Natl. Acad. Sci. USA.* 2009; 106(17):7233-8.

[82] Yamamoto S, Matsunaga N, Hitaka T, et al: Design, synthesis, and biological evaluation of 4-phenylpyrrole derivatives as novel androgen receptor antagonists. *Bioorg. Med. Chem.* 2011.

[83] McGregor N, Patel L, Craig M, et al: AT-101 (R-(-)-gossypol acetic acid) enhances the effectiveness of androgen deprivation therapy in the VCaP prostate cancer model. *J. Cell Biochem.* 2010; 110(5):1187-94.

[84] Sonpavde G, Matveev V, Burke JM, et al: Randomized phase II trial of docetaxel plus prednisone in combination with placebo or AT-101, an oral small molecule Bcl-2

family antagonist, as first-line therapy for metastatic castration-resistant prostate cancer. *Ann. Oncol.* 2011.

[85] Zhang L, Altuwaijri S, Deng F, et al: NF-kappaB regulates androgen receptor expression and prostate cancer growth. *Am. J. Pathol.* 2009; 175(2):489-99.

[86] Voutsadakis IA, Papandreou CN: The ubiquitin-proteasome system in prostate cancer and its transition to castration resistance. *Urol. Oncol.* 2010.

[87] Kraft AS, Garrett-Mayer E, Wahlquist AE, et al: Combination therapy of recurrent prostate cancer with the proteasome inhibitor bortezomib plus hormone blockade. *Cancer Biol. Ther.* 2011; 12(2):119-24.

[88] Jathal MK, Chen L, Mudryj M, et al: Targeting ErbB3: the New RTK(id) on the Prostate Cancer Block. *Immunol Endocr Metab Agents Med Chem.* 2011; 11(2):131-149.

[89] Sharifi N, Gulley JL, Dahut WL: Androgen deprivation therapy for prostate cancer. *JAMA.* 2005; 294(2):238-44.

In: Androgens: Production, Functions and Disorders
Editors: B. F. Thompson and D. J. Robinson
ISBN: 978-1-62081-145-0

Chapter VII

Androgens and Male Fertility: Spermatogenesis and Contraception

Yang Sun and Eric Chung
Department of Urology, Princess Alexandra Hospital,
Brisbane QLD, Australia

Abstract

Androgens are critical steroid hormones that determine the expression of the male phenotype, including the outward development of secondary male sexual characteristics as well as the initiation and maintenance of spermatogenesis and male fertility. Their actions are mediated by the androgen receptors. Several mechanisms involved in testosterone regulation of spermatogenesis include activation of gene transcription through the translocation of the androgen receptor, phospholipase and calcium influx, as well as specific protein receptors and kinases in Sertoli cells. On the other hand, the principle of male hormonal contraception is based on the suppression of luteinizing hormone and follicle stimulating hormone and the replacement of testosterone to maintain androgenicity. Novel contraceptive methods for men are currently not available but testosterone alone or in combination with progestin shows promising leads. The reasons for slow progress in hormonal male contraception are partly biological and partly practical and economical. This book chapter evaluates the role of androgen receptors in male fertility, as well as classical and non-classical mechanisms of actions of testosterone and spermatogenesis. It will also review the main principles of hormonal male contraception, results of clinical trials and potential innovations in this field.

Introduction

In both sexes, the gonads have a dual function: the production of germs cells (gametogenesis) and the secretion of sex hormones. The androgens are the steroid sex hormones that are masculinising in their action; the oestrogens are those that are feminising. Both types of hormones are normally secreted in both sexes.

The testes secrete large amounts of androgens, principally testosterone, but they also secrete small amounts of oestrogens. Androgens are also secreted from the adrenal cortex in both sexes, and some of the androgens are converted to oestrogens in fat and other extra-gonadal and extra-adrenal tissues. In both sexes, the gonads secrete other polypeptides, including inhibin B, a polypeptide that inhibits follicle stimulating hormone (FSH) secretion.

The secretory and gametogenic functions of the gonads are both dependent on the secretion of the anterior pituitary gonadotropins, FSH and luteinizing hormone (LH). The sex hormones and inhibin B feed back to inhibit gonadotropin secretion. In males, gonadotropin secretion is non-cyclic, in contrast to in post-pubertal females, where an orderly, sequential secretion of gonadotropins is necessary for the occurrence of menstruation, pregnancy and lactation (Ganong, 2009, p. 411).

Androgens

Androgens are hormones that exert masculinising effects, promoting protein anabolism and growth. They determine the expression of the male phenotype and the development of male reproductive organs. Furthermore, androgens are needed for puberty, male fertility and male sexual function.

Testosterone is required for the maintenance of spermatogenesis and for inhibition of germ cell apoptosis (Singh, O'Neill, & Handelsman, 1995). Testosterone from the testes is the most active androgen. The adrenal androgens have less than 20% of the activity of gonadal testosterone. Secretion of the adrenal androgens is controlled acutely by adrenocorticotropic hormone (ACTH) and not by gonadotropins. However, the concentration of dehydroepiandrosterone sulphate (DHEAS) increases until it peaks at about 225mg/dL in the early twenties, and falls to very low values in old age.

All but about 0.3% of the circulating DHEA is conjugated to sulphate (DHEAS). The secretion of adrenal androgens is quite similar between castrated males and females, and normal males, so it is clear that these hormones exert very little masculinising effect when secreted in normal amounts. However, they can produce appreciable masculinisation when secreted in excessive amounts. In adult males, excess adrenal androgens merely accentuate existing characteristics, but in pre-pubertal boys that can cause precocious development of the secondary sex characteristics without testicular growth (precocious pseudo-puberty). In females they cause female pseudo-hermaphroditism and the adrenogenital syndrome. Some health practitioners recommend injections of DHEA to combat the effects of aging, but the results to date are controversial.

Role of Androgens in Male Sexual Development

The embryology of the gonads is summarised in Figure 1. Male sexual development starts between the 7th and 12th week of gestation. In the 7th week, the embryo has both male and female primordial genital ducts. The undifferentiated gonads develop into a foetal testis by expression of the sex-determining region of the Y chromosome (SRY), a gene complex located on the short arm of the Y chromosome. In a normal male foetus, the Wolffian duct system on each side develops into the epididymis and vas deferens. The external genitalia are similarly bi-potential until the eighth week. Thereafter, the urogenital slit disappears and male genitalia form, or, alternatively, it remains open and female genitalia form.

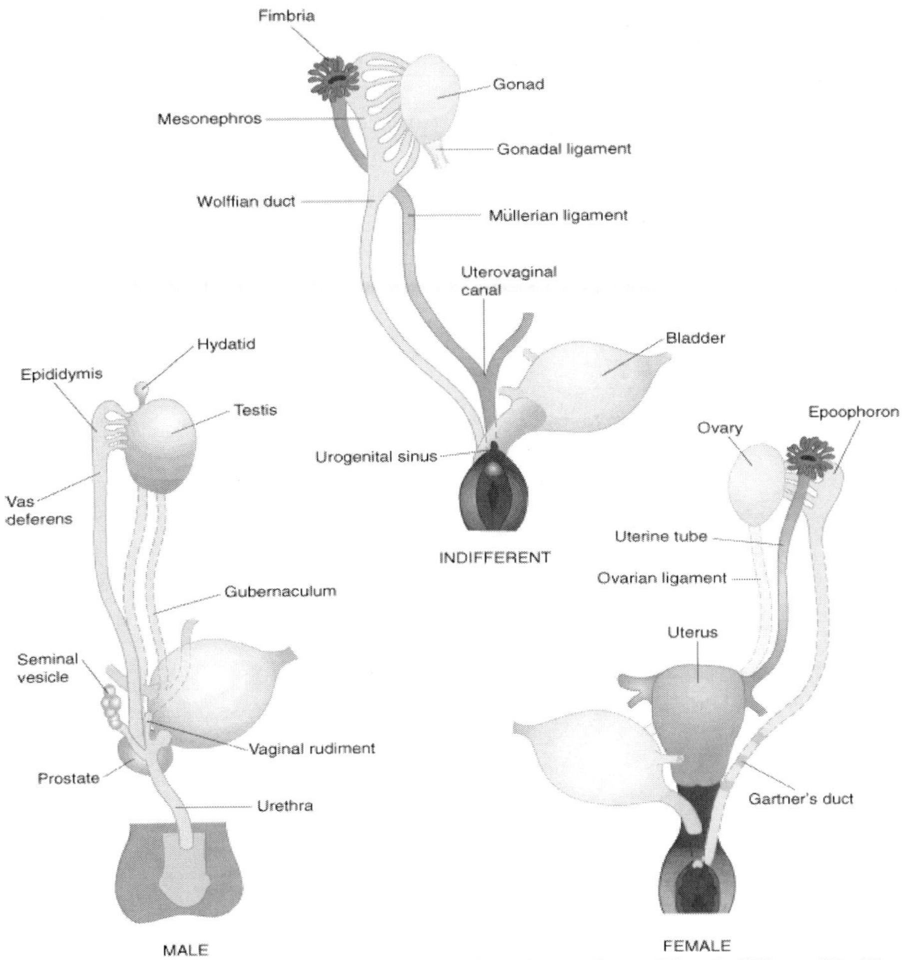

After Corning HK, Wilkins L *Williams Textbook of Endocrinology*, 7th ed. Wilson JD, Foster DW [editors]. Saunders,1985.

Figure 1. Embryonic differentiation of male and female internal genitalia (genital ducts) from wolffian (male) and mullerian (female) primordia.

When the embryo has functional testes, male internal and external genitalia develop. The Leydig cells of the foetal testis secrete testosterone, and the Sertoli cells secrete müllerian-inhibiting substance (MIS). MIS is a 536-amino-acid homodimer that is a member of the TGFβ superfamily of growth factors, which includes inhibins and activins. In their effects on the internal as opposed to the external genitalia, MIS and testosterone act unilaterally. MIS causes regression of the müllerian ducts by apoptosis on the side on which it is secreted, and testosterone fosters the development of the vas deferens and related structures from the Wolffian ducts. The testosterone metabolite dihydrotestosterone induces the formation of male external genitalia and male secondary sex characteristics [Figure 2].

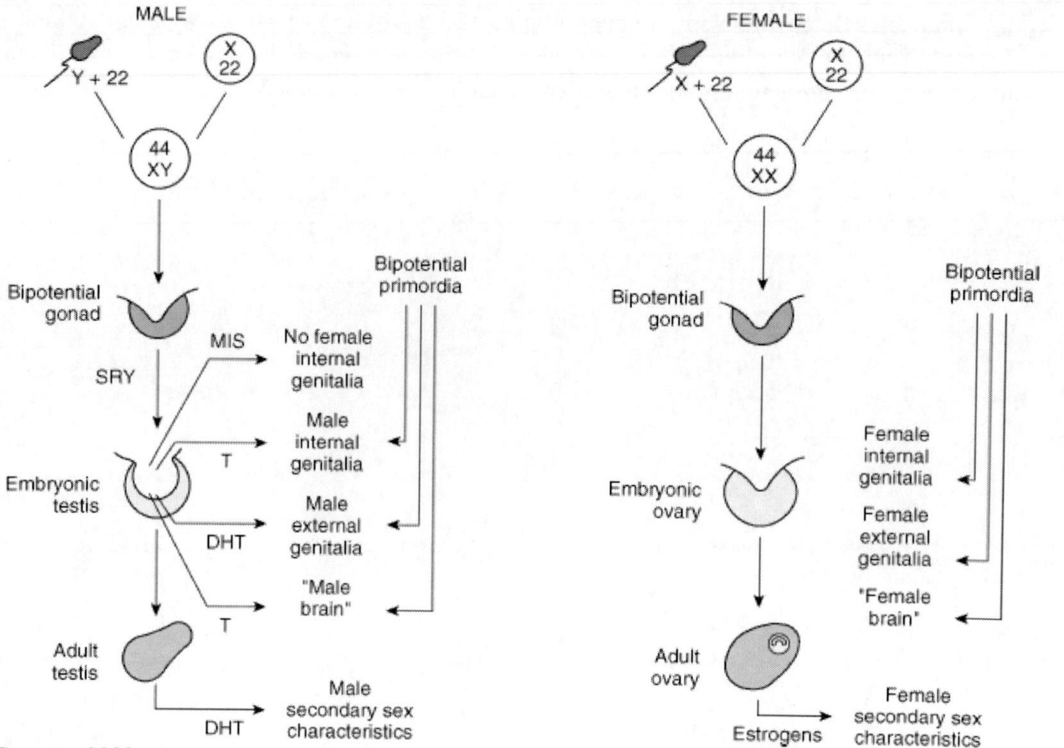

Ganong, 2009.

Figure 2. Diagrammatic summary of normal sex determination, differentiation, and development in humans. MIS, mullerian inhibiting substance; T, testosterone; DHT, dihydrotestosterone.

Spermatogenesis

Spermatogonia, the primitive germ cells located next to the basal lamina of the seminiferous tubules, mature into primary spermatocytes (Figure 3). This process begins during adolescence.

The primary spermatocytes undergo meiotic division, thereby reducing the number of chromosomes. In this two-stage process, they divide into secondary spermatocytes and then into spermatids, which contain the haploid number of 23 chromosomes. The spermatids mature into spermatozoa (sperms). As a single spermatogonium divides and matures, its descendants remain tied together by cytoplasmic bridges until the late spermatid stage. This apparently ensures synchrony of the differentiation of each clone of germ cells. The estimated number of spermatids formed from a single spermatogonium is 512. In humans, it takes an average of 74 days to form a mature sperm from a primitive germ cell by this orderly process of spermatogenesis.

Each sperm is an intricate motile cell, rich in DNA, with a head that is made up mostly of chromosomal material. Covering the head like a cap is the acrosome, a lysosome-like organelle rich in enzymes involved in sperm penetration of the ovum and other events involved in fertilization. The motile tail of the sperm is wrapped in its proximal portion by a sheath holding numerous mitochondria. The membranes of late spermatids and spermatozoa contain a special small form of angiotensin-converting enzyme called germinal angiotensin-converting enzyme. The function of this enzyme in the sperms is unknown, although male mice in which the function of the angiotensin-converting enzyme gene has been disrupted have reduced fertility.

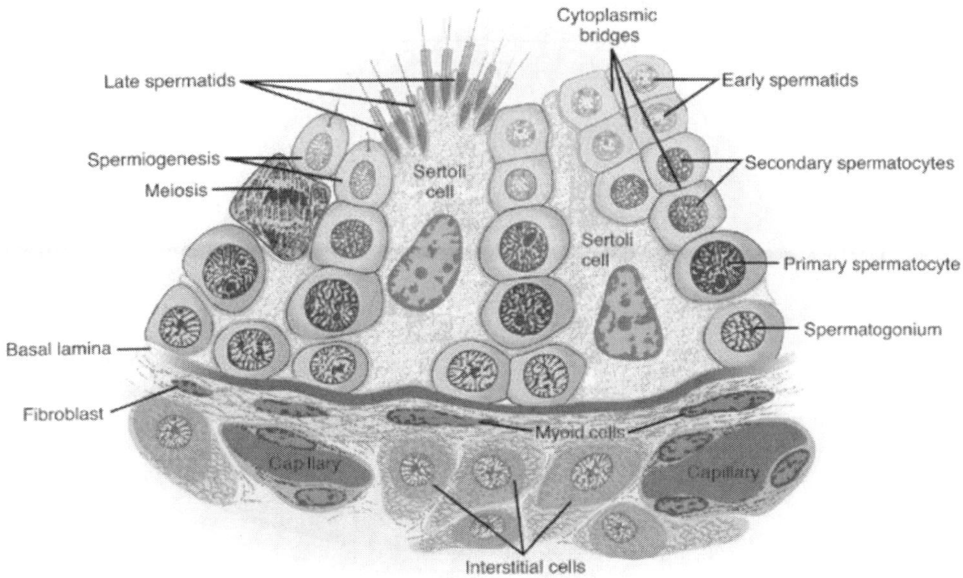

From Junqueira LC, Carneiro J: *Basic Histology: Text & Atlas*, 10th ed. McGraw-Hill, 2003.

Figure 3. Seminiferous epithelium. Note that maturing germ cells remain connected by cytoplasmic bridges through the early spermatid stage and that these cells are closely invested by Sertoli cell cytoplasm as they move from the basal lamina to the lumen.

The spermatids mature into spermatozoa in deep folds of the cytoplasm of the Sertoli cells. Mature spermatozoa are released from the Sertoli cells and become free in the lumens of the tubules. The Sertoli cells secrete androgen-binding protein (ABP), inhibin, and MIS. They do not synthesize androgens, but they contain aromatase (CYP19), the enzyme responsible for conversion of androgens to oestrogens, and they can produce oestrogens. ABP probably functions to maintain a high, stable supply of androgen in the tubular fluid. Inhibin inhibits FSH secretion.

FSH and androgens maintain the gametogenic function of the testis. After hypophysectomy, injection of LH produces a high local concentration of androgen in the testes, and this maintains spermatogenesis. The stages from spermatogonia to spermatids appear to be androgen-independent. However, the maturation from spermatids to spermatozoa depends on androgen acting on the Sertoli cells in which the developing spermatozoa are embedded. FSH acts on the Sertoli cells to facilitate the last stages of spermatid maturation. In addition, it promotes the production of ABP.

An interesting observation is that the oestrogen content of the fluid in the rete testis is high, and the walls of the rete contain numerous ERα oestrogen receptors. In this region, fluid is reabsorbed and the spermatozoa are concentrated. If this does not occur, the sperm entering the epididymis are diluted in a large volume of fluid, and infertility results.

Hypthalamic-Pituitary-Gonadal Axis

The hypothalamic–pituitary–gonadal axis (also HPG axis) refers to the effects of the hypothalamus, pituitary gland, and gonads as if these individual endocrine glands were a single entity as a whole. This axis controls development, reproduction, and ageing in animals. The hypothalamus produces gonadotropin-releasing hormone (GnRH). The anterior portion of the pituitary gland produces LH and FSH, and the gonads produce oestrogen and testosterone.

The hypothalamus is located in the brain and secretes GnRH (Millar, Lu, Pawson, Flanagan, Morgan, & Maudsley, 2004). GnRH travels down the anterior portion of the pituitary via the hypophyseal portal system and binds to receptors on the secretory cells of the adenohypophysis (Charlton, 2008) (Vadakkadath Meethal & Atwood, 2005). In response to GnRH stimulation these cells produce LH and FSH, which travel into the blood stream (Vadakkadath Meethal & Atwood, 2005).

These two hormones play an important role in communicating to the gonads. In males LH stimulates the interstitial cells located in the testes to produce testosterone, and FSH plays a role in spermatogenesis. Only small amounts of oestrogen are secreted in males. Recent research has shown a neurosteroid axis exists, which help the cortex to regulate the hypothalamus's production of GnRH (Meethal, et al., 2009).

In males, the production of GnRH, LH, and FSH are similar, but the effects of these hormones are different (Veldhuis, Keenan, Liu, Iranmanesh, Takahashi, & Nehra, 2009). FSH stimulates sustentacular cells to release androgen-binding protein, which promotes testosterone binding. In contrast, LH binds to the interstitial cells, causing them to secrete testosterone. Testosterone is required for normal spermatogenesis and inhibits the hypothalamus. Inhibin is produced by the spermatogenic cells, and is responsible to inactivate activin and acts on the hypothalamus. After puberty these hormones levels remain relatively constant.

Role of the Androgen Receptor

Testosterone exerts its function through androgen receptors, which are located in the cytoplasm and nucleus of target cells. Androgen binding results in DNA activation and transcription of several proteins, with different intercellular modifying activities. Testosterone in the foetal testis increases the number of androgen receptors in two ways, by increasing the number of cells with androgen receptors and also by increasing the number of androgen receptors in each individual cell (Bentvelsen, McPhaul, Wilson, & George, 1994).

The AR-gene is found on the X chromosome (Xq 11-12) and consists of eight exons. Defects and mutations of the AR gene can results in male sexual maldevelopment, which may cause testicular feminisation or low virilisation. Less severe mutations of the AR gene may cause mild forms of androgen resistance and male infertility (Brinkmann, et al., 1995).

Classical Testosterone Signaling Pathways

In the regulation of spermatogenesis, testosterone imparts direct actions on Sertoli cells, which involve both non-classical and classical signaling pathways (Walker & Cheng, 2005).

Within the seminiferous tubules, only Sertoli cells have receptors for testosterone, and therefore they are the major transducer of testosterone signals that are required to support germ cell survival and development (Walker, Non-classical actions of testosterone and spermatogenesis, 2010).

Testosterone is known to act through the AR to regulate gene expression. The classical mechanism of testosterone action begins with the diffusion of testosterone through the plasma membrane, and binding to AR that is sequestered in the cytoplasm by heat shock proteins. When bound by ligand, AR undergoes a change in structural conformation allowing it to be released from the heat shock protein complex. AR then translocates to the nucleus where it binds to specific DNA sequences called androgen response elements (AREs), resulting in the recruitment of co-activators, and the regulation of gene expression. This classical pathway of testosterone action requires at least 30 – 45 min to alter gene expression and hours to produce nascent proteins (Shang, Myers, & Brown, 2002).

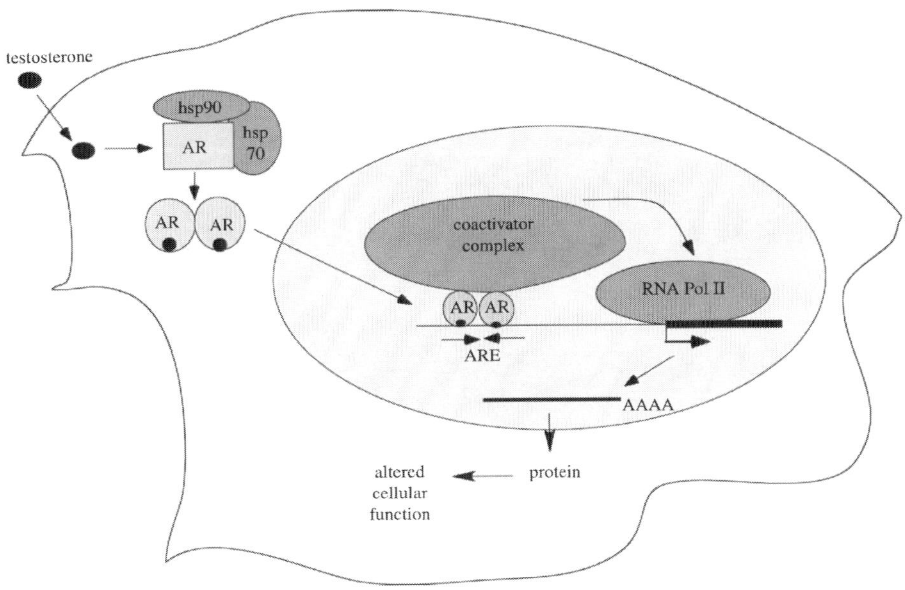

Walker, Non-classical actions of testosterone and spermatogenesis, 2010.

Figure 4. The classical testosterone signaling pathway. Testosterone diffuses through the plasma membrane and binds with the AR. The AR undergoes an alteration in conformation allowing it to be released from heat shock proteins in the cytoplasm. AR then is able to translocate to the nucleus where it binds to specific DNA sequences called androgen response elements (AREs). AR binding to an ARE allows the recruitment of co-activator and co-repressor proteins that alter the expression of genes to alter cellular function.

Non-Classical Testosterone Pathways in Sertoli Cells

Contemporary observations suggest that testosterone acts in Sertoli cells through alternative pathways to regulate spermatogenesis. Firstly, testosterone is present in the testis at levels much greater than that needed to regulate transcription via AREs. Second, relatively few genes in Sertoli cells are regulated by testosterone and even fewer are known to be regulated through AREs. Third, a large percentage of testosterone-regulated genes are inhibited in the presence of AR. Finally, intracellular calcium levels are elevated in primary Sertoli cells within seconds of androgen stimulation and thus cannot be dependent on AR-DNA interactions and the initiation of gene expression (Walker, Non-classical actions of testosterone and spermatogenesis, 2010).

Thus far, testosterone has been found to act via two non-classical pathways in Sertoli cells. First, testosterone stimulation can depolarize the Sertoli cell and cause calcium influx into Sertoli cells. Second, testosterone can also activate a series of kinases resulting in activation of the MAP kinase cascade and phosphorylation of the CREB transcription factor. At present, it is not yet known to what extent the two pathways are interlinked.

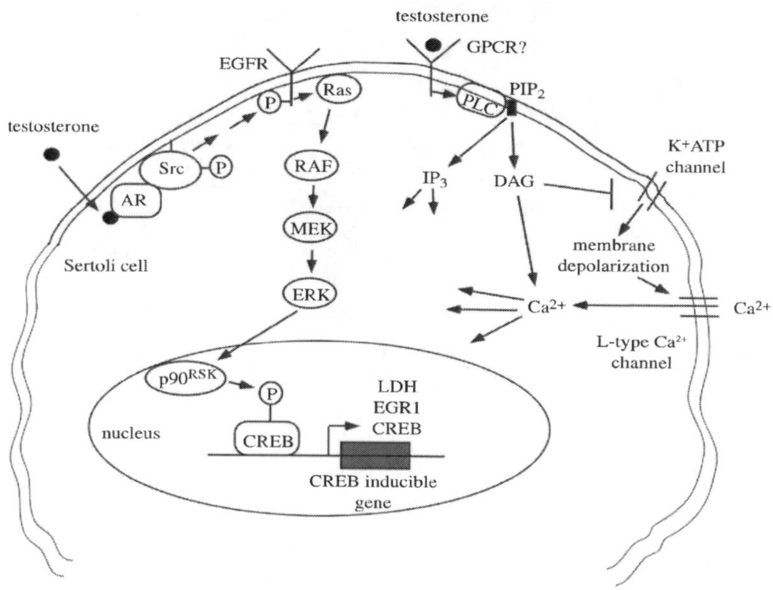

Walker, Non-classical actions of testosterone and spermatogenesis, 2010.

Figure 5. Non-classical testosterone signaling pathways. Left side: testosterone interacts with the classical AR which then is able to recruit and activate Src, which causes the activation of the EGF receptor via an intracellular pathway. The EGF receptor then activates the MAP kinase cascade most probably through Ras, resulting in the sequential activation of RAF and MEK and then ERK that activates p90-kinase, which is known to phosphorylate CREB on serine 133. As a result, CREB-regulated genes such as lactate dehydrogenase A (LDH-A) and early growth response 1 (EGR1) and CREB can be induced by testosterone. Right side: Testosterone interacts with a receptor in the plasma membrane that has characteristics of a Gq coupled G-protein coupled receptor (GPCR). Phospholipase C (PLC) is activated to cleave PIP2 into IP3 and DAG. Lower concentrations of PIP2 inhibit KþATP channels causing membrane depolarization and Ca2þ entry via L-type Ca2þ channels.

Male Hormone Contraception

In recent years there has been an interest to develop new approaches for male contraception in order to provide additional means to reduce over-population or control population growth, particularly in developing countries, so that limited global resources can sustain the world's population. Traditional male methods of contraception such as periodic abstinence or coitus interruptus are associated with relatively high rates of unwanted pregnancies and also disturbances to the sexual activity. Due to the shortcomings of the existing methods of male contraception, efforts have been made to develop an alternative hormonal contraceptive for men analogous to that of the oral contraceptive pill used by women. The notion of taking a pill to prevent paternity and pregnancy is therefore very popular and highly desired.

In the last decade, hormonal birth control for men may be a step closer to reality. Testosterone functions as a contraceptive by suppressing the secretion of the pituitary gonadotropins LH and FSH levels. Low levels of these hormones decrease endogenous or intra-testicular testosterone secretion from the testis and deprive developing sperm of the signals required for normal maturation. Interference with sperm maturation causes a decline in sperm production and male infertility. After 2-3 months of treatment, low levels of pituitary gonadotropins lead to markedly decreased sperm counts and effective contraception in the majority of men. Treatment with exogenous testosterone has proven not to be associated with serious adverse effects and is well tolerated by men. In addition, sperm counts uniformly normalize when testosterone is discontinued. Thus, male hormonal contraception is safe, effective, and reversible; however, spermatogenesis is not suppressed to zero in all men, meaning that some diminished potential for fertility persists (Amory, 2005; Amory, 2008; Nieschlag, 2011).

There are various contraceptive regimes that are commercially available (Nieschlag, 2011). Androgen only regimes include testosterone enanthate, testosterone implants, testosterone undecanoate, transdermal testosterone preparations and 7α-Methyl-19-nortestosterone (Amory, 2008; Nieschlag, 2011). The androgen-only approach is certainly attractive but it is not always effective and the effects of long-term androgen-related side effects remain a significant disincentive. Combination testosterone and other substances to suppress pituitary gonadotropin secretion has been trialled in efforts to improve its contraceptive efficacy. Progestin/androgen combinations such as depot medroxyprogesterone acetate, cyproterone acetate, levonorgestrel, desogesterel, and norethisterone enanthate, appear to be more effective than testosterone alone in suppressing spermatogenesis (Wu, 2006). This combined approach is more suited to Caucasians who is less responsive to steroid suppression. GnRH antagonist like Nal-Glu, in combination with testosterone could be used for rapid spermatogenic suppression but is associated with high cost, short half-life and the need for frequent subcutaneous injection (Wu, 2006).

However the current models of male contraception does not work in all men. Several possible reasons include the ethic differences in the response to sex steroid suppression of spermatogenesis between Asian (more responsive) compared to Caucasians men (less responsive), genetic differences in the number of CAG repeats in AR or polymorphisms in the CYP3A4 gene, as well as differential production of other enzymes or hormones such as 5α–reductase activity and peptide hormone insulin-like factor 3 level (Wu, 2006; Amory,

2008; Nieschlag, 2011). Therefore future research into the field of male contraception should be focused on improving the efficacy of hormonal drugs while minimizing the side effects profile. An ongoing large multicenter trial launched by World Health Organization and the US CONRAD Programme will decide on the future of male hormonal contraception. Another potential avenue to explore is to target the sperm and prevent sperm survival beyond testis and seminal emission in the male genital tract. Indeed, the promise of male hormonal contraception remains elusive.

References

Amory JK. Male hormonal contraceptives: current status and future prospects. *Treat. Endocrinol.* 2005;4(6):333-41

Amory JK. Progress and prospects in male hormonal contraception. *Curr Opin Endocrinol Diabetes Obes.* 2008;15(3):255-60)

Bentvelsen, F. M., McPhaul, M. J., Wilson, J. D., & George, F. W. (1994). The androgen receptor of the urogenital tract of the fetal rat is regulated by androgen. *Mol Cell Endocrinol , 105* (1), 21-26.

Brinkmann, A. O., Jenster, G., Ris-Stalpers, C., van der Korput, J. A., Bruggenwirth, H. T., Boehmer, A. L., et al. (1995). Androgen receptor mutations. *J Steroid Biochem Mol Biol , 53* (1-6), 443-8.

Charlton, H. (2008). Hypothalamic control of anterior pituitary function: a history. *J. Neuroendocrinol , 20* (6), 641-6.

Ganong, W. F. (2009). Chapter 25. The Gonads: Development & Function of the Reproductive System. In W. F. Ganong, & W. F. Ganong (Ed.), *Review of Medical Physiology* (23rd Edition ed., pp. 411-453). New York: The McGraw-Hill Companies, Inc.

Meethal, S. V., Liu, T., Chan, H. W., Ginsburg, E., Wilson, A. C., Gray, D. N., et al. (2009). Identification of a regulatory loop for the synthesis of neurosteroids: a steroidogenic acute regulatory protein-dependent mechanism involving hypothalamic-pituitary-gonadal axis receptors. *J. Neurochem , 110* (3), 1014-27.

Millar, R. P., Lu, Z. L., Pawson, A. J., Flanagan, C. A., Morgan, K., & Maudsley, S. R. (2004). Gonadotropin-releasing hormone receptors. *Endocr. Rev. , 25* (2), 235-75.

Nieschlag E. The struggle for male hormonal contraception. *Best Pract Res Clin Endocrinol Metab* 2011;25(2):369-75

Shang, Y., Myers, M., & Brown, M. (2002). Formation of the androgen receptor transcription complex. *Mol. Cell , 9*, 601-10.

Singh, J., O'Neill, C., & Handelsman, D. J. (1995). Induction of spermatogenesis by androgens in gonadotropin-deficient (hpg) mice. *Endocrinology , 136* (12), 5311-21.

Vadakkadath Meethal, S., & Atwood, C. S. (2005). The role of hypothalamic-pituitary-gonadal hormones in the normal structure and functioning of the brain. *Cell. Mol. Life Sci. , 62* (3), 257-70.

Veldhuis, J. D., Keenan, D. M., Liu, P. Y., Iranmanesh, A., Takahashi, P. Y., & Nehra, A. X. (2009). The aging male hypothalamic-pituitary-gonadal axis: pulsatility and feedback. *Mol. Cell. Endocrinol. , 299* (1), 14-22.

Walker, W. H. (2010). Non-classical actions of testosterone and spermatogenesis. *Phil. Trans. R. Soc. B* , *365*, 1557-69.

Walker, W. H., & Cheng, J. (2005). FSH and testosterone signalling in Sertoli cells. *Reproduction* , *130*, 15-28.

Wu, FCW. Hormonal approaches to male contraception: Approaching reality. *Mol Cell Endocrinol.* 2006;250:2-7.

In: Androgens: Production, Functions and Disorders
Editors: B. F. Thompson and D. J. Robinson
ISBN: 978-1-62081-145-0

Chapter VIII

Androgens, Cognition and Social Behavior in Children*

Aitziber Azurmendi, Aizpea Sorozabal and J. R. Sánchez-Martín†
Area of Psychobiology, Faculty of Psychology,
University of the Basque Country, San Sebastian, Spain

Introduction

We currently know that androgens act in the brain during its development, affecting both its structure and its neural function. In other words, in addition to the activating effects of adolescence, these hormones also have diverse organizational effects that, since they mold the central nervous system (CNS), influence both cognitive processes and the subject's behavior. Thus, a wide variety of data gathered over recent decades show that gonadal hormones affect the cognitive abilities of both men and women. Furthermore, recent studies carried out with normal populations instead of selected groups with hormonal abnormalities, have provided evidence that associates endogen hormone levels with behavior.

This chapter reviews existing research into both the relationship between androgen levels and diverse cognitive abilities, and the relationship between androgen levels and social behavior during childhood. First of all, it presents an overview of the principal androgens and their ontogenetic development, their main action mechanisms and their influence in the sexual differentiation of the brain and behavior. Next, it reviews the existing literature on the relationship between androgen levels and diverse cognitive abilities, paying special attention to social cognition (theory of mind, etc.) in children, before reviewing those research projects

* A version of this chapter also appears in Advances in Psychology Research. Volume 58, edited by Alexandra M. Columbus, published by Nova Science Publishers, Inc. It was submitted for appropriate modifications in an effort to encourage wider dissemination of research.
† Correspondence: José Ramón Sánchez-Martín. Area de Psicobiología. Facultad de Psicología. Universidad del País Vasco. Av/ Tolosa, 70. 20018 San Sebastián (Spain). Phone. 0034 943 01 5729; Fax: 0034 943 015670; E-mail: joseramon.samchez@ehu.es.

focusing on the relationship between androgens and social behavior in children, paying special attention to behaviors of dominance and aggression. Finally, the principal conclusions are discussed from an evolutionist or ultimate causes perspective.

Androgens

Androgens have diverse effects at both a physiological and behavioral level. These steroid hormones are present right from the first weeks of gestation and, in addition to playing a key role in the development of the male sex organs (both internal and external), they are also involved in the development of the CNS in both males and females, and therefore influence subjects' subsequent behavior.

The principal androgens found in the human body are testosterone, androstenedione, dehydroepiandrosterone (DHEA) and its sulfate (DHEA-S). Testosterone is the principal product of the testicles, while DHEA and DHEA-S are predominantly of adrenal origin and androstenedione is produced by both the adrenal glands and the gonads [1]. The following is a brief overview of the most important characteristics of the main androgens.

Testosterone

Testosterone has androgenic effects in the organism, being responsible for virilizing its internal structures, while one of its most active metabolites, dihydrotestosterone, is responsible for virilizing the external genitals. Another important metabolite of testosterone is estradiol. The conversion of testosterone to estradiol (the principal female hormone), a process which takes place in the CNS thanks to the enzyme aromatase, is considered important for the early development of the 'masculine brain'. Aromatase is mainly concentrated in cerebral areas related to the sexual differentiation of the CNS, such as the hypothalamus. Many sexual differences in behavior, especially in aggression, are, in all probability, the result of the different extent to which the CNS is exposed to testosterone and estradiol in the development of males and females [2].

There are three periods in the life of human males in which the plasmatic concentration of testosterone is relatively high: during embryonic development (the phase during which sexual differentiation is established), during the neonatal period and throughout the whole of the adult sexual life. During fetal development, by around 10-12 weeks gestation, testosterone levels are higher in males than in females, reaching their peak concentration between 12 and 18 weeks. This sexual difference is the result of the fact that circulating testosterone is produced by the testicles, since no evidence exists to suggest that human fetal ovaries synthesize testosterone, and fetal adrenal production in both sexes is very low. From weeks 22-28 onwards, sexual differences are not so apparent. A few days after birth, a rapid increase occurs in the testosterone levels of boys, reaching maximum levels (comparable to those observed in adolescents and approximately half of that found in adult males) between 20 and 60 days after delivery. From hereon, the levels start to drop and no sexual differences are observed again until puberty. In girls, testosterone levels at birth are similar to those found in adult women, although they begin to decrease after two weeks of life and then remain unchanged throughout childhood. During puberty, between 12 and 18 years of age, testosterone levels in boys increase by a factor of 20 to 30. In adulthood, testosterone concentrations in men are 15 times higher than in women. Finally, a gradual decrease has been observed in the levels of this hormone in men from the age of 60 onwards. In postmenopausal women, testosterone levels are also significantly lower than during their reproductive life [1].

Total testosterone levels are subject to diurnal and seasonal variations, reaching their daily peak between 06:00-08:00 h, and then dropping slowly down to 35% over the course of the day, before starting to rise again at midnight. This daily peak is tempered in old age [3]. As regards seasonal variations, it is in the months of November and December that adult men experience their highest annual peaks [4]. Furthermore, we should not forget that diverse environmental factors may also influence day-time testosterone concentrations; for example, levels may fluctuate from minutes up to hours depending on the subject's sexual activity and experiences, as well as his experiences of competitive success or failure [5].

Androstenedione

The main biological actions of androstenedione (which works together with DHEA and DHEA-S) are anabolic. This hormone stimulates erythropoiesis (the production of red corpuscles in the hemopoietic - blood producing - organs) and the development and maintenance of underarm and pubic hair. Under normal conditions, its virilizing action in the human male is weak, due to the high concentrations of testosterone present in the male body. Nevertheless, it can virilize human females if secreted in excess. Furthermore, androstenedione plays other key roles in both sexes, regulating lipids, reducing susceptibility to atherosclerosis and osteoporosis and fostering longevity [6].

During the fetal period, androstenedione levels are relatively low and there are no sexual differences. Although there is evidence that this hormone is produced by the testicles, the majority of circulating fetal androstenedione is of adrenal origin and, as in the case of testosterone, there is no evidence to suggest that fetal ovaries synthesize androstenedione. Postnatal changes in androstenedione levels are more modest than in the case of testosterone. In both sexes, levels decrease rapidly during the first week of life, but in boys they stop dropping after the first month, whereas in girls, they continue to decrease until the third month. This is the period in which sexual differences can be observed in the concentrations of this hormone. During childhood, the levels remain very low until they begin to increase with adrenarche (around the age of 8), and then continue rising gradually until they reach double or even triple their initial level during puberty. In young adults, levels stabilize and then remain the same throughout the rest of the subject's life (not including natural variations caused by the menstrual cycle) [1].

Androstenedione concentrations follow a circadian rhythm, with the highest levels being reached between 05:30 and 09:30 h and the lowest concentrations between 19:00 and 02:30 h. Furthermore, there is a concentration peak that coincides with each meal (lunch and dinner) [7].

Dehydroepiandrosterone (DHEA)

In addition to the aforementioned biological actions that it enables alongside androstenedione and DHEA-S, DHEA also influences the working of diverse systems, such as the immune system, the cardiovascular system, the endocrine system and the central nervous system [8]. We know that as well as the gonads and the adrenal glands, the brain also synthesizes DHEA and DHEA-S, causing these hormones to act as neurosteroids [9]. For example, in the CNS, DHEA stimulates axonal growth, while DHEA-S stimulates dendritic growth [2].

Human and non-human primates are the only creatures whose adrenal glands secrete large quantities of DHEA and, above all, DHEA-S, which can be converted into powerful androgens and/or estrogens in the peripheral tissues. Indeed, plasmatic levels of DHEA-S in adult men and women are between 100 and 500 times as high as testosterone levels and between 1000 and 10000 times as high as estradiol levels; this provides a large reserve pool for the conversion of androgens and/or estrogens in peripheral tissues which contain the enzymatic machinery required to transform DHEA into active sex steroids [10].

In the fetal stage, DHEA levels increase significantly during the final trimester of gestation, and do not show any sexual differences. In the neonatal stage, levels are also high (the same ranges as those found in adults) and again, no sexual differences are observed [1]. Later on, from the first month of life to the age of 5, DHEA concentrations decline. From the age of 7 in girls and 9 in boys, the levels start to increase rapidly, reaching their maximum concentration at the age of 20 in girls and between 20 and 30 in boys. In adulthood, women experience another concentration peak at the age of 40, after which time DHEA levels begin to drop in both sexes [11].

Normally, plasmatic levels of DHEA and DHEA-S follow a modest circadian rhythm. In humans, the highest concentrations are reached during the day, regardless of whether the hormones are produced by the adrenal gland or the brain [9].

Androgen-Secreting Glands

Suprarenal Glands

Adrenal or suprarenal glands owe their name to their anatomical location at the upper pole of each kidney [6]. In many mammalian species, including humans, each suprarenal gland consists of two embryologically and biochemically different organs: the external layer of the gland, the adrenal cortex, consisting of three different zones (glomerulosa, fasciculata and reticularis); and the inside of the gland, the adrenal medulla, which is actually a part of the autonomic nervous system [12] [13].

In order to enable them to carry out their function, the suprarenal glands are well supplied with arterial blood from diverse branches of the aorta, the inferior phrenic artery and the renal arteries, thus obtaining the highest rate of blood per gram of tissue in the whole body [14].

The outermost layer of the adrenal cortex, the zona glomerulosa, is the thinnest and its cells are arranged in rounded groups. It is in this zone that mineralocorticoids (such as, for example, aldosterone) are produced. The middle layer, known as zona fasciculata, is the largest and its cells are arranged radially in parallel cords. It is in this zone that glucocorticoids (such as, for example, cortisol) are produced. The innermost later, the zona reticularis, is only slightly thinner than the fasciculata. Its most important function is the secretion of adrenal androgens (DHEA, DHEA-S and androstenedione), and, in much lesser quantities, of estrogens [6]. The zona reticularis is smaller in infants and only begins to grow between the ages of 6 and 7, its development being parallel to the increase in adrenal androgens in plasma during the prepubertal period [15].

As regards the regulation of the quantity of circulating cortical adrenal hormones, it is the adrenocorticotropic hormone (ACTH) of the pituitary gland that promotes its synthesis. Adrenal steroids have a negative feedback effect on the release of ACTH. Thus, as the level of hormones secreted by the adrenal cortex rises, the secretion of ACTH drops, thus reducing the rate of hormonal release in the adrenal cortex. When adrenal steroid levels drop, the secretory cells of the pituitary gland are freed from their inhibition and the concentration of ACTH in the blood rises. This in turn results in an increase in the release of hormones from the adrenal cortex [13].

Gonads

Testicles

The testicle is an endocrine gland that fulfils two different functions: the production and secretion of sex steroids (especially testosterone) by the Leydig cells, and spermatogenesis in the seminiferous tubules, which provides the fertilizing capacity. The testicle also secretes small quantities of estradiol, although the majority (approximately 75% to be exact) of the estradiol circulating around the male body comes from the peripheral aromatization of testosterone in diverse tissues, particularly the adipose [16].

The production and release of testosterone is regulated by a hormone of the anterior pituitary gland known as Luteinizing hormone (LH) [13]. During infancy, the pituitary gland secretes LH sporadically, causing the testicle to remain in a state of 'functional repose' practically until the onset of puberty. During this stage, due to mechanisms we still do not really understand, the biological clock is activated and as a result, the testicle begins gradually to mature, both functionally and morphologically [16].

Ovaries

The ovaries also produce mature gametes (ovules) and hormones. Given the correct conditions, i.e. the presence of the required enzymes, the progesterone produced in the ovary is transformed into androgens and estrogens [17].Note that the hormonal activities of the ovaries are more complex than those of the testicles and their production, in humans, is carried out in cycles that last around four weeks [13].

The ovary has three functional subunits: the follicles, each of which contains a developing egg or ovule; the corpora lutea, structures that evolve from the follicles when the ovule is released; and the support tissue or stroma. Around the follicular cells, although separated from them by a membrane, are the cells of the theca interna and the theca externa [12]. The stroma of the ovaries secretes androstenedione, usually in small quantities, although it can lead to virilization in diseases in which the ovarian stroma is overly developed (polycystic ovary syndrome) [17].

It is mainly the theca cells that produce the highest concentrations of androgens, particularly DHEA and androstenedione [18], which are then used for synthesizing estrogens, the principal sex steroids in females.

As with the testicles, the production of hormones in the ovaries is controlled by LH [13].

Biosynthesis of Steroid Hormones

The adrenal glands, testicles and ovaries share the same steroidogenic capacity, although each one specializes in the production of one specific group of steroids. The fact that these three pairs of glands have the same hormone synthesis mechanism is the result of their sharing the same embryonic origin (all three derive from areas near to the neural crest).

The precursor of all steroid hormones in vertebrates is cholesterol, which in turn is produced in large quantities from acetate in our livers [12]. Through a complex enzymatic process of hydroxylation, which takes place in the mitochondria, cholesterol is converted to pregnenolone. It is during this phase that the aforementioned specific regulation occurs, in the adrenals through ACTH and in the gonads through LH [17]. From hereon, both in the gonads and the suprarenal glands, specific enzymes convert pregnenolone into androgens [12].

Demolase transforms 17-hydroxypregnenolone into DHEA, which can be esterified reversibly into DHEA-S. Alternatively, the 3β-hydroxyl group of DHEA can be oxidized, thereby producing androstenedione. This product can also be obtained directly from 17-hydroxiprogesterone. The reduction of androstenedione produces testosterone [17]. Finally, the reduction of testosterone by the 5α-reductase enzyme results in dihydrotestosterone.

Action Mechanisms of Androgens

Since steroid hormones are fat soluble, they move easily through cellular membranes. However, at the same time, they also have the disadvantage of not being very water soluble, which makes their movement around the circulatory system somewhat complicated. Therefore, steroid hormones tend, in general, to bind with water soluble carrier proteins which increase their solubility and carry them through the bloodstream to the target tissues (tissues whose cells contain specific receptors for a specific hormone) [12].

Thus, steroid hormones are transported in the bloodstream in two fractions, one bound to plasmatic proteins and the other, much smaller, dissolved and free [19]. DHEA and androstenedione bind mainly with albumin, unlike testosterone and estrogens which combine with great affinity with one specific globulin, sex hormone binding globulin (SHBG). However, only free hormones are able to act and penetrate cells in order to bind to specific cytoplasmic receptors [19]. Although hormones can influence target cells in many different ways, they hardly ever change their function, altering rather the rhythm of the normal cellular function, stimulating growth and cellular [20] [12].

Once inside the target tissues, the steroid hormone dissociates itself from the carrier molecule (if it was bound to it) and spreads through the cellular membrane, where it binds to the cytoplasmic receptors [12]. Next, the steroid-receptor complex is transported to the nucleus of the cell where it regulates the synthesis of specific proteins, mainly altering the transcription rate of specific genes [20].

The brain contains receptors for all kinds of steroid hormones, including androgens, and these receptors are similar to those found in any other part of the body [19]. Thus, steroids can also affect cerebral functions, both during the fetal-perinatal period and during adulthood, since the central nervous system contains many places to which steroids can bind. Of these, the most extensively studied are the classic intracellular receptors mentioned above (steroid-receptor complexes); however, more recently, a number of specific places have been described in the membrane also, where some families of steroid hormones can bind. This may be very important for explaining the extremely quick and fleeting effects induced by steroid hormones, effects which cannot be accounted for by slower genomic actions. Finally, steroid hormones and their metabolites may interact with the receptors for some neurotransmitters [21].

The cellular action mechanism of androgens frequently includes a local metabolism in the target cells. For example, in specific organs (prostate, seminal vesicles, skin, etc.) testosterone is reduced by the enzyme 5α-reductase, located in the endoplasmic reticulum, thus turning into dihydrotestosterone (DHT) and binding to the receptors. As a result, in these organs testosterone is a prohormone, and DHT is the true active component. In other organs (kidneys, muscles), the amount of DHT formed is very low, and testosterone binds directly to the receptor. In these organs, testosterone is the active hormone. In the hypothalamus, on the other hand, part of the testosterone is aromatized into estradiol by the aromatase enzyme. Thus, some authors are inclined to believe that estradiol or one of its derivatives is the active intermediary of the specific effects of testosterone in diverse cerebral structures [17].

Sexual Differentiation of the Brain and Behavior

Sexual Differentiation

The process which takes place during development that turns an individual into either a male or a female is known as sexual differentiation [22]. The chain of events which occur during differentiation is well known in placental mammals such as humans. Sexual differentiation begins with the establishment of chromosomal sex in fertilization, which is followed by the development of gonadal sex and culminates in the development of secondary sexual characteristics, commonly called masculine and feminine phenotypes [23].

Genetic Sex

During early development, both XX and XY individuals have gonads that have still not turned into testicles or ovaries, and which are simply called "non-differentiated gonads". One crucial task carried out by at least one gene of the Y chromosome is the transformation of these non-differentiated gonads into testicles. This gene of the Y chromosome is called the sex determining region of the Y gene, or SRY [22]. SRY possesses the testicular determination factor (TDF) which induces the testicular development of the non-differentiated gonad and consequently, the development of the Leydig cells, which are responsible for secreting testosterone and androstenedione around the eighth week of development [24]. If the cells of the non-differentiated gonad contain a Y chromosome with the SRY gene, testicles will develop. In the absence of the SRY gene, the gonad will develop as an ovary. Thus, genetic sex (i.e. whether an individual is XX or XY) determines gonadal sex (whether the individual will have ovaries or testicles). After this, the rest of the sexual differentiation process is not determined directly by genetic sex, but rather by the hormones secreted by the gonads [22].

Gonadal Sex

As mentioned above, the differentiation of the gonads into testicles begins in humans around the eighth week after conception; if the gonads are to become ovaries, this process begins later, around the twelfth week. Testicles are differentiated earlier than ovaries because the androgens produced by the testicles are necessary for the subsequent phases of male development. In this phase (7/8-12 weeks after conception) each embryo has two primitive duct systems: the Müllerian duct system, which may evolve into feminine reproductive structures (oviduct, uterus and upper part of the vagina) and the Wolffian duct system, which may evolve into masculine structures (epididymis, vas deferens and seminal vesicles). It is the presence or absence of testicles which determines which of these duct systems evolves. Subsequently, the testicles promote the differentiation of the external genitals in the masculine form, through the secretion of the androgen DHT. If the androgens do not act on these tissues, the external genitals will be differentiated as female [13].

Hormonal Sex

Once the testicles are formed, they begin to produce sex hormones, particularly testosterone. As for the ovaries, they mainly secrete estrogens and progesterone. The important thing here is to note that the difference between males and females does not lie in the type of sex hormones they produce, but rather in the relative proportion of the set of sex hormones produced by each sex. Women do produce testosterone (in this case in the suprarenal glands, ovaries and through peripheral conversion), but in much lower quantities than those produced by the testicles in human males. Similarly, men also produce estrogens and progesterone, but only in very small amounts [24].

Therefore, hormonal sex is determined by the quantity or level of hormones produced in the gonads. If there is enough testosterone, then masculine differentiation will occur, otherwise, the embryo will follow the female course regardless of chromosomal or gonadal sex [24]. Even if a fetus is female, and therefore does not possess testicles, its genitals may assume a masculine form under the influence of androgens in certain clinical conditions such as congenital adrenal hyperplasia: the adrenal cortex normally produces small amounts of androgens, but a hyperactive adrenal or maternal gland may produce sufficient androgens during the critical period of fetal development to cause the genitals to adopt a masculine form in genetically female subjects [13].

Effects of Organization and Activation

The sexual differentiation of the nervous system and, consequently, of behavior, is guided by the same steroids that cause the sexual differentiation of the body [22]. Thus, in the development and differentiation of the brain, a series of extremely long and complex events occur, beginning during gestation and continuing, at least in some species, during the early postnatal period [25].

The action of steroid hormones in the central nervous system can be classified as organizational and activating. Organizational effects refer to the steroids' ability to sculpt the structure of the central nervous system during development. Structural organization is permanent, remaining long after the period of development during which it is exposed to the hormone [26]. It is believed that this occurs because the hormones direct basic neural development processes, thus determining which nervous cells live and which die, with which other cells they should connect automatically and which neurotransmitters they should use [27]. Activating effects, on the other hand, refer to the steroids' ability to modify the activity of target cells so as to facilitate behavior in a specific social context. Activating effects are temporary, they come and go according to the presence or absence of the hormones in question and are typically associated with the action of steroids during adulthood [26].

We know that androgens act on the developing brain, producing sexual differences in both the neural structure and function. All the cells of the sexually dimorphic regions of the brain are rich in androgen receptors, and their development is mainly affected by testosterone, both during early fetal life and later on [28].

Many sexual differences become more evident after puberty as the result of the activation effects of steroid hormones [26]. However, as mentioned above, the organizational influence is already present during early fetal development, organizing the nervous system in such a way that the processing of information for specific aspects with adaptive implications is different for each sex. Furthermore, due to the early organizational effects of hormones, major sexual differences may also occur in an individual's sensitivity to sex steroids from a very early age. In fact, evidence exists to suggest that individuals differ not only with regard to their levels of circulating hormones, but also with regard to their sensitivity to said levels [29]. According to Baron-Cohen et al.[28],there seems to be a very close relationship between sexual differences in the brain and androgen levels, since the sexually dimorphic regions of the brain (amygdala, corpus callosum, etc.) contain numerous androgen receptors and their development seems to be influenced by androgens during the fetal stage and subsequent development phases.

For example, studies carried out with clinical populations in humans have found that prenatal exposure to atypically high levels of androgens, as in the case of women suffering from congenital adrenal hyperplasia (CAH), masculinizes both behavior and cognitive skills [30]. Moreover, studies carried out with healthy populations have found that inter-individual hormonal variations in prenatal levels of androgens are associated with subsequent behavior typical of the specific sex [28].

As we will see below, the hormones responsible for the sexual differentiation of the organism are also involved in the sexual dimorphism of behavior and diverse cognitive abilities, since because they have an organizational effect on the CNS, they help mold both cognition and adult behavior.

Relation of Androgens to Psychological Processes

Although there is no difference between human males and females as regards general intelligence, sexual differences have been observed in relation to specific cognitive tasks. For example, men perform better in mental rotation and spatial navigation tests, while women score higher in emotional recognition, social sensitivity and verbal fluency tests. Moreover, girls begin to speak before boys [28]. A number of different biological and socio-cultural factors have been used to explain these differences; one such biological factor proposed is related to the role played by hormones in early development [31].

Based on the studies carried out in other species, it is reasonable to assume that the presence of high levels of masculinizing hormones (androgens) during critical periods of cerebral development cause masculinization or defeminization in cognitive development, while low levels of the same hormones result in demasculinization or feminization. Thus, it has been hypothesized that hormone levels atypical of the sex to which the individual belongs are associated with cognitive patterns also atypical of that sex. In this model of sexual differences in cognition, high levels of androgens (as observed in typical males and females with an atypical exposure to androgens) would be associated with high scores in spatial ability, and relatively low scores in verbal fluidity, perceptual rapidity and memory; on the other hand, low levels of androgens (such as those found in typical women and men with reduced levels of androgens) would be associated with relatively poor spatial abilities and high scores in verbal fluidity, perceptual rapidity and memory.

Studies have also been carried out on the possible activating influences of hormones on human cognition. Many of these works suggest a non-linear (inverted U-shaped) relationship between circulating testosterone levels and various cognitive abilities, mainly visuospatial abilities in adult men. These studies suggest the existence of an optimum level of testosterone for the correct performance of cognitive tests. Thus, an increase in testosterone would not necessarily mean greater spatial skills, in fact, just the opposite [32]. In a study carried out by Gouchie and Kimura [33], in which women were included in the study sample, the authors found that men with low levels of T performed better in spatial tests and mathematics than those with high levels of T. However, women with high levels of T performed better at said tests than men with low levels. It is also worth noting that a number of other studies failed to find any association between androgens and the cognitive function [34]. Nevertheless, the idea that hormones have an activating influence on the human cognitive function is widely accepted [35].; although the precise nature of this influence is not yet clear [36].

In studies carried out with elderly individuals who were administered exogenous hormones for different reasons, replacement testosterone was observed to have diverse beneficial effects on a number of cognitive abilities, including spatial cognition [37] and the visual working memory [38]. Studies taking into account endogenous testosterone levels found a positive relationship between verbal memory and testosterone in older women, and a negative one between verbal fluidity and testosterone in older men [39]. Some studies have also analyzed the effects of administering DHEA on cognition, although results are inconsistent [40] [41]. Hirshman et al. [42] suggest that the influence of DHEA on diverse cognitive abilities may be mediated by its estrogenic and androgenic metabolites. A study carried out with postmenopausal women found that those taking replacement DHEA had varying levels of circulating estrogens and androgens (thus confirming previous studies regarding the rapid metabolism of DHEA); it was therefore impossible for the authors to observe the direct effect of DHEA on diverse cognitive tests, since its estrogenic and androgenic metabolites also influenced subjects' performance.

The idea that emerges from the data provided by diverse sources is that prenatal hormone levels affect the configuration of cognition differently depending on the sex of the subject (for example, better visuospatial abilities in boys and better verbal skills in girls), and that the normal daily and monthly fluctuations in the levels of these hormones in healthy human adults correlate with subjects' performance in diverse cognitive tests [43].

Nevertheless, could androgens influence those cognitive abilities for which no sexual differences have been observed? It seems that the answer is yes, at least in the case of fluid intelligence. This kind of intelligence is related to a person's adaptability and flexibility in coping with unexpected situations during problem solving, and is typically assessed through tasks containing matrixes, series and classifications, etc. Tan and Tan [44] found that very low and very high levels of testosterone may be disadvantageous for fluid intelligence in women, while in men, very low or medium levels of testosterone have a positive relationship with fluid intelligence, and very high levels of testosterone have a negative relationship with this type of cognition.

The relationship between androgens and cognitive abilities in prepubertal children has not been researched in any great depth, given that, as we have seen, the majority of studies have focused on young people and adults. We can only highlight three studies in this field. The first was carried out by Jacklin et al.[45], who analyzed the relationship between perinatal testosterone levels (obtained from the umbilical cord) and performance in spatial tasks at the age of 6. The authors found that girls with higher levels of testosterone scored lower in said tests, while no relationship was observed for boys. In this sense, it is worth highlighting the fact that hormone levels obtained from the umbilical cord do not necessarily reflect the levels to which the developing fetus has been exposed. The second study, carried out by Finegan et al.[46], found a negative non-linear relationship between prenatal testosterone levels (obtained from the amniotic fluid) and performance in a spatial visualization test in 4-year-old girls, while no relationship was observed for boys. Here also we should point out that, as in the previous case, hormone levels taken from the amniotic fluid may not reflect the levels that have influenced the fetus. Finally, Grimshaw et al. [47] observed a positive correlation between testosterone levels (obtained from the amniotic fluid) and the mental rotation rate of 7-year-old girls, while these correlations were not significant for boys, although they did suggest a positive relationship.

Another very interesting aspect of cognition in children, that is also relevant from the point of view of subjects' social adaptation, is that related to theory of mind skills. Between the ages of 3 and 5, children acquire theory of mind, i.e., the ability to understand others' mental states, including beliefs, emotions, desires, etc. [48] [49]. Over recent years, a wide variety of tests have been developed to assess different traits of theory of mind. Thus, various authors have found sexual differences in relation to some of these abilities, such as the understanding of emotions, for example, in which girls tend to score better than boys [50] [51]. Geary [52] suggests that endocrine correlates may underlie these sexual differences, although this has not received any experimental attention to date. Nevertheless, it is worth noting that other researchers have failed to discover any sexual differences in the performance of diverse theory of mind tests [53] [54].

Our research team recently analyzed the relationship between circulating levels of androgens (testosterone, DHEA and androstenedione) and performance in diverse cognitive tests in 5-year-old children [55].

As seen earlier, in prepubertal children, adrenal androgens such as DHEA, DHEA-S and androstenedione play a very important role (as much as or more than testosterone). Given the technological progress of recent years, which enables us to analyze androgens in saliva samples gathered in a non-invasive manner, we decided to measure also testosterone levels, DHEA levels and androstenedione levels (as far as we know, no kits are available which measure DHEA-S in saliva, since this molecule is too large to enter the oral cavity and is therefore not found in saliva, unless, of course, there is blood contamination).

In our opinion, another very important aspect of our study is that we selected a sample of 129 healthy normal children, 60 boys and 69 girls. As we know, many studies in this field have focused on subjects who have developed in atypical hormonal environments. Without wishing to detract from the importance of all these studies, which have provided much useful information, it is nevertheless necessary to carry out studies with healthy subjects in order to see how normal inter-individual variations in hormone levels are related to diverse cognitive abilities.

The data for this study were gathered over the course of two months, during March and April 2004. Two saliva samples were obtained (one each month) for subsequent analysis of hormone levels. The levels of salivary androgens were measured in duplicate using an enzyme immunoassay kit. The two values for each hormone were averaged, as they presented a positive correlation, with the result serving as a base line for each androgen in each subject. This base line was then used in the subsequent statistical analyses.

Cognitive abilities were assessed over a series of sessions deliberately kept short in order to avoid tiring the subjects, thus maintaining their motivation to perform. Cognitive tests were administered to all subjects between 09:00 and 10:00 h in a room adjacent to the classroom in each of the schools. The Kaufman intelligence test or K-BIT was used to measure intelligence. This test provides three IQ measures, assessing fluid intelligence through the Matrices subtest and crystallized intelligence through the Vocabulary subtest, as well as giving an IQ composite or global intelligence measure, which is the combination of the two previous subtests. Four further tests were used to assess theory of mind or subjects' understanding of others' mental states (beliefs, emotions, intentions, etc.): a) False belief, which measures whether or not the child is capable of attributing to others beliefs that are different from their own; b) Emotional labeling, which assesses the child's ability to recognize emotions; c) Appearance-reality distinction, which assesses whether or not the child understands the critical distinction between the appearance of an emotion and the reality of that same emotion; and finally, d) Rules of expression, which assesses whether or not the child comprehends the coexistence of contradictory feelings towards another person in the same situation.

As regards the results, despite the fact that no sexual differences were found for any of the cognitive abilities assessed or in the levels of testosterone and androstenedione (in the case of DHEA, girls' levels were significantly higher than boys'), the correlation analysis revealed a positive relationship between fluid intelligence and testosterone levels in boys, a negative relationship between crystallized intelligence and androstenedione levels in girls and a negative relationship between emotional labeling and androstenedione levels in boys. Furthermore, a multiple regression analysis indicated that androstenedione was one of the best predictors for some of the cognitive abilities assessed; in specific terms, it predicted negatively both emotional labeling in boys and global intelligence for the whole sample group (both sexes combined).

It is worth highlighting that almost all the relationships observed in our study between hormone measures and cognitive performance were negative. In other words, the higher the androgen levels, the poorer the cognitive performance (except in the case of fluid intelligence in boys).

Relationship between Androgens and Social Behavior

Over recent decades, much progress has been made in the study of the relationship between hormones and behavior. Recent studies focusing on normal populations instead of selected groups with hormonal abnormalities have established a relationship between the circulating levels of various androgens (mainly testosterone) and social behavior in humans.

Although the specific mechanisms through which hormones affect behavior are still unknown, Susman and Ponirakis [56] offer a detailed description of the four main models that have been developed in order to try and explain hormone-behavior interactions. The first model proposes that hormones have a causal influence on behavior, and is based on research carried out with animal models. These studies support the organizational and activating effects of hormones on behavior. The second model postulates that it is behavior which has a causal influence on hormone concentrations. According to this model, the environment in which an organism develops affects both its structures and biological functions. Many studies have observed hormone changes resulting from diverse behaviors and experiences. The third model focuses on hormone-behavior interactions and assumes that any changes in one dominion will influence changes in the other. Raine et al. [57] call this sequential biosocial effect. Finally, the fourth model takes context into account also. According to this model, the role of contextual factors, such as psychological and social factors, influence both behavior and hormone levels. In other words, such factors may moderate or mediate hormone-behavior interactions.

In addition to sexual behavior, there are also three main types of social behavior for which relationships have been established with androgen levels: aggression [58], dominance [59] and prosocial behavior [60]. The following is a summary of the different models or hypotheses which have been developed to explain the relationship between androgens and the display of aggression or dominance behaviors in humans. The reader should have no trouble identifying similarities with the models described above. There are three main models or hypotheses:

1. *Mazur and Booth's* [58] *biosocial hypothesis of status.* This model postulates a bidirectional relationship between testosterone levels and dominance. Thus, in the same way that high testosterone levels can encourage behavior aimed at dominating others, an increase in the experience of dominance can also increase testosterone levels.

2. *The challenge hypothesis*. This hypothesis suggests that aggression and testosterone correlate in moments of social instability or when an individual is challenged by a conspecific [61]. In the case of humans, evidence obtained to date indicates a low (but inconsistent) correlation between aggression and testosterone levels, and a higher and more consistent association between dominance and testosterone levels [62].
3. *The multivariate association model between dominance and testosterone levels*. Nyborg [63] [64] proposed a multivariate theoretical model of association between testosterone and dominance (the general trait covariance model), which integrates testosterone, dominance and intelligence. According to this author, individuals with a high IQ and low levels of testosterone may have high levels of formal dominance and enjoy a high status in areas in which their analytical abilities combine favorably with sensitivity. For their part, individuals with a high IQ and high levels of testosterone may also possess formal dominance and enjoy a high status, although in areas in which a combination of high intelligence and a certain degree of aggression and insensitivity are valued.

An association has also been found between androgen levels and affiliation and prosocial behavior, although the existing literature is a lot more scarce than in the case of aggression and dominance. In specific terms, a negative correlation has been found between testosterone and prosocial personality in adult men and women [60]. Furthermore, a negative correlation has also been found between testosterone levels and the emission of smiles and kindness/pleasantness [65] [66].

The majority of studies on the relationship between social behavior and androgen levels have focused on pubescent or post-pubescent subjects (especially males), and the few that were carried out with prepubertal individuals tend to focus on the relationship between hormones and aggression and between hormones and dominance (for a review on hormones and aggression in childhood and adolescence, see [67]).

Not enough studies have focused on the question of whether or not prenatal androgen levels influence human aggression, i.e. whether they have organizational effects [68]. Therefore, there is no reliable evidence regarding the androgen-aggression relationship in children in this respect.

As regards the possible relationship between circulating androgens and aggressive behavior, only a very few studies have focused on this question in preschool children. Of these, the majority have focused on testosterone. Constantino et al. [69] failed to find any relationship between testosterone levels and aggression in children aged between 4 and 10. Van Goozen et al. [70] also failed to find any significant relationships between testosterone levels and aggression and delinquency scores in the CBCL test. However, Sánchez-Martín et al. [71] found a significant positive correlation between testosterone levels and the aggression observed in social interactions in healthy normal four-year-old boys. The results obtained in the study carried out by Chance et al. [72] do nothing to clarify things: the authors found a positive association between testosterone and aggression in children aged between 9 and 11, but not in children aged between 5 and 8.

Even more scarce are studies which focus on the relationship between adrenal androgens and aggressive behavior in prepubertal children. Van Goozen et al. [70], found that DHEA-S levels in boys with conduct disorder were significantly higher than in the control group and were closely related to aggression and delinquency scores. In another study carried out by the same research group (Van Goozen et al., 2000b) the authors also found that boys with oppositional-defiant disorder had significantly higher DHEA-S levels than normal boys.

In light of these data, it is not unreasonable to assume that adrenal androgens, which are characteristic of childhood, contribute to initiating and maintaining human aggression [71] [74] [75]. It is important to remember that androgens are at their lowest levels during the preschool period [1], and the majority of circulating androgens in prepubertal children are produced by the adrenal glands. We therefore believe that it is very important to take adrenal androgens into account when studying the development of aggression in children.

Recently, our research team explored the potential relationship between social behavior (aggression, dominance and affiliation) and androgen levels (testosterone, DHEA and androstenedione). We also aimed to analyze whether or not intelligence played a moderating or mediating role in said potential relationship [76].

Given that there is evidence to suggest that the data obtained from behavioral observation enable researchers to establish a more consistent relationship between androgen levels and aggressive and dominance behavior than data obtained through self-reports [77], we decided to study children's social behavior through the systematic observation of their behavior during interaction with their peers in contexts of free play.

Subjects' social interactions with their peers were recorded on video daily throughout an entire school year, in the context of free play in the school playground. We obtained 30 minutes of recorded behavior per child, of which 15 minutes were assessed using the behavioral analysis software Observer 4.1 (Noldus).

The behavioral patterns considered were grouped into 3 main categories: aggression, governance and affiliation, with each category containing behaviors with the same functional area. Within the categories selected, we opted for more compact variables. To this end, we calculated the principal components of the behavioral traits, and the solutions obtained were as follows: victimization, offensiveness and provocation (for the Aggression behavioral category); subordination and dominance (for the Governance category); and linking, prosociality and affectivity (for the Affiliation category). Given that in the section on the relationship between androgens and cognitive abilities we present part of our work, we will not repeat here how the hormonal and cognitive data were collected.

Despite the fact that there were no sexual differences in the social behavior displayed by the subjects of our sample, we did find androgen-behavioral relationships specific to each sex. The correlation analysis showed that, in boys, androstenedione levels are associated negatively with victimization and positively with provocation; while in girls, testosterone levels are related positively with affectivity.

Next, we carried out various regression analyses in order to see whether or not intelligence acted as a moderator in those hormone-behavior relationships which were found to have significant effects in the correlation analyses. One regression analysis was carried out for each social behavior category (Provocation, Victimization and Affectivity), with one behavioral factor being introduced into each regression as a dependent variable, and one hormone, IQ and the interaction between the two as predictors. We found that, in the case of boys, the androstenedione-IQ interaction negatively predicted provocation; while for girls, the testosterone-IQ interaction positively predicted affectivity. Subsequently, in those cases in which the interaction of the variables assessed was significant, the association between hormones and behavioral factors was analyzed through a regression analysis for both levels of intelligence, high and low. Thus, we found that it was in boys with a high IQ that androstenedione predicted provocative behavior, and it was in girls with a low IQ that testosterone predicted affective behavior.

Conclusion

The existence of sexual differences in diverse spheres of human behavior is a fact supported by a large body of evidence [78]. From a phylogenetic or ultimate causes perspective, the existence of differentiated reproductive strategies in males and females in our species is associated, among other things, with sexual dimorphism in the behavior favored by sexual selection [79]. Evolutionist Psychology has also proposed that diverse evolutionary-based information processing mechanisms (such as, for example, those related to theory of mind and spatial abilities, etc.), have been molded by sexual selection [80].

From the perspective of proximate causes, the hormones known as sex hormones (a group which includes androgens) are among the principal elements responsible, at a biological level, for these different cognitive and behavioral manifestations [31]. These hormonal influences are even more evident both in those stages in which their organizational effects are more pronounced, and in those in which circulating levels in males and females are markedly different.

The results obtained from our research support the idea that, as suggested by other authors, even in stages in which hormone levels (in this case androgen levels) are low and present hardly any sexual differences, a relationship can be observed between hormones and cognition and hormones and behavior, precisely in those aspects of cognition and behavior in which, later on in the life cycle, sexual differences can be observed between individuals.

Firstly, it is likely that early organizational effects result in each sex being particularly sensitive to specific hormones (or at least to certain levels of such hormones). And secondly, as Archer [67] points out, it is probable that hormones (in this case androgens) are associated with the lifetime trajectories of specific subjects (for example, certain levels of androgens will be associated with specific levels of cognitive abilities and/or aggressive behaviors).

Finally, the biopsychosocial approach [82] [83] suggests that the hormone-behavior relationship is moderated by other factors also, both those that are strictly psychological in nature (cognitive, emotional, etc.) and social ones (family ecology, peers, etc.). Our work supports the moderating effect of intelligence on the relationship between hormones and behavior (both aggressive and affective) even during very early stages of development.

References

[1] Forest, M. G., (1989). Physiological changes in circulating androgens. *Pediatr. Adolesc. Endocrinol.,* 19, 104-129.

[2] Pfaff, D. W., Phillips, M. I. y Rubin, R. T., (2004). Principles of hormone/behavior relations. Elsevier Academic Press, London.

[3] Mooradian, A. D., Morley, J. E., y Korenman, S. G., (1987). Biological actions of androgens. *Endocrine Reviews,* 8(1), 1-28.

[4] Dabbs, J. M., (1990). Age and seasonal variation in serum testosterone concentration among men. *Chronobiol. Int,* 7(3), 245-249.

[5] Dabbs, J. M., (1990). Salivary testosterone measurements: Reliability across hours, days, and weeks. *Physiology and Behavior,* 48, 83-86.

[6] Jara, A. y González, G., (2001). *Fisiopatología de la corteza suprarrenal.* In: A. Jara (Ed.), Endocrinología. Editorial Médica Panamericana, Madrid.

[7] Goldman, J., Wajchenberg, B. L., Liberman, B., Nery, M., Achando, S. y Germek, O. A., (1985). Contrast analysis for the evaluation of the circadian rhythms of plasma cortisol, androstenedione, and testosterone in normal men and the possible influence of meals. *Journal of Clinical Endocrinology and Metabolism,* 60, 164-167.

[8] Granger, D. A., Schwartz, E. B., Booth, A., Curran, M. y Zakaria, D., (1999). Assesing dehydroepiandrosterone in saliva: A simple radioimmunoassay for use in studies of children, adolescents and adults. *Psychoneuroendocrinology,* 24, 567-579.

[9] Majewska, M. D., (1995). Neuronal actions of dehydroepiandrosterone. Possible roles in brain development, aging, memory, and affect. *Annals New York Academy of Sciences,* 774, 111-120.

[10] Labrie, F., Luu-The, V., Bélanger, A., Lin, S. X., Simard, J., Pelletier, G. y Labrie, C., (2005). Is dehydroepiandrosterone a hormone? *Journal of Endocrinology,* 187, 169-196.

[11] Kroboth, P. D., Salek, F. S., Pittenger, A. L., Fabian, T. J. y Frye, R. F., (1999). DHEA and DHEA-S: A Review. *Journal of Clinical Pharmacology,* 39, 327-348.

[12] Nelson, R. J., (1996). *Psicoendocrinología.* Ariel Psicología, Barcelona.

[13] Rosenweig, M. R. y Arnold, L., (1994). Psicología fisiológica. McGraw-Hill, Madrid.

[14] Kaplan, N. M., (1992). The adrenal glands. In: J. E. Griffin y S. R. Ojeda (Eds.), Textbook of Endocrine Physiology. Oxford University Press, New York.

[15] Bertrand, J. y Saez, M., (1992). *Adrenal steroids: Biochemistry and physiology.* In: J. Bertrand, R. Rappaport, and P. C. Sizonenko (Eds.), *Pediatric Endocrinology.* Williams and Wilkins, Baltimore.

[16] Miralles, J. M., (2001). *Testículo. Exploración. Fisiopatología.* In: A. Jara (Ed.), Endocrinología. Editorial Médica Panamericana, Madrid.

[17] Milgrom, E., (1990). *Steroid Hormones.* In: E. Baulieu and P. A. Kelly (Eds.), *Hormones.* Chapman and Hall, London.

[18] Chiva, L. y Chiva de Agustín, L., (2001). *Ovario. Fisiopatología. Exploración.* In: A. Jara (Ed.), Endocrinología. Editorial Médica Panamericana, Madrid.

[19] Donovan, B. T., (1985). Hormones and human behavior. Cambridge University Press, New York.

[20] Mendelson, C. R., (1992). Mechanisms of hormone action. In: J. E. Griffin y S. R. Ojeda (Eds.), *Textbook of Endocrine Physiology.* Oxford University Press, New York.

[21] Negri-Cesi, P., Poletti, A., Martini, L. y Piva, F., (1999). *Steroid metabolism in the brain: Role in sexual differentiation.* In: A. Matsumoto (Ed.), Sexual differentiation of the brain. CRC Press, Boca Raton, Florida.

[22] Breedlove, S. M. y Hampson, E., (2002). *Sexual differentiation of the brain and behavior.* In: J. B. Becker, S. M. Breedlove, D. Crews, and M. M. McCarthy (Eds.), *Behavioral Endocrinology.* MIT press, London.

[23] George, F. W., (1992). Sexual differentiation. In: *Textbook of Endocrine Physiology* (Ed.), Griffin, J.E.

[24] Colom, R. y Zaro, M. J., (2004). *La psicología de las diferencias de sexo*. Biblioteca Nueva, Madrid.

[25] Cooke, B., Hegstrom, D., Villeneuve, L. S. y Breedlove, S. M., (1998). Sexual differentiation of the vertebrate brain: Principles and mechanisms. *Frontiers in Neuroendocrinology,* 19, 323-362.

[26] Sisk, C. L. y Zehr, J. L., (2005). Pubertal hormones organize the adolescent brain and behavior. *Frontiers Neuroendocrinology,* 26, 163-174.

[27] Hines, M., (1999). *Gonadal hormones and sexual differentiation of human behavior: Effects on psychosexual and cognitive development*. In: A. Matsumoto (Ed.), *Sexual differentiation of the brain.* CRC Press, Boca Raton, Florida.

[28] Baron-Cohen, S., Knickmeyer, R. C. y Belmonte, M. K., (2005). Sex differences in the brain: Implications for explaining autism. *Science,* 310(4), 819-823.

[29] Cohen-Bendahan, C.C., van den Beek, C. y Berenbaum, S. A., (2005). Prenatal sex hormona effects on child and adult sex-typed behavior: Methods and findings. *Neuroscience and Biobehavioral Reviews,* 29, 353-384.

[30] Collaer, M. L. y Hines, M., (1995). Human behavioral sex differences: A role for gonadal hormones during early development. *Psychological Bulletin,* 118(1), 55-107.

[31] Berenbaun, S. A., Korman, K. y Leveroni, C., (1995). Early hormones and sex differences in cognitive abilities. *Learning and Individual Differences,* 7(4), 303-321.

[32] Christensen, K. y Knussman, R., (1987). Sex hormones and cognitive functioning in men. *Neuropsychobiology,* 18, 27-36.

[33] Gouchie, C. y Kimura, D., (1991). The relationship between testosterone levels and cognitive ability patterns. *Psychoneuroendocrinology,* 16(4), 323-334.

[34] McKeever, W. F., Rich, D. A., Deyo, R. A. y Conner, R. L., (1987). Androgens and spatial ability: Failure to find a relationship between testosterone and ability measures. *Bulletin of the Psychonomic Society,* 25, 438-440.

[35] Kimura, D. y Hampson, E., (1994). Cognitive pattern in men and women is influenced by fluctuations in sex hormones. *Current Directions in Psychological Science,* 3, 57-61.

[36] Hines, M., (2004). Brain gender. Oxford University Press, New York.

[37] Janowsky, J. S., Oviatt, S. K. y Orwoll, E. S., (1994). Testosterone influences spatial cognition in older men. *Behavioral Neuroscience,* 108(2), 325-332.

[38] Janowsky, J. S., Chavez, B., y Orwoll, E., (2000). Sex steroids modify working memory. *Journal of Cognitive Neuroscience,* 12, 407-414.

[39] Wolf, O. T. y Kirschbaum, C., (2002). Endogenous estradiol and testosterone levels are associated with cognitive performance in older women and men. *Hormones and Behavior,* 41, 259-266.

[40] Wolf, O. T., Naumann, O., Hellhammer, D., Geiben, A. C., Strasburger, C. J., Dressendörfer, R. A., Pirke, K. M. y Kirschbaum, C., (1997). Effects of a two-week physiological dehydroepiandrosterone substitution on cognitive performance and well-being in healthy elderly women and men. *Journal of Clinical Endocrinology and Metabolism,* 82(7), 2363-2367.

[41] Wolf, O. T., Naumann, O., Hellhammer, D. y Kirschbaum, C., (1998). Effects of dehydroepiandrosterone replacement in elderly men on event-related potential, memory and well-being. *Journal of Gerontology: Medical Sciences,* 53, M385-M390.

[42] Hirshman, E., Merrit, P., Wang, C., Wierman, M., Budescu, D., Kohrt, W., Templin, J. y Bhasin, S., (2004). Evidence than androgenic and estrogenic metabolites contribute to

the effects of dehydroepiandrosterone on cognition in postmenopausal women. *Hormones and Behavior,* 45, 144-155.

[43] Halpern D. F. y LaMay, M. L., (2000). The smarter sex: A critical review of sex differences in intelligence. *Educational Psychology Review,* 12(2), 229-246.

[44] Tan, Ü. y Tan, M., 1998. Curvilinear correlation between total testosterone levels and fluid intelligence in men and women. *International Journal of Neuroscience,* 95, 77-83.

[45] Jacklin, C. N., Wilcox, K. T. y Maccoby, E. E., (1988). Neonatal sex-steroid hormones and cognitive abilities at six years. *Developmental Psychology,* 21, 567-574.

[46] Finegan, J. K., Niccols, G. A. y Sitarenios, G., (1992). Relations between prenatal testosterone levels and cognitive abilities at 4 years. *Developmental Psychology,* 28, 1075-1089.

[47] Grimshaw, G. M., Sitarenios, G. y Finegan, J. K., (1995). Mental rotation at 7 years: Relations with prenatal testosterone levels and spatial play experiences. *Brain and Cognition,* 29, 85-100.

[48] Astington, J., (1993). The child's discovery of the mind. Harvard University Press, Cambridge, MA.

[49] Wellman, H., (1990). The child's theory of mind. MIT Press, Cambridge, MA.

[50] Brown, J. y Dunn, J., (1996). Continuities in emotion understanding from three to six years. *Child Development,* 67, 789-802.

[51] Dunn, Y., Brown, J., Slomkowski, C., Tesla, C. y Youngblade, L., (1991). Young children's understanding of other people's feelings and beliefs: Individual differences and their antecedents. *Child Development,* 62, 1352-1366.

[52] Geary, D., (1999). *Male, female. The evolution of human sex differences.* American Psychological Association, Washington, DC.

[53] Roazzi, A. y Santana, S., (1999). Theory of mind: the effect of age, sex, and the use of animate and inanimate "actors" in inferring mental states. *Psicologia: Reflexao e Critica,* 12(2), 307-330.

[54] Simpson, D. L., (2003). The role of culture and gender in children's development of display rules and theory of mind. *Dissertation Abstracts International: Section B: The Sciences and Engineering,* 63(7-B), 3498.

[55] Azurmendi, A., Braza, F., Sorozabal, A., García, A., Braza, P., Carreras, M. R., Muñoz, J. M., Cardas, J. y Sánchez-Martín, J. R., (2005). Cognitive abilities, androgen levels, and body mass index in 5-year-old children. *Hormones and Behavior,* 48, 187-195.

[56] Susman, E. J. and Ponirakis, E., (1997). Hormones-context interactions and antisocial behavior in youth. In: Raine, A., Brennan, P.A., Farrington, D.P., and Mednick, S.A. *Biosocial bases of violence.* New York: Plenum.

[57] Raine, A., Brennan, P., and Farrington, D.P. (1997). Biosocial bases of violence: Conceptual and theoretical issues. In Raine, A., Brennan, P.A., Farrington, D.P. and Mednick, S.A. (Eds). *Biosocial bases of violence* (pp. 1-20). New York: Plenum

[58] Archer, J., (1991). The influence of testosterone on human aggression. *British Journal of Psychology,* 82, 1-28.

[59] Mazur, A. y Booth, A., (1998). Testosterone and dominance in men. *Behavioral and Brain Sciences,* 21, 355-397.

[60] Harris, J. A., Rushton, J. P., Hampson, E. y Jackson, D. N., (1996). Salivary testosterone and self-report aggressive and pro-social personality characteristics in men and women. *Aggressive Behavior,* 22, 321-331.

[61] Wingfield, J. C., (2005). A continuing saga: The role of testosterone in aggression. *Hormones and Behavior,* 48, 253-255.

[62] Archer, J., (2006). Testosterone and human aggression: an evaluation of the challenge hypothesis. *Neuroscience and Biobehavioral Reviews*, 30, 319-345.

[63] Nyborg, H., (1994). *Hormones, sex, and society: The science of physicology*. Praeger, Westport, CT.

[64] Nyborg, H., (2004). Multivariate modelling of testosterone-dominance associations. *Behavioral and Brain Sciences,* 27(1), 155-159.

[65] Dabbs, J. M., Hargrove, M. F. y Heusel, C., (1996). Testosterone differences among college fraternities. Well-behaved vs. rambunctious. *Personality and Individual Differences,* 20(2), 157-161.

[66] Dabbs, J. M., (1997). Testosterone, smiling, and facial appearance. *Journal of Nonverbal, Behavior* 21, 45-55.

[67] Ramirez, J., (2003). Hormones and aggression in childhood and adolescence. *Aggression and Violent Behavior,* Nov-Dec 8(6), 621-644.

[68] Van Goozen, S., (2005). Hormones and the developmental origins of aggression. In Tremblay, R.E., Hartup, W.W., and Archer, J. (Eds.) *Developmental origins of aggression.* Guilford Publications, New York.

[69] Constantino, J. N., Grosz, D., Saenger, P., Chandler, D. W., Nandi, R. y Earls, F. J., (1993). Testosterone and aggression in children. *Journal of The American Academy of Child and Adolescent Psychiatry,* 32, 1217-1222.

[70] Van Goozen, S. H. M., Matthys, W., Cohen-Kettenis, P., Thijssen, J. H. H. y van Engeland, H., (1998). Adrenal androgens and aggression in conduct disorder prepuberal boys and normal controls. *Biological Psychiatry,* 43(2), 156-158.

[71] Sánchez-Martín, J. R., Fano, E., Ahedo, L., Cardas, J., Brain, P. F. y Azpíroz, A., (2000). Relating testosterone levels and free play social behavior in male and female preschool children. *Psychoneuroendocrinology,* 25, 773-783.

[72] Chance, S. E., Brown, R. T., Dabbs, J. M. y Casey, R., (2000). Testosterone, intelligence and behavior disorders in young boys. *Personality and Individual Differences,* 28, 437-445.

[73] Van Goozen, S., Van den Ban, E., Matthys, W., Cohen-Kettenis, P.T., Thijssen, J. and Van Engeland, H., (2000). Increased adrenal androgen functioning in children with oppositional defiant disorder: A comparison with psychiatric and normal controls. *Journal of the American Academy of Child and Adolescent Psychiatry,* 39(11), 1446-1451.

[74] Chance, S.E., Brown, R.T., Dabbs, J.M., and Casey, R., (2000). Testosterone, intelligence and behavior disorders in young boys. *Personality and Individual Differences*, 28, 437-445.

[75] Scerbo, A. y Kolko, D. J., (1994). Salivary testosterone and cortisol in disruptive children: Relationship to aggressive, hyperactive, and internalizing behaviors. *Journal of The American Academy of Child and Adolescent Psychiatry,* 33(8), 1174-1184.

[76] Azurmendi, A., Braza, F., García, A., Braza, P., Muñoz, J. M. y Sánchez-Martín, J. R., (2006). Aggression, dominance, and affiliation: their relationships with androgen levels and intelligence in 5-year-old children. *Hormones and Behavior,* 50(1), 132-140.

[77] Archer, J., Graham-Kevan, N., y Davies, M., (2005). Testosterone and aggression: A reanalysis of Book, Starzyk and Quinsey´s (2001) study. *Aggression and Violent Behavior,* 10, 241-261.

[78] Rose, A. J. y Rudolph, K. D., (2006). A review of sex differences in peer relationships processes: Potential trade-offs for the emotional development of girls and boys. *Psychological Bulletin,* 132(1), 98-131.

[79] Geary, D., (2006). Sex differences in social behavior and cognition: Utility of sexual selection for hypothesis generation. *Hormones and Behavior*, 49, 273-275.

[80] Buss, D. M., 1995. Psychological sex differences. Origins through sexual selection. *American Psychologist,* 50, 164-168.

[81] Archer, J., (1996). Sex differences in social behaviour: Are the social role and evolutionary explanations compatible? *American Psychologist,* 51, 909-917.

[82] Booth, A., Carver, K. y Granger, D. A., (2000). Biosocial perspectives on the family. *Journal of Marriage and Family*, 62, 1018-1034.

[83] Granger, D. A. y Kivlighan, K. T., (2003). Integrating biological, behavioral, and social levels of analysis in early child development: Progress, problems, and prospects. *Child Development* 74(4), 1058-1063.

In: Androgens: Production, Functions and Disorders
Editors: B. F. Thompson and D. J. Robinson
ISBN: 978-1-62081-145-0

Chapter IX

Androgens, Androgen Suppression, and Prostate Cancer: Recent Preclinical and Clinical Findings and Implications*

***Jeanny B. Aragon-Ching*[1], *Nima Sharifi*[1,2] *and James L. Gulley*[3]**
[1]Medical Oncology Branch, Center for Cancer Research,
National Cancer Institute, National Institutes of Health, Bethesda, MD, US
[2]Cancer Stem Cell Section, Laboratory of Cancer Prevention,
Center for Cancer Research, National Cancer Institute,
ational Institutes of Health, Bethesda, MD, US
[3]Laboratory of Tumor Immunology and Biology,
Center for Cancer Research, National Cancer Institute,
National Institutes of Health, Bethesda, MD, US

Abstract

Prostate cancer is the leading cause of malignancy among American men and the third leading cause of cancer deaths. Approximately 42% of men will progress following local definitive therapy, with either rising PSA or metastatic disease. Androgen-deprivation therapy (ADT) is the primary treatment for metastatic prostate cancer. Although it is increasingly used in different disease stages and patient populations, there is still controversy regarding its use in patients whose sole determinant of disease progression is rising PSA. Various treatment modalities have been studied in combination with ADT to determine their overall efficacy and benefit in terms of progression-free and overall survival, leading to the use of ADT in selected patient populations. However, ADT is not without side effects. Different dosing strategies, such as intermittent ADT and rapid hormonal cycling, are being employed to determine how to administer ADT for maximum benefit and minimal toxicity. For the many patients who will eventually develop incurable, often debilitating metastatic castrate-resistant disease, strategies for

* A version of this chapter also appears in Trends in Testosterone Research, edited by Raymund E. Veritos, published by Nova Science Publishers, Inc. It was submitted for appropriate modifications in an effort to encourage wider dissemination of research.

directing novel therapies at the androgen-receptor signaling pathway and apoptotic genes hold future promise. This chapter discusses the clinical applications, side effects, and overall effects of ADT, clinical studies that support or controvert its use in selected patient populations and stages of prostate cancer, and the biology of castrate-resistant prostate cancer and novel approaches to its treatment.

Introduction

Since Huggins and Hodges established a link between prostate cancer and androgens in 1941 [1], androgen-deprivation therapy (ADT) in the form of chemical or surgical castration has become a cornerstone of treatment for metastatic prostate cancer. The rationale behind the use of ADT in prostate cancer derives from the interdependent actions of various hormones and their effects on the prostate gland (figure 1).

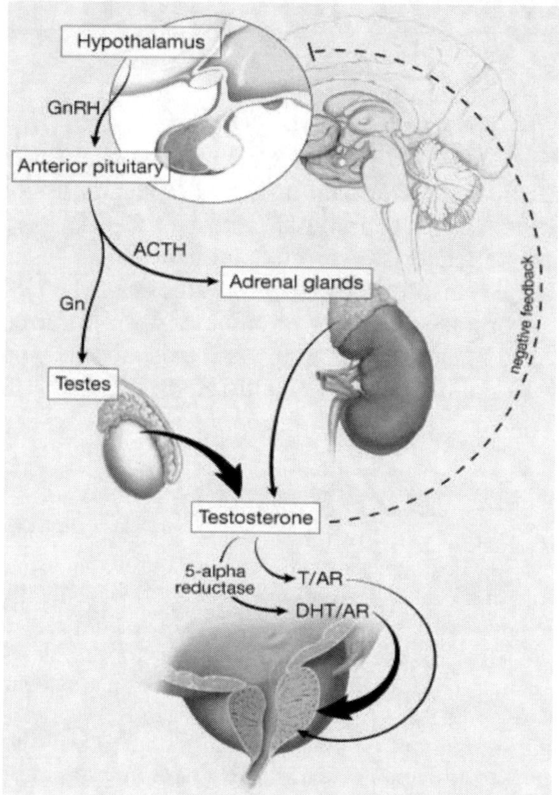

Figure 1. Hypothalamic-pituitary-gonadal axis. The hypothalamus/pituitary/testes endocrine loop is shown above. Gonadotropin-releasing hormone (GnRH) stimulates the secretion of gonadotropic hormone (Gn) from the anterior pituitary, and Gn in turn stimulates the production of testosterone. Circulating testosterone (and estrogens) acts in a negative feedback loop to down-regulate the expression of GnRH. Adrenocorticotropic hormone (ACTH), also made by the pituitary, stimulates androgen synthesis in the adrenal gland. Testosterone (T) and dihydrotestosterone (DHT) bind to the androgen receptor (AR), causing increased expression of androgen-responsive genes and leading to cell growth.

The hypothalamus initiates a series of events that lead to the secretion of testosterone, a hormone required by the prostate to develop into and remain a functional gland. Gonadotropin-releasing hormone (GnRH), also known as luteinizing hormone-releasing hormone (LHRH), is produced in the neurosecretory cells of the hypothalamus in a physiologically pulsatile manner. These neurons terminate in the hypothalamic-hypophysial portal system, a unique vascular bed that shuttles GnRH directly to the anterior lobe of the pituitary. In response to GnRH, the anterior lobe of the pituitary secretes luteinizing hormone, perhaps more accurately known as gonadotropic hormone (Gn), which enters the circulation and exerts an effect on the testes.

The testes contain a network of seminiferous tubules that produce sperm. In between these tubules are the interstitial cells of Leydig, which have high-affinity receptors on their plasma membranes to which the Gn binds. Leydig cells produce testosterone as a result of enzymatic conversion of cholesterol upon stimulation by Gn. Testosterone then enters the blood stream and the cells of the prostate, where it may be converted by 5-alpha reductase into dihydrotestosterone (DHT). Both DHT and testosterone can directly bind to the cytoplasmic androgen receptor (AR) and initiate growth, although DHT binds with greater affinity. The adrenal cortex plays a lesser role in prostatic growth, producing only a small proportion (about 10% to 15%) of total testosterone. These androgens, which are essential for normal glandular growth, are also implicated as growth factors in the development of prostate cancer. As in other endocrine systems, negative feedback control is fundamental. When testosterone (or estrogen) levels in the blood stream increase, the hypothalamus reduces the secretion of GnRH, which in turn inhibits the secretion of Gn from the pituitary gland. The overall effect is a net decrease in the amount of Gn that can stimulate the Leydig cells, thus decreasing testosterone production.

Table 1. FDA-approved hormonal therapy agents and their dosing

Hormonal Therapy	Dosing	Manufacturer
GnRH agonists		
Lupron Depot® (leuprolide acetate for depot suspension)	7.5 mg s.c. q mo 22.5 mg s.c. q 3 mos 30 mg s.c. q 4 mos 45 mg s.c. q 6 mos	TAP Pharmaceuticals, Inc.
Zoladex® (goserelin acetate implant)	3.6 mg s.c. q mo 10.8 mg s.c. q 3 mos	Astra-Zeneca Pharmaceuticals
Viadur® (leuprolide acetate implant)	yearly implant	ALZA Corporation under license to Bayer Corporation
GnRH antagonist		
Plenaxis® (abarelix)	100 mg i.m. days 1,15, 29, q 4 wks thereafter	Praecis Pharmaceuticals
Antiandrogens		
Casodex® (bicalutamide) tablets	50 mg daily	Astra-Zeneca Pharmaceuticals
Eulexin® (flutamide) capsule	250 mg p.o. t.i.d.	Schering-Plough Pharmaceuticals
Nilandron® (nilutamide) tablets	150 mg p.o. daily	Aventis

s.c. = subcutaneous; i.m. = intramuscular; q = every; mo = month; p.o. = per orem; t.i.d. = 3x per day.

Androgens are the hormones that primarily regulate the growth, proliferation, and differentiation of prostate cancer cells. Since AR is expressed in almost all prostate cancers, this expression is the basis of the initial response of prostate cancer to ADT. AR is a 919 amino acid member of the steroid receptor transcription factor superfamily and has incorporated within its protein structure an N-terminal regulation domain, a central DNA binding domain, and a C-terminal domain, which includes the ligand binding domain [2]. In the absence of stimulation, AR resides in the cytoplasm. Once activated, it localizes to the nucleus within a few minutes [3], where it regulates the transcription of numerous genes [4]. When androgens bind to AR, AR protein is stabilized and levels increase [5]. AR is critical for normal male development, and its role in the physiology and health of the adult male is best illustrated by the adverse effects of ADT described later in this chapter. The role of AR in development is probably best illustrated by some loss-of-function AR mutations, which can result in complete androgen insufficiency syndrome, including a superficial female habitus and lack of a prostate [6].

ADT, also known as androgen-suppression therapy, derives from the basic principle that androgens stimulate the growth of prostate cancer. Androgen deprivation is achieved by various methods, including surgical castration with bilateral orchiectomy, or medical castration in the form of GnRH agonists (GnRH-A) or antagonists, estrogens, androgen-receptor antagonists or antiandrogens, ketoconazole, and aminoglutethimide (see table 1 for FDA-approved hormonal agents and their dosing). The first large-scale study on the effectiveness of orchiectomy showed that patients with advanced prostate cancer had a 73% one-year survival and a 35% 5-year survival following orchiectomy, compared to 66% and 20% respectively for placebo [7]. Although orchiectomy is a relatively simple procedure with minor operative risks and low physical morbidity, it has fallen out of favor because of its psychological effect on the patient, and because medical alternatives to achieving androgen suppression are now available [8, 9].

Diethylstilbestrol (DES) is a synthetic estrogen that blocks Gn and follicle-stimulating hormone (FSH) via a negative feedback loop (figure 1), resulting in complete testosterone suppression. DES was the first alternative to orchiectomy and was commonly used for androgen suppression through the late 1970s. Although estrogen treatment resulted in improved cancer-specific survival compared to placebo, as established in the Veterans Administration Cooperative Urological Research Group Study, it was accompanied by side effects including nausea, vomiting, fluid retention, gynecomastia, and most significantly, serious cardiovascular events [7, 10, 11]. In 1971, Schally et al. successfully isolated and identified the decapeptide GnRH, which has a unique primary structure that is able to regulate production and secretion of Gn and FSH [12, 13]. However, a shortcoming of this form of GnRH is its half-life of 2 to 4 minutes, which precludes its therapeutic usefulness. Subsequent cloning studies were undertaken with single amino acid substitutions that identified compounds with longer half-lives. Substituting the amino acid positions 6 and 10 led to the development of superagonists 50 to 100 times more potent than the native peptide [14].

GnRH-A supplants the effect of physiologically pulsatile endogenous GnRH, producing steady-state levels of GnRH and desensitizing the pituitary. This causes down-regulation of receptors in the pituitary and arrests pituitary secretion, thereby decreasing testicular function.

Testosterone production in the gonads decreases within 3 weeks, with a subsequent decrease in the growth of androgen-dependent tissues [15-17]. Available formulations can be administered every 1, 3, 4, 6, or 12 months. An initial increase in the level of testosterone causes an acute stimulation of prostate cancer growth and may result in a "flare" reaction of increased bone pain in patients with metastatic prostate cancer. This effect can be alleviated by androgen-receptor antagonists [18]. It is recommended that an androgen antagonist be administered prior to initiation of treatment with a GnRH-A and continued for 2 to 4 weeks to block the effect of the testosterone surge on peripheral AR [19]. In clinical practice, this regimen is often reserved for patients with significant metastatic disease that may become more symptomatic with the testosterone surge.

As an alternative method of medical castration, a GnRH antagonist has the benefit of obviating the acute testosterone surge. However, abarelix, the only GnRH antagonist approved in the United States, caused an immediate-onset systemic allergic reaction in 1.1% (15/1,397) of patients [20, 21], and the cumulative risk of this adverse event increased with the duration of treatment. As a result, GnRH antagonists are no longer available in the U.S. for new patients.

A number of studies have examined the efficacy of medical vs. surgical castration for advanced prostate cancer, including 10 randomized trials comparing GnRH-A and orchiectomy. A published meta-analysis of all of these trials found equivalence between GnRH-A and orchiectomy in terms of survival, progression-related outcomes, and time to treatment failure [22]. Given a choice between bilateral orchiectomy and GnRH-A therapy, only 22% (32/147) of patients in 2 clinical studies elected to undergo bilateral orchiectomy [23].

Other classes of agents used clinically to block the effects of androgen are AR antagonists such as flutamide, bicalutamide, and nilutamide, which are often used alone or in combination with castration to block the effects of adrenal androgens. Other adrenal-ablating drugs, such as ketoconazole, inhibit cytochrome P450 enzymes that are required for synthesis of androgens and other steroids. Another class of drugs is the 5-alpha reductase inhibitors, which inhibit the enzyme responsible for the conversion of testosterone to DHT, the potent activator of AR. While 5-alpha reductase inhibitors currently have no role in the treatment of advanced prostate cancer, they may have a role in prevention [24].

Regardless of the means, the goal of androgen blockade is to achieve the lowest possible serum testosterone level in order to minimize stimulation of prostate cancer cells. Castration is generally achieved when the serum testosterone level falls to less than 50 ng/dL [25]. However, since most men who undergo surgical castration achieve levels below 20 ng/dL, reassessment of the definition of castrate levels has been proposed to reflect this threshold [26]. This chapter will address the indications and clinical settings for ADT, and the studies and issues that support or controvert its use, as well as the biology of castrate-resistant prostate cancer (CRPC).

Prevention Studies

5-Alpha Reductase Inhibitors

5-alpha reductase enzymatically converts testosterone into DHT, the most active androgen in the prostate. Two 5-alpha reductase enzymes have been isolated: type 1 is the predominant enzyme located in extraprostatic tissues such as the skin and liver; type 2 is predominantly expressed in the prostate [27]. Recent findings that may have important implications in the choice of drug treatment suggest that there is a shift in isozyme expression from type 2 to type 1 in recurrent prostate cancer, as well as in androgen-stimulated benign prostate and prostate cancers [28]. Two FDA-approved 5-alpha reductase inhibitors are currently available: finasteride and dutasteride. Finasteride, a selective inhibitor of type 2 5-alpha reductase, decreases levels of serum DHT by about 70%, reducing gland size by 20% to 30% and improving urinary symptoms for patients with benign prostatic hypertrophy [29]. The Prostate Cancer Prevention Trial (PCPT) demonstrated that 7 years of prophylactic finasteride decreased the incidence of prostate cancer by 6% (24.4% in the placebo group vs. 18.4% in the finasteride group). However, there was a statistically significant 1.3% absolute increase in the incidence of high-grade tumors (Gleason score 7 to 10) in the finasteride group, with 6.4% of men in the finasteride group and 5.1% of men in the placebo group having this pathology [24, 30]. One possible explanation for the increase in high-grade tumors in the finasteride group is that finasteride causes shrinkage of the normal prostate gland [29] that may be out of proportion to the tumor, and may cause more shrinkage in lower-grade tumors than in more aggressive-appearing tumors. This could lead to relative oversampling of tumors in the finasteride-treated patients. In the PCPT, the prostates of finasteride-treated men were about 25% smaller than the prostates of placebo-treated men. This effect of finasteride may have increased sampling bias by increasing the likelihood of detecting prostate cancer by needle biopsy. Another possible explanation for the absolute increase in high-grade cancer in the finasteride group is that increased sensitivity of the PSA test used in the finasteride group led to greater detection of prostate cancer in this group [31].

Dutasteride, another 5-alpha reductase inhibitor, has also been used in a chemoprevention trial. Whereas finasteride primarily inhibits type 2 5-alpha reductase, dutasteride inhibits both 5-alpha reductase isoenzymes and reduces DHT by > 90% [32]. Retrospective analysis of data from several early trials of dutasteride in benign prostatic hyperplasia indicated a lower incidence of prostate cancer with the use of this agent [33]. The Reduction by Dutasteride of Prostate Cancer Events (REDUCE) trial, a 4-year, international, multicenter, randomized, double-blind, placebo-controlled study, will evaluate the efficacy and safety of oral dutasteride 0.5 mg/day in men aged 50 to 75 who are at increased risk of developing prostate cancer. The target enrollment is 8,000 men, who will be randomized 1:1 to receive dutasteride or placebo for 4 years. Enrollment of approximately 4,000 participants per treatment group will provide over 90% overall power to detect the superiority of dutasteride 0.5 mg treatment vs. placebo after 4 years. Eligibility criteria include PSA of between 2.5 and 10 ng/mL for men aged 50 to 60, or between 3.0 and 10 ng/mL for men older than 60, and a free PSA of 25% or lower at initial screening. Enrollees must have no histological evidence of prostate cancer, based on a negative 6 to 12 core biopsy within 6 months prior to enrollment. Prostate biopsies will be performed at 2 and 4 years to evaluate for prostate cancer. PSA, genetic, and protein markers will also be obtained. The primary endpoint is biopsy-determined prostate cancer after 2 and 4 years of treatment. Several secondary endpoints will also be assessed, including overall survival [34].

A large chemoprevention trial that does not utilize a testosterone-lowering approach is the Selenium and Vitamin E Cancer Prevention Trial (SELECT) [35]. This clinical trial will attempt to confirm the findings of previous studies that suggest that selenium and vitamin E may reduce the risk of developing prostate cancer [36, 37]. SELECT closed to accrual in 2004 with an enrollment of over 35,000 men from various ethnic groups.

Treatment Studies

Role of ADT in Early Prostate Cancer

ADT is most commonly used to treat metastatic prostate cancer. However, there is emerging evidence that using ADT as a neoadjuvant, or as an adjuvant combined with local definitive therapy, in the treatment of localized disease can provide clinically significant outcomes in selected patient populations. Neoadjuvant therapy is therapy administered after diagnosis but before definitive local treatment. Surgery and radiotherapy (RT) are generally effective in treating localized disease. However, as a cure for patients with apparently localized disease, surgery may fail if the cancer is incompletely resected due to understaging of the tumor or the existence of micrometastases prior to surgery. The addition of systemic therapy is therefore appealing.

ADT Combined with Radiation

Combining RT and hormonal therapy has the potential benefits of decreasing prostate volume, diminishing the number of cancer cells, improving tumor oxygenation, and eliminating micrometastases. One of the first trials to indicate the benefit of combining RT with ADT was the Radiation Therapy Oncology Group trial (RTOG 83-07) [38]. Comparison to a historical control group showed that 2 different hormonal strategies—megestrol acetate and DES—had equivalent efficacy in patients with clinical stages T2B or T3 prostate cancer. Although the primary objective of this trial was to evaluate the relative effectiveness and toxicity of megestrol acetate vs. DES as neoadjuvant therapies to RT, this initial combination of hormonal agents with RT led to succeeding trials that examined the overall survival benefit of adjuvant ADT with RT.

Several randomized phase III trials have shown an overall survival benefit of RT plus ADT compared to RT alone for patients with locally advanced extracapsular or node-positive disease. (See table 2 for a list of selected studies combining ADT with RT.) A trial by the European Organisation for Research and Treatment of Cancer (EORTC) showed an overall survival benefit following 70 Gy RT plus 3 years of ADT compared to 70 Gy RT alone for patients with locally advanced and high-grade clinically localized prostate cancer [39, 40]. In the EORTC trial, 412 analyzable patients with T1 to T2 tumors of World Health Organization (WHO) class 3 (Gleason score 8 to 10) or T3 to T4, N0 to N1 M0 tumors were randomly assigned to receive RT alone (50 Gy delivered to the pelvis over 5 weeks, and 20 Gy over 2 weeks as a prostatic boost) or RT plus immediate treatment with goserelin (3.6 mg subcutaneously every 4 weeks beginning on the first day of irradiation, with 150 mg oral cyproterone acetate (CPA) given for 1 month starting 1 week before the first goserelin injection). ADT was continued for 3 years in the combined arm. The primary endpoint was clinical disease-free survival. At 5 years, overall survival was 78% for the combined-treatment group and 62% for the RT-alone group ($p = 0.0002$). Five-year disease-free survival was 74% for the combined-treatment group and 40% for the RT-alone group ($p = 0.0001$).

Table 2. Selected studies utilizing ADT with RT

Study	Study design	No. of patients	Population	Study intervention	Median follow-up (mos)	Results
Bolla et al. [40]	Randomized phase III	415	T1/T2 high-grade, T3/T4 any grade, N0-N1, M0	Definitive RT vs. RT + goserelin on day 1 x 3 yrs	66	Benefit in RT + ADT: 5 yr OS 78% vs. 62% (p = 0.0002); 5 yr DFS 74% vs. 40% (p = 0.0001)
Lawton et al. [42]	Randomized phase III	977	Node-positive or clinical stage T3	Definitive RT vs. RT + goserelin on last week of RT	96	No OS benefit at 8 yrs: 49% vs. 47%; improved local control with RT + ADT (failure rate 23% vs. 37%)
Pilepich et al. [49]	Randomized phase III	471	T2-T4, node-negative or –positive	Definitive RT vs. RT + CAB 2 mos prior and 2 mos during RT	96	Improvement with ADT in all endpoints: local control, DFS, cause-specific mortality, biochemical DFS, and OS (70% vs. 52%; p = 0.015) at 8 yrs only in patients with Gleason score 2–6
Hanks et al. [268]	Randomized phase III	1,500	T2C-T4	CAB 2 mos prior and during RT followed by 2 yrs CAB vs. no CAB	58	No difference in OS between LTAD-RT vs STAD-RT (80.0% vs. 78.5% at 5 yrs; p = 0.73), but improved in Gleason 8 to 10 group (80% vs. 69%; p = 0.02)
D'Amico et al. [269]	Retrospective cohort	1,586	T1-T2c	1,310 men got RT but no ADT; 276 got ADT 2 mos prior, 2 mos during, and 2 mos after RT	23-30	Decreased risk of biochemical failure for ADT + RT (88% vs. 62%) in intermediate-risk and 68% vs. 43% in high-risk group
D'Amico et al. [46]	Prospective randomized	206	T1b-T2b	70 Gy 3D-CRT vs. 70 Gy 3D-CRT + 6 mos GnRH-A + flutamide	54	Higher survival (p = 0.04), lower prostate cancer-specific mortality (p = 0.02), higher survival free of salvage ADT (p = 0.002)

RT: radiotherapy; OS: overall survival; DFS: disease-free survival; ADT: androgen-deprivation therapy; CAB: combined androgen blockade; 3D-CRT: 3-dimensional conformal radiation therapy.

The most compelling results from clinical trials combining hormonal therapy with RT are seen in patients at high risk for recurrent disease (table 3). In RTOG 85-31, patients with T3 or T1-2 with N1 disease were randomized to RT with adjuvant hormonal therapy, or RT alone with hormones initiated at relapse [41, 42]. Among 945 analyzable patients, improvement in overall survival favored the ADT arm, with an estimated 10-year absolute survival of 53% vs. 38% ($p < 0.0043$) [43]. When these results were updated, they still showed an absolute survival rate at 10 years of 49% vs. 39% ($p = 0.002$) favoring the adjuvant vs. RT-alone arm [44]. All other parameters, including local failure rate, incidence of distant metastases, and disease-specific mortality, favored the adjuvant arm. Further subset analysis showed ADT improved survival in patients with involved regional lymph nodes [45]. A prospective randomized controlled trial by D'Amico et al. that looked at 70 Gy 3-dimensional conformal RT (3D-CRT) with or without 6-month ADT using GnRH reported evidence of extraprostatic disease, as measured by a PSA of at least 10 ng/mL [46]. Patients enrolled in the trial had clinically localized prostate cancer with high-risk features, including Gleason score of at least 7. After a median follow-up of 4.52 years, patients randomized to receive 3D-CRT plus ADT had a significantly higher survival rate of 88% vs. 78% for the 3D-CRT-alone group ($p = 0.04$), lower prostate cancer-specific mortality ($p = 0.02$), and higher survival free of salvage ADT of 82% vs. 57% ($p = .002$). A similar suggestion of benefit was observed in a cohort of 241 men with clinically localized prostate cancer and pretreatment PSA velocity of > 2 ng/mL/year (which is increasingly identified as another feature of high risk) who underwent 6 months of ADT with RT vs. RT alone [47]. Treatment with RT and ADT compared to RT alone was associated with a longer time to PSA recurrence (adjusted hazard ratio [HR] = 0.22; 95% confidence interval [CI]: 0.14 to 0.35; $p < 0.001$), prostate cancer-specific mortality (HR = 0.23; 95% CI: 0.09 to 0.64; $p = 0.005$), and all-cause mortality (HR = 0.30; 95% CI: 0.16 to 0.58; $p < 0.001$). Conversely, the RTOG 86-10 trial of 471 men with stages cT2-4 prostate cancer indicated no difference in 5-year overall survival, but improved progression-free survival (36% in ADT plus RT vs. 15% in RT alone [$p < 0.001$]), as well as local control (46% in ADT plus RT vs. 71% in RT alone [$p < 0.001$]) [48]. In this trial, ADT was given 2 months prior to and throughout the duration of RT. Updated results at 8 years of follow-up showed an improvement in overall survival (70% vs. 52%; $p = 0.015$) favoring the short-term androgen ablation arm only for patients with Gleason scores of 2 to 6, and no benefit in overall survival or locoregional control in patients with Gleason scores > 8 [49]. Further secondary analysis of RTOG 86-10 showed that 54% (n = 247) developed relapse and required subsequent salvage hormonal therapy. Overall survival and disease-specific survival were compared with the extent of disease progression (manifested by presence or absence of distant metastasis) at the time salvage ADT was initiated. Both were reduced in patients with distant metastasis, but a statistically significant increase in disease-specific survival ($p = 0.045$) was observed among 143 patients with no known distant metastasis but in whom PSA was < 20 ng/mL, compared with those in whom PSA was ≥ 20 ng/mL at the time of initiation of salvage ADT [50]. Although the optimal duration of ADT in these studies has not been determined, evidence suggests that hormonal therapy in combination with RT is beneficial. However, it is important to note that none of these adjuvant studies involving radiation included an arm using hormonal therapy alone [51].

Table 3. Risk categories for PSA failure after definitive treatment

	Low	Intermediate	High
Stage	T1c, T2a	T2b	T2c
PSA	< 10	10-20	> 20
Gleason	≤ 6	7	≥ 8
Qualifier	and	or	or
5-year risk of biochemical failure	< 25%	25% to 50%	> 50%

* Adapted from Gulley JL, Dahut WL. Novel schedules of hormone therapy for prostate cancer. In: Devita V, Hellman S, Rosenberg S, eds. *Progress in Oncology*. Boston: Jones and Bartlett; 2004:236-253.

Table 4. Selected neoadjuvant ADT followed by radical prostatectomy (RP)

Study	Study design	Population	No. of patients	Study intervention	Median follow-up (mos)	Results
Gleave [53]	Nonrandomized	T1c-T3a	156	32 wks CPA + DES or CAB followed by RP	54	PSA recurrence rate 12.2%
Schulman [270]	Prospective randomized	cT2-cT3 PSA<100	402	12 wks CAB followed by RP	48	PFS 26.4% vs. 32.5% (p=0.19)
						Decrease in positive margins in cT2 (p<0.01) and cT3 (p=0.01)
Soloway [52]	Prospective randomized	cT2b, PSA<50	282	12 weeks CAB followed by RP	60	5 yr DFS 67.6%(RP only) vs. 64.8% (p=0.663)
Aus [271]	Prospective randomized	T1b-T3a	126	3 wks CPA + 12 wks GnRH-A followed by RP	82	7 yr DFS 51.5% (RP) vs. 49.8% (p=0.588)
Gleave [54]	Prospective randomized	T1c-T2	500	12 wks vs. 32 wks CAB followed by RP	24	Positive margin rate 12% (8-mo group) vs. 23% (3-mo group), p=0.016
						Undetectable PSA rate before RP 75% (8-mo group) vs. 43% (3-mo group), p=0.0001

CPA: cyproterone acetate; RP: radical prostatectomy; CAB: combined androgen blockade; DFS: disease-free survival, equivalent to biochemical recurrence rate; PFS: progression-free survival, equivalent to PSA progression-free survival; PSA: prostate-specific antigen; GnRH: gonadotropin-releasing hormone agonist; DES: diethystilbestrol.

ADT Combined with Surgery

Neoadjuvant ADT has been used to downstage primary tumors and to potentially improve outcome by decreasing tumor burden. However, data indicate that neoadjuvant ADT prior to radical prostatectomy (RP) provides no long-term overall or disease-free survival benefit (table 4). In a multicenter phase III study involving 282 patients, there was an apparently significant decrease in positive surgical margins after 3 months of neoadjuvant ADT prior to RP (18% in the neoadjuvant group vs. 48% in the RP-alone group [$p < 0.001$]). However, this did not translate to improved biochemical recurrence rate or disease-free survival rate at 5 years of follow-up (64.8% in the neoadjuvant group vs. 67.6% in the RP-alone group [$p = 0.663$]) [52]. In a study of 156 nonrandomized patients comparing shorter and longer durations of neoadjuvant ADT prior to RP, there was an overall PSA recurrence rate of 12% at 5 years of follow-up after 8 months of neoadjuvant ADT for clinical T3 disease [53]. This risk was proportional to known adverse risk factors such as higher pretreatment PSA, higher Gleason score, and higher pathologic stage. A randomized multicenter trial by the Canadian Uro-Oncology Group has reported interim data on 500 patients with T1c-T2 tumors, equally stratified to 3 months vs. 8 months of neoadjuvant complete androgen blockade prior to RP [54]. The percentage of patients with undetectable PSA prior to surgery

(3-month: 75% vs. 8-month: 43%; $p < 0.0001$) and the rate of positive margins (3-month: 23% vs. 8-month: 12%; $p = 0.016$) favored the 8-month group. There was a higher incidence of hot flashes and newly reported adverse events in the 8-month neoadjuvant group. However, follow-up was too short to draw any definitive conclusions. Because there are insufficient data to support either overall or disease-free survival, there is currently no definitive role for neoadjuvant ADT prior to RP, unless micrometastasis is suspected.

The role of ADT after definitive curative surgery is also controversial. However, for patients with the highest risk of recurrence, such as those with positive lymph nodes, ADT seems strongly indicated [55]. A prospective randomized trial by the Eastern Cooperative Oncology Group (ECOG) examined the effect of early ADT on outcomes for men with pelvic lymph node involvement who had undergone RP. Fifteen percent (7/47) of men in the early-ADT arm died after a median follow-up of 7.1 years vs. 35% (18/51) of men in the observation arm ($p = 0.02$). Furthermore, 36 men (77%) in the early-ADT arm were alive and disease-free with undetectable PSA levels vs. 22 men (43%) in the observation arm ($p < 0.001$) [56]. In a retrospective study from a single institution, data from 3,663 patients who underwent RP and bilateral pelvic lymphadenectomy for prostate cancer were examined. Patients with multiple regional lymph node metastases had increased risk of death from disease, whereas patients with single lymph node involvement appeared to have a more favorable prognosis after RP and immediate adjuvant ADT [57].

ADT for Patients Ineligible for Local Definitive Therapy

For patients deemed ineligible for definitive local therapy because of increased comorbidities, decreased life expectancy, advanced local tumor stage or other existing malignancies, or who refuse local therapy for any reason, ADT remains a viable treatment option. A prospective randomized Swiss trial (SAKK 08/88) with a total of 197 patients attempted to determine if immediate ADT had an advantage over deferred treatment in newly diagnosed asymptomatic prostate cancer patients who were not candidates for curative local treatment [58]. Although the study showed no statistically significant difference in overall survival between the 2 arms ($p = 0.09$), it was underpowered to conclude equivalence. Recently, EORTC 30891, a randomized trial involving 985 patients with newly diagnosed prostate cancer, demonstrated statistically significant improvements in subjective and objective disease progression and in overall survival of men with nonmetastatic prostate cancer who were not candidates for local therapy but who were treated with immediate ADT, compared to men not treated until they were symptomatic [59].

Interestingly, the significant survival benefit was apparently due to fewer nonprostate cancer deaths and not to prostate cancer-related deaths. It may be that patients with less aggressive cancers, who might therefore respond better to early ADT, have a better chance of disease-specific survival.

Alternatively, the finding of increased cardiovascular mortality in the control arm may be attributed to increased thrombotic events secondary to more advanced cancer [60]. It is important to note that, in this study, early ADT did not lead to earlier symptomatic or radiographic progressive disease with CRPC. The finding that there was no difference in time to progression after CRPC further allays concerns that earlier ADT would likely result in earlier development of hormone-refractory disease. This study therefore provides persuasive evidence that, for patients with nonmetastatic prostate cancer who are otherwise not good candidates for local therapy, immediate ADT provides meaningfully improved overall survival, with no decline in quality of life and no earlier time to progressive CRPC [61].

Role of Continuous ADT in Biochemical Recurrence of Prostate Cancer

Adoption of routine PSA testing after primary local therapy with surgery or RT has helped to identify a greater number of patients with PSA-only disease, also referred to as serologic or biochemical recurrence. The definition of biochemical recurrence is dependent on the treatment modality, since detectable PSA post-surgery indicates residual disease [62], while a PSA nadir post-RT can last as long as 18 months [63]. Although biochemical recurrence is considered a legitimate endpoint for defining treatment outcome, its clinical significance remains unclear [64]. One retrospective study of 1,997 men, of whom 315 (15%) developed biochemical recurrence after RP without hormonal therapy, revealed a median actuarial time to metastasis of 8 years after PSA elevation was noted, and a median of 5 years from metastasis to death [65]. However, of the subset of patients with Gleason scores > 8 and biochemical recurrence within 2 years of prostatectomy, only 53% remained free of metastasis at 3 years. A retrospective study with a larger cohort (379 analyzable patients out of 5,096 who underwent RP) and a longer median follow-up of 10 years showed that clinical parameters such as faster PSA doubling time (< 3 months), higher Gleason score (> 8), and shorter time from surgery to biochemical recurrence (< 3 years) were significant predictive risk factors for time to prostate-specific mortality [66]. [Note: PSA doubling time can be calculated by the natural log of 2 (0.693) divided by the slope of the relation between the log of PSA and the time of PSA measurement [67].] Several studies have examined various PSA dynamics, including doubling time, PSA response to salvage ADT, and PSA velocity, as indicators of which patients would develop clinically significant disease following biochemical recurrence after primary surgery or RT [68]. A PSA doubling time of < 3 months after RP or RT has consistently been shown to be associated with higher prostate cancer-specific mortality [69, 70]. Numerous treatment options are available for men with biochemical recurrence, including salvage RT or surgery (depending on the primary modality used), ADT, cryotherapy, brachytherapy, or observation. However, patients who present with biochemical recurrence may not be suitable candidates for salvage local therapy because of associated metastatic disease. ADT has been used in this treatment setting, but its value is unknown because there have been no randomized trials specifically for men with biochemical recurrence following local therapy. Researchers can only extrapolate from data on the use of ADT in patients with more advanced disease or high-grade disease, which demonstrate a survival advantage for adjuvant ADT.

Early versus Late Therapy

The question of whether to begin ADT early or late in the course of disease is in constant debate. The issue of timing stems from 2 hypotheses regarding the sensitivity of prostate cancer to hormonal ablation. The first hypothesis (the adaptation model) assumes that androgen-resistant clones of a homogeneous cell population develop through adaptive transformation after castration [71]. It would therefore be prudent to withhold androgen therapy for as long as possible. The other hypothesis (the clonal selection model) assumes that prostate cancer becomes heterogeneous soon after malignant transformation because of inherent genetic instability. Therefore, combining definitive local therapy (either surgery or radiation) with adjuvant ADT destroys the primary tumor when the percentage of androgen-resistant clones is still small and susceptible to eradication [72]. These and other hypotheses will be discussed further in the section on castrate resistance.

Several randomized trials have studied the use of early vs. delayed ADT in men with locally advanced or metastatic disease, but none of these trials has included patients with true PSA-only or biochemical recurrence. A retrospective, observational, multicenter database analysis of 5,382 men who underwent primary RP demonstrated that early hormonal therapy in patients with biochemical recurrence after RP was an independent predictor of delayed clinical metastasis only for high-risk cases. This delay in clinical metastasis was seen in patients with a pathological Gleason score > 7 or a PSA doubling time of ≤ 12 months (HR = 2.12; p = 0.01) [73].

Table 5. Selected studies comparing combined androgen blockade (CAB) with conventional ADT

Study	Study design	No. of patients	Tumor stage	Study intervention	Max/med follow-up (mos)	Results
Crawford et al. [83]	Randomized double blind	603	D2	Leuprolide + placebo vs. leuprolide + flutamide	48	Improved PFS and length of survival in CAB group (16.5 vs. 13.9 mos; p = 0.039 and 35.6 vs. 28.3 mos; p = 0.035, respectively)
Denis et al. [272]	Randomized phase III	327	M1	Orchiectomy alone vs. goserelin + flutamide	84	Improvement with CAB in OS (p = 0.04), time to death from malignant disease (p = 0.008), time to first progression (p = 0.009), and PFS (p = 0.02)
Janknegt et al. [85]	Randomized double-blind	457	M1	Orchiectomy alone vs. orchiectomy + nilutamide	35–54	Improvement with CAB in PFS (20.8 vs. 14.9 mos; p = 0.005) and median time to death (30.0 vs. 37 mos)
Eisenberger et al. [87]	Randomized	1387	M1	Orchiectomy + flutamide vs. orchiectomy + placebo	84	No difference in OS (p = 0.14)
Tyrrell et al. [273]	Randomized double-blind	571	M0 and M1	Goserelin vs. goserelin + flutamide	56.2	No difference in PFS and OS
Iversen et al. [274]	Randomized	264	M0 and M1	Orchiectomy vs. goserelin + flutamide	57	No difference in PFS (16.8 vs. 16.5 mos) and OS (27.6 vs. 22.7 mos)
Boccardo et al. [275]	Randomized	304	C and D	Goserelin vs. goserelin + flutamide	24	No difference in PFS (24 vs. 17 mos; p = 0.09) or OS

PFS: progression-free survival; CAB: combined androgen blockade; OS: overall survival.

Table 6. Advantages and disadvantages of nonsteroidal AA monotherapy over GnRH-A

Advantages	Disadvantages
Less loss of physical capacity	Not FDA-approved
Less sexual dysfunction	Significant gynecomastia and mastodynia
No loss of BMD	4% to 10% withdrawal due to adverse events
Equivalent outcomes in D0 disease	Worse outcomes in metastatic D2 disease
Fewer hot flashes	Visual and respiratory disturbance and alcohol intolerance (nilutamide)

The British Medical Research Council conducted a randomized trial of early vs. late ADT in more than 900 patients with M0/M1 locally advanced or asymptomatic metastatic disease [74]. Early reports showed a survival benefit for M0 patients in the immediate-treatment arm (71% vs. 62%; p = 0.001), but the benefit disappeared after follow-up report [75]. Immediate treatment was associated with disease-specific survival and decreased incidence of ureteric obstruction, spinal cord compression, and transurethral resection of the prostate, but also with an increased number of deaths from nonprostate cancers. Immediate ADT was not associated with an increase in cardiovascular deaths. The reduction in overall survival on long-term follow-up may be a result of prolonged hormonal treatment contributing to greater nonprostate cancer-related mortality. The study did not report on patient quality of life or bone pain, either of which could influence the decision to begin ADT early or late in the course of disease.

ADT should be used judiciously, taking care not to overtreat patients. Disease progression after biochemical recurrence can be long and variable, and the 8-year median time to metastasis may be even longer in patients treated with ADT. Therefore, any potential benefits of early ADT should be weighed against the adverse effects of long-term ADT. Such caution is especially important for patients with rising PSA but no radiographic evidence of metastatic disease, for whom the potential harm from long-term ADT may be substantially greater. A large systematic review of 4 trials involving 2,167 patients was conducted by Cochrane Database. The review showed an overall survival favoring the early-treatment group that was significant only at 10 years, when few patients had survived. Although these randomized controlled trials varied in study design, stage of cancer, subjects enrolled, and specific treatment utilized, the available information suggests that early androgen suppression for treatment of advanced prostate cancer reduces disease progression and its accompanying complications, and may provide a small but statistically significant improvement in overall survival at 10 years (odds ratio (OR) = 1.5; 95% CI: 1.04 to 2.16) [76].

Peripheral Androgen Blockade

Peripheral hormonal therapy for biochemical recurrence of prostate cancer has been tested in small studies. In a recent study, 98 evaluable patients with rising PSA after definitive local therapy were put on an oral regimen of finasteride 5 mg/day and flutamide 250 mg 3 times/day [77]. PSA was reduced by > 80% in 91 of 94 (97%) patients, with a median time of 1 month to a PSA level of < 1 ng/mL; 76 patients achieved a median time of 3.2 months to

PSA nadir of < 0.2 ng/mL. After 59 months of follow-up, 48 patients were still undergoing treatment. Of these patients, 15 had documented disease progression that was responding to further hormonal manipulation. Toxicity was reportedly mild, with few reported grade 3 or 4 adverse events. Peripheral androgen blockade using finasteride and flutamide has fewer side effects than conventional ADT, especially in terms of sexual potency and libido. However, the long-term efficacy of this approach is currently unknown.

Role of Continuous ADT in Advanced Disease

Huggins and Hodges established prostate cancer's dependence on hormones over 50 years ago [1, 78]. In the setting of metastatic disease, ADT produces clinical response rates of 70% to 80% [79]. Early randomized controlled trials comparing orchiectomy to DES have demonstrated equivalent efficacy [7] and overall survival [80]. A systematic review and meta-analysis of 10 trials involving 1,908 patients reported no significant difference in overall survival between ADT and orchiectomy. The hazard ratio showed GnRH-A to be essentially equivalent to orchiectomy (HR = 1.1262; 95% CI: 0.915 to 1.386) [22]. The 3 GnRH-A used in these studies were associated with similar overall survival, although CIs were wider for leuprolide (HR = 1.0994; 95% CI: 0.207 to 5.835) and buserelin (HR = 1.1315; 95% CI: 0.533 to 2.404) than for goserelin (HR = 1.1172; 95% CI: 0.898 to 1.390). Bilateral orchiectomy or chemical castration with GnRH are thus the recommended initial treatments for patients with metastatic prostate cancer. Different approaches to delivering ADT have been studied in the setting of advanced prostate cancer and will be reviewed here.

Combined Androgen Blockade

ADT remains the primary systemic treatment for advanced prostate cancer. Testosterone produced by the testes supplies around 90% of the body's androgens, with the rest produced by the adrenal glands. Labrie et al. introduced complete blockade of the effects of both testicular and adrenal androgens through the use of nonsteroidal antiandrogens (AAs), a practice now known as combined or complete androgen blockade (CAB) or maximal androgen blockade (MAB) [81]. The rationale is that low concentrations of androgens of adrenal origin can remain in the body after androgen suppression and exert a stimulatory effect on hormone-sensitive prostate cancer cells, and that any such residual effect can be reduced by the use of AAs.

There are 3 FDA-approved AAs in common use in the United States (table 1). These drugs competitively antagonize the binding of testosterone and DHT to the AR and thus block the androgens' growth signal. Since the early 1980s, numerous randomized trials have studied the efficacy of CAB [82]. Though the approach was initially promising and some trials suggested that CAB conferred a significant survival advantage [83-86], others showed no survival advantage [87, 88] (see table 5). Specifically, the largest prostate cancer trial (SWOG or INT-0105 trial), which compared 687 patients treated with orchiectomy plus placebo with 700 patients treated with orchiectomy plus flutamide, found no difference in overall survival ($p = 0.14$) [87]. CAB had no advantage over orchiectomy alone in terms of time to progression, even in patients with minimal disease. The only notable differences in the arm using CAB with flutamide were higher rates of diarrhea and anemia, and a higher rate of discontinuation from therapy because of this toxicity (5% vs. 1%).

The Prostate Cancer Trialists' Collaborative Group reviewed 27 trials of CAB vs. androgen suppression involving 8,275 men. The meta-analysis found that the addition of an AA to androgen suppression improved overall 5-year survival by about 2% to 3%, depending on whether the analysis included CPA, which conferred a slightly unfavorable outcome compared to nilutamide and flutamide [88]. A subsequent meta-analysis of 21 randomized trials with a total of 6,781 patients comparing ADT and CAB found no statistically significant difference in overall survival at 2 years (HR = 0.970; 95% CI: 0.866 to 1.087) [89]. However, based on more limited data, the meta-analysis found a statistically significant increase in overall survival at 5 years (total of only 10 trials, HR = 0.871; 95% CI: 0.805 to 0.942) for CAB minus CPA. This modest increase in overall survival should be weighed against the increased risk of adverse events that could affect the patient's overall quality of life. CAB with GnRH-A plus AA confers the highest estimated quality-adjusted survival, but at an incremental cost of over USD 1 million per quality-adjusted life-year over orchiectomy alone [8].

Antiandrogen Monotherapy

Nonsteroidal AAs are often used in conjunction with castration as a component of CAB. However, monotherapy with a nonsteroidal AA has also been explored as an alternative therapeutic approach. Nonsteroidal AAs act through competitive inhibition of androgen-binding at the receptor level, inhibiting the action of androgens from both the adrenal glands and the testes. Because of their mode of action, decreases in testosterone concentration are not seen after monotherapy with these agents. Instead, testosterone levels generally increase slightly, offering potential quality-of-life advantages over conventional ADT [90]. Flutamide has been used in AA monotherapy for the longest period of time. However, only 2 studies have reported survival data: one comparing flutamide monotherapy to orchiectomy (319 patients) and another comparing flutamide to CAB (104 patients). Neither study reported an overall survival or clinical progression-free advantage [91, 92].

Bicalutamide monotherapy has been most widely evaluated. Bicalutamide 150 mg/day was compared with flutamide 250 mg 3 times/day, in combination with an LHRH, in a study involving 220 patients. No statistically significant differences in overall or progression-free survival were noted between the 2 groups, although bicalutamide treatment fared better in terms of quality-of-life assessments, with fewer instances of diarrhea or withdrawal from study due to adverse events [93]. Lower doses of bicalutamide (50 mg/day) showed inferior overall survival compared to castration [94]. In PSA dose-response studies, bicalutamide at once-daily 150 mg was found to be an appropriate, well-tolerated dosage, prompting subsequent evaluation of bicalutamide in larger comparative trials. The clinical efficacy of bicalutamide 150 mg/day monotherapy for treatment of 480 patients with locally advanced M0 disease and 805 patients with M1 disease was studied in comparison with medical or surgical castration in 2 randomized phase III trials [95, 96]. Survival data from patients randomized to castration vs. bicalutamide after a follow-up of 1.9 years showed that bicalutamide was not as effective as castration. Further analysis of mature data at 6.3 years in patients with locally advanced nonmetastatic disease demonstrated no statistically significant difference in overall survival or time to progression [97]. An overall survival advantage of 6 weeks (42 days) was seen in favor of castration in patients with M1 disease (HR = 1.3; 95% CI: 1.04 to 1.65; p = 0.02). This advantage was observed only in patients with PSA levels of ≥ 400 ng/mL; for patients with M1 disease with PSA ≤ 400 ng/mL at baseline, bicalutamide 150 mg seemed just as effective as castration in terms of overall survival, which may have been due to differences in tumor burden in these groups of patients. Since AAs work by competitively inhibiting the AR binding site, it is conceivable that the total number of ARs present and the penetration of AA into the tumor, which vary according to tumor burden, dictate the clinical efficacy of AAs [98].

These trials further demonstrated that patients receiving bicalutamide had measurably greater physical capacity and sexual interest. This assessment, which covered most aspects of daily life, indicated that patients receiving bicalutamide monotherapy maintained a more normal lifestyle than those who underwent castration (table 6). In these trials, the most commonly reported adverse events with bicalutamide treatment were breast pain (40.1%) and gynecomastia (49.4%) (these figures were higher in the more recent Early Prostate Cancer (EPC) program involving bicalutamide [97, 99]), but withdrawal from therapy because of these adverse events was low. Prophylactic radiation at 12 to 15 Gy x 1 has been used to prevent this adverse effect [100]. Once gynecomastia has developed, estrogen antagonists and aromatase inhibitors may be effective complements to treatment, since gynecomastia is caused by an increase in the ratio of estrogen to androgen [101].

A major collaborative clinical trials program was set up to investigate the efficacy and tolerability of bicalutamide 150 mg combined with standard care (RP, RT, or watchful waiting) in patients with localized or locally advanced disease [102]. The bicalutamide EPC program consisted of 3 randomized, double-blind, placebo-controlled trials of similar design conducted in distinct geographical areas, including North America (Trial 23), Europe, South Africa, Mexico, Israel, and Australia (Trial 24), and Scandinavia (Trial 25). In the first analysis of the EPC program, after a median follow-up of 3 years, patients in the bicalutamide arm had 42% less objective disease progression compared with standard care alone (p < 0.001). Patients treated with watchful waiting had more disease-progression events than patients in the other groups. At the second analysis, after 5.4 years of follow-up, there were no overall survival differences, but improved progression-free survival with bicalutamide was seen in Trials 24 and 25 [103]. Of note were the differences in entry and selection criteria among the 3 trials. Trial 23 (n = 3,292) excluded patients with lymph node metastasis and had no watchful-waiting arm, while Trial 24 (n = 3,603) and Trial 25 (n = 1,218) included node-positive patients and those managed with watchful waiting. At a median follow-up of 7.7 years for Trial 23, no differences in the primary endpoints of objective progression-free survival and overall survival were seen between patients treated with bicalutamide and those treated with standard care alone, although once-daily bicalutamide 150 mg significantly improved time to PSA progression (HR = 0.80; 95% CI: 0.72 to 0.90; p < 0.001) [104]. In a combined analysis of all 3 trials at 7.4 years follow-up, adding bicalutamide to standard care for patients with localized disease offered no benefit in terms of progression-free survival and, in fact, trended toward decreased survival compared to patients undergoing watchful waiting alone (HR = 1.16; 95% CI: 0.99 to 1.37; p = 0.07) [99]. This was in contrast to patients with locally advanced disease, for whom bicalutamide significantly improved progression-free survival irrespective of standard care. Bicalutamide also improved overall survival in patients receiving RT (HR: 0.65; 95% CI: 0.66 to 1.01; p = 0.03), and appeared favorable to watchful waiting (HR = 0.81; 95% CI: 0.44 to 0.95; p = 0.06). The EPC program concluded that adjuvant bicalutamide provided no benefit to patients with localized prostate cancer, but did improve progression-free survival in patients with locally advanced disease, and appeared to increase overall survival in a selected group of patients. Another large trial conducted by the EORTC closed to accrual in February 2003. It is a randomized phase III step-up study on initial AA monotherapy compared to watchful waiting in asymptomatic T1 to T3 (any Gleason) N0 or NxM0 prostate cancer patients without local treatment with curative intent. Trial results have not been reported [105].

Based on current data, AA monotherapy, specifically bicalutamide 150 mg once daily, is an attractive alternative to castration for patients with locally advanced prostate cancer, especially those who wish to retain potency or those for whom quality-of-life issues are a major concern (table 6). Ongoing studies are needed to determine the exact role of AA monotherapy in the treatment of localized prostate cancer. Currently, the American Society of Clinical Oncology states that monotherapy with a nonsteroidal AA may be discussed with patients as an alternative to ADT, but steroidal AAs, which are currently not approved in the United States, should not be offered as monotherapy [8].

Antiandrogen Withdrawal

Another method of hormone manipulation in prostate cancer patients involves withdrawal of the use of AAs. This paradoxical response to flutamide withdrawal was recognized only recently, despite over 2 decades of use in patients with progressive prostate cancer [106]. Subsequent studies of this withdrawal syndrome found that it was not limited to flutamide, but was also observed with bicalutamide, another AA, and reported with nilutamide, ketoconazole, megestrol acetate, and CPA, suggesting that the phenomenon is more generalized than originally believed. The biochemical response can be achieved in approximately 15% to 20% of patients with CRPC on CAB upon AA therapy withdrawal [107]. In addition to PSA declines, tumor regression and resolution of tumor-related symptoms have also been documented, along with improvements in abnormal hematological parameters [108, 109]. It is believed that the mechanism of this withdrawal phenomenon is related to mutations in the hormone-binding domain on the AR. In addition, there may be a different binding site for each AA, turning the AA from an AR antagonist (AR corepressor) to an AR agonist (AR coactivator). In vitro studies have shown that the AR in prostate cancer cell line LNCaP contains a specific point mutation [110] that results in a paradoxical stimulation of growth following incubation of hydroxyflutamide, nilutamide, CPA, and progestins [111, 112].

Beyond the obvious advantages of lower toxicity and greater cost-effectiveness, AA withdrawal has important therapeutic implications for patients who are progressing on hormonal therapies despite castrate levels of testosterone [113], especially if they are enrolled in clinical trials. AA withdrawal should be considered before alternative strategies since some improvements, such as PSA levels, may be attributed to the study agent when, in fact, they are due to AA withdrawal. It is unknown at this point whether tumor-cell apoptosis in response to AA withdrawal is due to an in vivo step selection because of continual exposure to an AA, or whether it is a natural action of prostate cancer cells as they undergo androgen independence. Regardless of the mechanism of action, AA withdrawal may further delay progression of prostate cancer, allowing for consideration of alternative therapies.

Intermittent ADT

Intermittent ADT (or intermittent androgen suppression) is an alternative approach to treating hormone-sensitive prostate cancer. It involves administering hormonal agents for 2 to 3 months beyond "best response," then discontinuing hormonal therapy and restarting it at a predetermined time. This strategy evolved in response to concerns that early ADT could lead to androgen independence [114]. Studies using animal models have suggested that successive rounds of androgen withdrawal and replacement delay progression to androgen independence [115]. The underlying hypothesis is this: if prostate cancer cells that survive androgen deprivation are forced by androgen replacement to continue normal differentiation, then treatment-induced apoptosis may occur, androgen dependence may be prolonged, and progression to androgen independence may be delayed [116]. Other concerns about prolonged or continuous ADT include the side effects of prolonged hormonal ablation, as well as the cost of treatment as many more patients are being treated earlier in the course of disease [117].

Role of Intermittent ADT in Biochemically Recurrent Prostate Cancer

Intermittent ADT in the setting of biochemical recurrence has uncertain long-term efficacy. A 31-month interim follow-up of the European EC 507 phase III trial of intermittent vs. continuous ADT (RELAPSE study) for men with PSA relapse after RP showed no difference in progression-free survival between the 96 patients who received intermittent ADT and the 78 patients who received continuous ADT (1,234 days vs. 1,010 days; $p = 0.9852$). Progression was reported in 8.3% of those receiving intermittent ADT vs. 6.4% of those receiving continuous ADT (p = not significant) [118]. Although these results suggest equivalent efficacy between continuous and intermittent ADT in patients with biochemical recurrence, definite conclusions cannot be made because of early follow-up. One ongoing phase III trial sponsored by the National Cancer Institute of Canada is the PR.7 trial (available through Southwest Oncology Group (SWOG), Medical Research Council, and Cancer Trials Support Unit), which has the primary objective of comparing overall survival for intermittent vs. continuous ADT in men with biochemical progression after RT.

Understanding the kinetics of testosterone after GnRH-A therapy is an important tool in planning subsequent treatment and timing of intermittent ADT. In a randomized prospective phase III analysis of 80 men with PSA-only biochemically recurrent disease, treatment consisted of 6 months of either leuprolide 22.5 mg or goserelin 10.8 mg, after which patients received either thalidomide or placebo [119]. When PSA rose to $\geq$ 5 ng/mL or baseline without evidence of metastatic disease, patients received another 6-month course of GnRH-A and crossed over to the opposite oral therapy. Testosterone and DHT levels were obtained every 3 months on GnRH-A therapy and monthly on oral study drug; analyses of these levels were secondary endpoints of this trial. The median time to an increase of testosterone $\geq$ 50 ng/dL and DHT $\geq$ 212 ng/dL was 12.9 weeks and 16.6 weeks, respectively. Levels remained elevated and did not return to normal until a median of 14.9 weeks and 16.6 weeks, respectively. Understanding the kinetics of androgen normalization following GnRH-A therapy may help guide treatment decisions, especially where discontinuation of GnRH-A therapy is employed, as in the case of intermittent ADT.

The use of a predetermined PSA trigger point or threshold level as a basis to initiate therapy has also been studied in a prospective, noncomparative, multicenter phase II trial of 103 men with localized prostate cancer who had biochemical recurrence after external-beam radiation. Study subjects underwent cyclic ADT after 36 weeks of GnRH-A treatment [120]. If serum PSA was not < 4.0 ng/mL at 24 weeks and 32 weeks of treatment, patients were taken off study. Serum PSA and testosterone levels were monitored thereafter and ADT was resumed when PSA rose to ≥ 4.0 ng/mL. After a median follow-up of 4.2 years, the time off treatment averaged 53% of the total cycle time, but progressively decreased with each succeeding cycle, ranging from 63.7 weeks in cycle 1 to 25.6 weeks in cycle 5. The good response rates seen in this study again demonstrated the feasibility of cyclic ADT. Although the proportion of the total length of each cycle spent in the off-treatment phase became progressively shorter with time, the off-treatment interval remained appreciable, ranging from 65% in cycle 1 to 46% in cycle 5.

Role of Intermittent ADT in Locally Advanced or Metastatic Prostate Cancer

Data from a number of phase II trials have provided the basis for the current use of intermittent ADT (table 7). One of the first trials involved 12 patients with localized disease and 8 with metastatic disease who received DES and flutamide for a median of 10 months prior to halting treatment [121]. Disease progression was noted after a median of 8 months. In addition, 9 out of 10 patients who were rendered impotent by ADT resumed sexual activity within 3 months after going off treatment, and all patients who relapsed had a rapid response upon reinitiation of treatment. The use of PSA level to define an optimal trigger point or threshold was first described in a trial where 47 patients were treated with combined ADT for 24 to 32 weeks, when a PSA nadir in the normal range (≤ 4.0 ng/mL) was obtained [122]. Patients went off treatment until PSA rose to a mean value between 10 ng/mL and 20 ng/mL. The first 2 cycles were of 73 weeks and 75 weeks duration, with mean percentages of time off therapy of 41% and 45%, respectively. The mean time to achieve a PSA nadir was 20 weeks in cycle 1 and 18 weeks in cycle 2. Serum testosterone returned to normal pretreatment levels at a median of 8 weeks after stopping treatment. This normalization of serum androgens has been confirmed in a prospective analysis of a phase III trial utilizing ADT, where the median time to a testosterone level of ≥ 212 ng/dL was 16.6 weeks [119].

Several factors have been shown to predict the time off ADT in patients undergoing intermittent ADT [123]. Patients who had an undetectable PSA (< 0.05 ng/mL) within one year had a median off phase of 29 months vs. 8.5 months for those whose PSA did not become undetectable within one year ($p = 0.00002$). For patients who had biochemical recurrence without radiographic evidence of disease, median duration of the off phase was twice as long as it was for patients who had no prior definitive therapy (24 vs. 12 months; $p = 0.028$), which may be related to a higher disease burden in the former group. Finally, those patients whose testosterone recovered within 4 months after discontinuing ADT had a shorter median time in the off phase than those whose PSA did not recover within 4 months (10 vs. 25 months; p = 0.0022).

To address the effects of intermittent vs. continuous institution of CAB in patients with locally advanced or metastatic prostate cancer, a large phase III trial conducted on behalf of the South European Urooncological Group demonstrated no association between intermittent CAB and higher rates of subjective or objective disease progression (p = 0.32), and no evidence that intermittent ADT leads to a significantly elevated risk of death (p = 0.79) compared to continuous ADT, after a median follow-up of 4.3 years [124]. In this study, 766 treatment-naïve patients with advanced-stage or metastatic prostate cancer were treated for 3 months with hormonal therapy consisting of 2 weeks of CPA at 200 mg/day, followed by monthly depot injections of decaptyl plus CPA. Patients who achieved a PSA of < 4 ng/mL, or reduced their PSA level by 80%, were further randomized to continuous (n = 312) or intermittent (n = 314) CAB. In the intermittent arm, patients underwent treatment only when their PSA level was ≥ 10 ng/mL with symptoms, or ≥ 20 ng/mL without symptoms, or when they reached ≥ 20% of the nadir value. About 50% of patients were off treatment for a year after the first cycle of hormonal therapy; of those who had a nadir PSA < 2 ng/mL, the time off treatment was 74 weeks. After these patients returned to therapy, they had a median of 14 weeks of treatment, followed by a second period off treatment for a median of 70 weeks. The results of this study suggest the feasibility and equivalent outcome of intermittent CAB compared to continuous CAB in men with locally advanced or metastatic disease.

Table 7. Published reports of use of intermittent ADT with at least 50 patients

Author	Year	No. of patients	Treatment	Stage of disease (L/BF/M)	Median time on therapy (mos)	Median time off therapy (mos)
Bruchovsky [125]	1998	110	GnRH + AA	0/110/0	9	9
Bracarda [276]	1998	55	GnRH + AA	44/0/11	7	6
Theyer [277]	1998	52	GnRH + AA	15/0/37	6	16
Gleave [278]	1998	70	GnRH + AA	41/0/29	9	9
Crook [279]	1999	54	GnRH + AA	11/4/39	8	9
Strum [123]	2000	52	GnRH + AA	24/19/9	19	24
Grossfeld [280]	2001	61	GnRH + AA	34/27/0	9	6
Prapotnich [281]	2003	233	GnRH + AA	NR ^	6 *	13 *
Tangen [282]	2003	663	GnRH + AA	0/0/663	NR	NR

L: localized; BF: biochemical failure without radiographic evidence of disease; M: metastatic; AA: antiandrogen; NR: not reported.

^ All patients had either biochemical failure, PSA > 30 (without local therapy), or metastatic disease.

* Median for first cycle. Treatment was stopped when PSA was < 4 ng/mL and restarted when PSA was > 20 ng/mL.

** Adapted from Gulley JL, Dahut WL. Novel schedules of hormone therapy for prostate cancer. In: Devita V, Hellman S, Rosenberg S, eds. *Progress in Oncology*. Boston: Jones and Bartlett; 2004:236-253.

While there are definite benefits to the use of intermittent ADT, especially improved quality of life [125, 126], the overall survival benefit of intermittent ADT over CAB cannot be fully ascertained [127]. This is particularly true for patients with advanced prostate cancer who exhibit poor prognostic features and should not be considered for intermittent therapy [128]. An ongoing phase III trial that will attempt to address this issue is SWOG 9436, which is currently available via various cooperative groups [129]. Recent data published from SWOG 9436 and INT-0162 showed that D2 prostate cancer patients with a PSA nadir between 0.2 ng/mL and 4 ng/mL had one third less risk of death than those with a PSA nadir ≥ 4 ng/mL, and those with PSA nadir ≤ 0.2 ng/mL had less than one fifth the risk of death of patients with a PSA nadir of ≥ 4 ng/mL ($p < 0.001$). This translated to a median survival of 75 months in the ≤ 0.2 ng/mL PSA group compared to 44 months for the group with PSA between 0.2 ng/mL and 4 ng/mL, and only 13 months for the PSA > 4 ng/mL group [130]. All 1,395 eligible patients received 7 months of induction ADT using a GnRH-A and bicalutamide, of whom 81% (1,134) achieved a PSA of ≤ 4 ng/mL on months 6 or 7 and were randomly assigned to continuous vs. intermittent ADT on month 8. Four statistically significant ($p < 0.05$) risk factors predicting postinduction survival are performance status, Gleason score, bone pain, and PSA upon study entry. Although this study concluded that achieving a PSA of ≤ 4 ng/mL after 7 months of ADT is a strong predictor of survival, the primary objective of SWOG 9436 to compare overall survival between patients undergoing intermittent ADT vs. continuous ADT has not been reported.

Individual Dosing Based on Testosterone Levels

Taking an individualized approach to androgen suppression is an evolving strategy that appears to be safe and cost-effective [131]. For many men, testosterone suppression continues well beyond the scheduled readministration of GnRH-A [132], especially for those with a long history of GnRH use [133]. Thus, delaying readministration of a GnRH-A until testosterone suppression has abated to noncastrate levels might lead to significant cost savings, with no negative effect on outcomes. The feasibility of this approach was demonstrated by a small phase I study composed of 32 patients with stage T3N± M± or greater treated with a 3-month depot of leuprolide injection [134]. Serum testosterone and PSA levels were obtained every month beginning 3 months after the last leuprolide injection. The median duration of castrate-level testosterone was 6 months. Another study prospectively enrolled 22 patients in a control-crossover-designed trial of dosing depot GnRH-A based on serum testosterone [132]. The median duration of castrate testosterone was 5.5 months (range 3.5 to 10), but the primary outcome of the study was to examine the health-related quality of life of patients with prostate cancer undergoing this treatment strategy. Short-term overall health-related quality of life and patient satisfaction were significantly improved over baseline measurements, and the yearly cost of care was significantly decreased using the method of redosing based on serum testosterone (USD 3,567.90 vs. USD 7,135.80). A similar finding of castrate levels of testosterone lasting for 6 months after the last GnRH-A dosing was seen in a study involving 80 patients, all of whom were treated with 6 months of GnRH-A therapy, with testosterone levels obtained monthly [119]. These early results suggest that using serum testosterone levels as a guide in redosing GnRH-A may be a novel, cost-effective approach to administering intermittent ADT.

Rapid Hormonal Testosterone Cycling

Another evolving strategy being explored in patients with locally advanced or metastatic prostate cancer combines hormonal cycling and chemotherapy, the rationale being that cycling, rather than prolonged androgen deprivation, will induce multiple cycles of apoptosis [135]. In a trial consisting of 23 patients (18 with metastatic disease and 5 with biochemical recurrence), treatment involved a 3-month induction of GnRH-A monthly plus bicalutamide for a month, followed by monthly cycles of androgen repletion with topical testosterone for 7 days and a GnRH-A monthly [136]. No disease progression was observed, and 6 of the 15 evaluable patients achieved the study's primary endpoint of successively declining PSA troughs after androgen deprivation. To further improve responses with rapid hormonal cycling, addition of cytotoxic chemotherapy using docetaxel was studied in a total of 63 patients, 25 with biochemical recurrence and 38 with metastatic prostate cancer [137]. Six treatment cycles were given, each lasting for 4 weeks, consisting of GnRH-A plus docetaxel 75 mg/m^2 given on day 1 of week 1, followed by testosterone gel given for 7 days beginning on day 1 of week 4. The primary endpoint of the study was to determine the number of patients who achieved the following PSA levels: $\leq$ 0.05 ng/mL in those who had undergone prostatectomy; $\leq$ 0.5 ng/mL in those who had undergone radiation therapy; and $\leq$ 2.0 ng/mL in those with metastatic prostate cancer without local therapy. Results showed that 4 of 11 evaluable patients (36%) in the biochemical-recurrence group and 9 of 26 evaluable patients (35%) in the metastatic-disease group achieved the primary endpoint. A notable toxicity was grade 3 and 4 neutropenia, which occurred in 82% of the biochemical-recurrence group and 88% of the metastatic-disease group. These studies show that rapid hormonal cycling plus docetaxel is feasible. Longer follow-up is necessary to compare this treatment approach to standard hormonal therapy in patients with biochemical recurrence after primary therapy for prostate cancer. A phase II study of rapid hormonal cycling with testosterone and leuprolide in combination with docetaxel in patients with recurrent or noncastrate metastatic prostate cancer has closed to accrual with results pending [138].

Other Agents

After first-line ADT has failed, usually in the form of orchiectomy, GnRH-A therapy, CAB, and AA withdrawal, alternative second-line ADT may be considered. Second-line ADT has not been shown to provide an overall survival advantage, but it can decrease symptomatic disease and cause radiographic responses and PSA declines.

Ketoconazole

Ketoconazole, an oral imidazole antifungal agent, suppresses testicular and adrenal steroidogenesis and is associated with significant PSA declines when used as a second-line hormonal agent [139-141]. In vitro data with androgen-independent cell lines PC-3 and DU-145 suggest that ketoconazole may have direct cytotoxic effects as well [142]. A decline in

PSA of at least 50% in 11 of 20 patients (55%) and a median duration of PSA response of 9 months was seen in one trial [143]. Castrate levels of testosterone may be seen after only 48 hours of therapy, making ketoconazole a treatment of choice in cases requiring urgent ADT, such as spinal cord compression or active disseminated intravascular coagulation [140, 144, 145]. In another study of 16 patients with D2 prostate cancer despite ADT in whom pain was a prominent clinical feature, 13 patients (81%) noted improvement in pain for periods of 1 to 8 months (median 3 months) [146]. Despite its advantages, ketoconazole can cause toxicity at high doses and requires glucocorticoid (e.g., hydrocortisone) supplementation to prevent adrenal insufficiency even at lower doses [141].

Aminoglutethimide

Aminoglutethimide, originally developed as an anticonvulsant, blocks adrenocorticoid synthesis by inhibiting the conversion of cholesterol to pregnenolone. One study showed a > 50% PSA decline in 37% of patients with CRPC treated with 1,000 mg/day aminoglutethimide and 40 mg/day hydrocortisone, with a median duration of response of 9 months and median survival of 23 months [147]. However, changes in androgen levels do not seem to correlate with tumor response. Aminoglutethimide has also been found to ameliorate bone pain and is used as a second-line hormonal treatment in patients with advanced disease [148, 149]. Like ketoconazole, it must be administered with glucocorticoid replacement. Although aminoglutethimide has largely been replaced by ketoconazole, it remains an active available agent and is a reasonable alternative for patients requiring a secondary hormonal approach. Aminoglutethimide is started at a dose of 250 mg 3 times/day for 3 weeks, then increased to 4 times/day [150].

Abiraterone Acetate

Abiraterone acetate is an oral 17-alpha hydroxylase/C17,20-lyase inhibitor developed as a steroidal inhibitor following observations that nonsteroidal 3-pyridyl esters had improved selectivity for inhibiting testosterone synthesis [151]. Several phase I studies evaluating the effects of abiraterone acetate have shown that enhanced testosterone suppression is achieved in castrate men [151, 152].

ADT Combined with Other Modalities

ADT and Chemotherapy

As a prostate cancer treatment, chemotherapy alone has been shown to palliate symptoms of disease, but until the recent introduction of taxanes it has shown no survival advantage [153]. The success of chemotherapy noted in early trials by the National Prostatic Cancer Project (NPCP) in the 1980s caused enthusiastic speculation on the potential benefits of combining chemotherapy with ADT. It was suggested that multiply heterogeneous prostate cancer cells may be susceptible to both chemotherapy and hormonal therapy agents [154]. The first prospective trial by the NPCP randomized 246 newly diagnosed metastatic prostate cancer patients to orchiectomy or DES alone, DES plus cyclophosphamide, or cyclophosphamide plus estramustine phosphate (EMP) [155]. Overall response rates were similar for all 3 treatment arms, although there appeared to be significantly poorer survival in patients with painful metastasis compared to those who were asymptomatic. This led to a subsequent NPCP study (Protocol 1300) that randomized 319 men to ADT alone, a combination of ADT and chemotherapy (cyclophosphamide plus 5-fluorouracil), or EMP alone [156]. There were no significant differences between the groups in progression-free or overall survival. In another trial, 96 patients with locally advanced or metastatic prostate cancer were randomized to hormonal therapy with a GnRH-A and flutamide, with or without 4 cycles of adjuvant mitoxantrone [157]. Men with localized prostate cancer receiving adjuvant chemotherapy had a higher initial objective response rate (95% vs. 53%; p = 0.008) and median survival (80 vs. 36 months; p = 0.04) than patients treated with hormonal therapy alone. No advantage was observed with adjuvant chemotherapy in men with metastatic disease. The use of adjuvant chemotherapy in conjunction with other hormonal agents such as prednisone has been studied.

EMP, a synthetic drug with both nitrogen mustard and estrogenic properties, has been used in the treatment of prostate cancer since 1981. As a single-agent treatment for metastatic disease, radiographic improvement was noted in only 5% to 19% of patients [158, 159]. EMP is believed to target the microtubule-associated proteins alpha-tubulin and beta-tubulin. Preclinical data suggesting additive antitumor effects of combining EMP with microtubule-targeted agents such as taxanes, etoposide, and vinca alkaloids spurred clinical trials. Neoadjuvant chemotherapy, with or without EMP, has been examined [160-164]. In the 2 trials involving patients who were to undergo radical prostatectomy, combination chemotherapy with EMP and etoposide was given to 18 patients, and KAVE (ketoconazole plus adriamycin alternating with vinblastine and EMP) plus CAB was given to 27 patients. None of the patients had a pathologically complete response, although undetectable PSA was noted in all patients [160, 164]. Docetaxel and EMP combination was studied in 21 patients prior to local therapy with either RT or RP [165]. EMP was given over 3 days and therapy was repeated every 3 weeks. Clinical response was noted in all patients as measured by DRE, imaging, and PSA responses, and other pathologic endpoints, such as a negative margin rate of 70%, were favorable. All patients also achieved suppressed testosterone levels despite the short duration of EMP treatment. These improvements in radiographic and biochemical response rates in patients with advanced disease led to further phase III trials [166-168]. In the phase III SWOG 9916 trial that compared docetaxel and EMP with mitoxantrone and prednisone in patients with advanced CRPC, docetaxel and EMP improved median survival by nearly 2 months (p = 0.01) [169]. Patients in the docetaxel and EMP treatment groups also had a significantly higher PSA response rate (50% vs. 27%; p < 0.0001). There was, however, no statistically significant difference in bone-pain palliation or quality of life between the 2 treatment arms [170].

The TAX 327 trial compared docetaxel weekly, docetaxel every 3 weeks, and mitoxantrone plus prednisone every 3 weeks in men with CRPC. Compared with mitoxantrone plus prednisone, docetaxel every 3 weeks improved overall survival (p = 0.009), and both docetaxel regimens improved patient quality of life [171]. Because it has a similar survival advantage and fewer side effects than docetaxel and EMP, the combination of docetaxel and prednisone has become the standard of care for patients with metastatic CRPC [153]. An alternative treatment sequence utilizing cytotoxic chemotherapy with docetaxel followed by ADT has also been studied. Severe combined immunodeficient mice bearing human LNCaP prostate tumors were treated with docetaxel and/or surgical castration, applied singly, concurrently, or in different sequences. Initial observations showed that docetaxel followed by castration had the most potent antitumor effect [172]. The beneficial therapeutic effect of this treatment sequence is thought to be due to cytotoxic drugs known to be active in CRPC targeting the androgen-sensitive components of prostate cancer cells, since a subpopulation of tumor cells that are resistant to apoptosis, as demonstrated by an altered bax-to-bcl2 ratio, is believed to emerge within weeks of androgen ablation [173]. The feasibility of this treatment sequence was examined in 32 men with biochemical recurrence and 7 with metastatic disease [174]. When treatment was initiated with docetaxel at 70 mg/m^2 every 3 weeks for up to 6 cycles, followed by CAB for 12 to 20 months, 48.5% of patients (17/35) had a serum PSA decrease of $\geq$ 50%, and 20% (7/35) had a serum PSA decrease of $\geq$ 75%.

Taken together, these studies provide rationale and feasibility for the use of chemotherapy as an adjunct to localized treatment for a select group of high-risk patients. Ongoing studies will further evaluate clinical endpoints in patients treated with ADT alone vs. ADT with chemotherapy. TAX 3501 will randomize patients to observation, ADT, or ADT plus docetaxel every 3 weeks. SWOG 9921 will evaluate overall survival following hormone therapy alone and hormone therapy plus adjuvant chemotherapy with 6 cycles of mitoxantrone plus prednisone in men who have had RP for high-risk prostate cancer.

ADT and Vaccines

While numerous ADT regimens have been described, no standard therapy options involving ADT have been shown to prolong survival for prostate cancer patients with biochemical progression after localized therapy. The generally indolent nature of prostate cancer compared with other solid tumors and the low tumor burden seen with biochemical progression have spurred significant interest in the development of vaccines for this disease state, and in the potentially significant benefits of combining ADT with vaccines. Several lines of research have demonstrated the immunosuppressive effects of androgens. Both testosterone and DHT have been shown to increase levels of TGF-beta, which can negatively modulate the immune response [175-181]. Murine studies showed that androgen deprivation increased levels of T cells in peripheral lymphoid organs, and that T cells from these mice proliferated more vigorously with antigen stimulation [182]. There is also evidence of improved immunologic function following ADT. In a murine tumor-prevention model, Hsueh et al. showed that androgen ablation with an oral AR antagonist augmented immune responses, with a modest improvement in antitumor efficacy to a whole-tumor cell vaccine (B16) [183]. A similar augmentation of immune response was observed in a transgenic mouse model that stably expressed prostate antigen. Failure of immunotherapy usually stems from the rapid development of tolerance by prostate-specific T cells. This study found that androgen ablation brought about a decrease in tolerance of prostate-specific T cells [184].

These preclinical data linking castration to more potent immune response mirror findings from studies done in humans. Research on autoimmune disease in humans also supports the immunoinhibitory effects of androgens [185]. ADT induces profuse T-cell infiltration of benign glands and tumors in human prostates, which is readily apparent after 7 to 28 days of therapy [186]. Recruitment and activation of antigen-presenting cells in treated prostate tissues may also contribute to local T-cell activation. Androgen ablation therefore induces priming of T cells to prostate antigens. Blazar et al. showed that ADT increased circulating levels of T ($p \leq 0.01$) and NK ($p \leq 0.05$) cells in men [187]. In this study, 6 of 10 patients showed a substantial increase in absolute T-cell receptor rearrangement excision circles

(TRECs), indicating an increase in thymic function. A larger study analyzed peripheral blood mononuclear cells from 83 patients who underwent either autologous or allogeneic bone marrow transplant for hematologic malignancies. Patients received either ADT or no ADT, along with high-dose myeloablative chemotherapy, followed by bone marrow transplant. While complete data from this trial are not yet available, interim results from a subset of 20 adult patients showed a significant increase in naïve CD4+ T cells and CD4+ TRECs in patients treated with ADT [188]. Results in these animal and human studies suggest that immunotherapy for prostate cancer may be most effective when administered following androgen ablation—thus forming the rationale for combining ADT and vaccines.

Several phase I trials have evaluated the safety and biological effects of vaccinia virus expressing human PSA in patients with prostate cancer [189-191]. In some patients, serum PSA stabilized for 6 to 11 months [190]. In a phase II trial designed to analyze the toxicity, immunogenicity, and time to treatment failure using vaccine, AA therapy, or their sequential use, 42 patients with CRPC and increasing PSA levels without radiographic evidence of metastatic disease (stage D0.5) were randomized (21 in each arm) to receive vaccine vs. AA therapy with nilutamide [192]. Nilutamide was chosen because clinical and biochemical responses have been reported with nilutamide after flutamide and bicalutamide failure [193, 194]. Median time to treatment failure for the vaccine arm was 9.9 months vs. 7.6 months for the nilutamide arm ($p = 0.28$). Eight patients in the combined AA/nilutamide arm who were given vaccine at the time of PSA progression had a median time to treatment failure of 5.2 months; 12 patients in the vaccine arm who had nilutamide added at the time of PSA progression had a median time to treatment failure of 13.9 months. Based on the potential synergy between ADT and vaccines, a randomized, double-blind, phase II trial of ONY-P1 vaccine vs. placebo in men with D0 prostate cancer following limited androgen ablation is underway that will determine if the vaccine can lengthen the time to PSA progression. It is hoped that these trials will determine whether there is a role for combination therapy in this patient population for whom there is currently no standard of care.

Adverse Effects of ADT

Skeletal Changes and Loss of Bone Mineral Density

As the number of prostate cancer patients being treated with ADT has increased, significant declines in bone mineral density (BMD) [195-197] and increased risk of fracture [198-200] have been reported (figure 2). A recent study of over 50,000 men from the Surveillance, Epidemiology, and End Results program (SEER) and Medicare databases compared the risk of fracture in men with prostate cancer who were treated with ADT vs. those not receiving ADT [197]. The study found that for men treated with ADT, the risk of fracture increased one year after diagnosis and rose with the number of doses of GnRH-A received. Several randomized trials of bisphosphonates for the treatment of ADT-induced osteoporosis have been published. A randomized trial of 47 men with nonmetastatic prostate cancer compared GnRH-A alone with GnRH-A plus 60 mg pamidronate given every 12 weeks. After 48 weeks, men in the GnRH-A-alone arm had a mean decrease in lumbar, trochanter, total hip, and trabecular lumbar spine BMD of 3.3% ($p < 0.001$), 2.1% (p =

0.003), 1.8% ($p = 0.005$), and 8.5% ($p = 0.02$), respectively. Mean BMD in the GnRH-A-plus-pamidronate group did not change significantly [201]. In a randomized trial of 106 men with nonmetastatic prostate cancer, men who were beginning ADT were randomly assigned to receive zoledronic acid or placebo every 3 months for a year [202]. Mean lumbar spine BMD decreased in the placebo group by 2.2%, but increased by 5.6% in the zoledronic-acid group ($p < 0.001$). In another study, 44 men with nonmetastatic prostate cancer who were receiving a GnRH-A were assigned randomly to receive zoledronic acid (4 mg IV × 1) or placebo [203]. At the end of 12 months, mean (± SE) BMD of the posteroanterior lumbar spine increased by 4.0% ± 0.9% in men treated with zoledronic acid and decreased by 3.1% ± 0.9% in men who received placebo ($p < 0.001$ for between-group comparison). BMD of the total hip decreased by 0.7% ± 0.6% in men treated with zoledronic acid and decreased by 1.9% ± 0.7% in men who received placebo ($p = 0.005$). Although these data suggest the feasibility of once-yearly dosing of bisphosphonates for patients undergoing ADT, it is too soon to recommend routine use of bisphosphonates as standard treatment for all men receiving ADT.

At present, no systematic guidelines have been published on diagnosing or managing osteoporosis in patients receiving ADT [204]. Furthermore, no trials of other interventions, including nonpharmacologic therapies, calcium, or vitamin D, have been performed in this setting. Based on available data, it is recommended that patients have a baseline evaluation of BMD before starting ADT and periodic follow-up assessments depending on baseline findings, and that patients already on ADT for prostate cancer undergo annual assessment of BMD [205]. Men should be counseled about calcium and vitamin D intake and behaviors that reduce the risk of bone loss. They should also be screened for risks such as family history of osteoporosis, low body weight, prior fractures, excessive alcohol use, smoking, glucocorticoid use, low vitamin D levels, and other medical comorbidities.

Hot Flashes

Hot flashes, a troublesome adverse effect, can impair quality of life for men on ADT. They occur in up to 80% of patients undergoing treatment with GnRH-A [206]. In some studies, megestrol acetate reduced hot flashes in 74% of men on ADT vs. 20% for placebo [207, 208]. Other trials have studied the effectiveness of antidepressants [209, 210], but to date there have been no large-scale placebo-controlled trials of any of these interventions in patients with prostate cancer [211].

Metabolic Syndrome

Recently, male hypogonadism has emerged as an independent risk factor in the development of metabolic syndrome. Men with low testosterone levels have a higher prevalence of metabolic syndrome, after controlling for other risk factors [212]. Serum testosterone levels have a negative correlation with fat mass and a positive correlation with muscle mass. Several prospective studies have compared body composition and metabolism in cohorts of men with prostate cancer before ADT and 6 to 12 months after ADT [213-215], with the following results ($n = 35$): increased body mass index of 1.6% to 2.4%; increased fat body mass of 10% to 20% in 7 (20%) men, 20% to 50% in 8 (22.8%) men, and more than 50% in 5 (14.3%) men; decreased lean body mass of 2% to 5% in 8 (22.8%) men and > 5% in 7 (20%) men [214], as well as increases in total cholesterol and triglycerides, HDL, LDL, and, in one study, fasting glucose. All of these studies were comparisons of the same patient population before and after ADT and therefore lacked a true control group.

A cross-sectional study divided 58 men into 1) an ADT group consisting of 20 patients with prostate cancer undergoing ADT for at least 12 months, 2) a non-ADT group of 18 age-matched men with nonmetastatic prostate cancer who had rising PSA after local treatment, and 3) a control group of 20 age-matched men. Patients on ADT had significantly higher body mass indexes and lower total and free testosterone levels.

The prevalence of metabolic syndrome was higher in the ADT group compared with the non-ADT ($p < 0.01$) and control ($p = 0.03$) groups [216]. Among the components of metabolic syndrome, men on ADT had a higher prevalence of abdominal obesity and hyperglycemia. Androgen-deprived men also had elevated triglycerides compared with controls ($p = 0.02$).

The prevalence of hypertension and low HDL levels was similar. In a large observational population-based cohort study of 73,196 Medicare enrollees 66 years old or older who had locoregional prostate cancer, GnRH-A use was associated with an increased risk of coronary heart disease (adjusted HR = 1.16; $p = 0.004$), myocardial infarction (adjusted HR = 1.11; $p = 0.03$), sudden cardiac death (adjusted HR = 1.34; $p < 0.001$), and diabetes (adjusted HR = 1.44; $p < 0.001$).

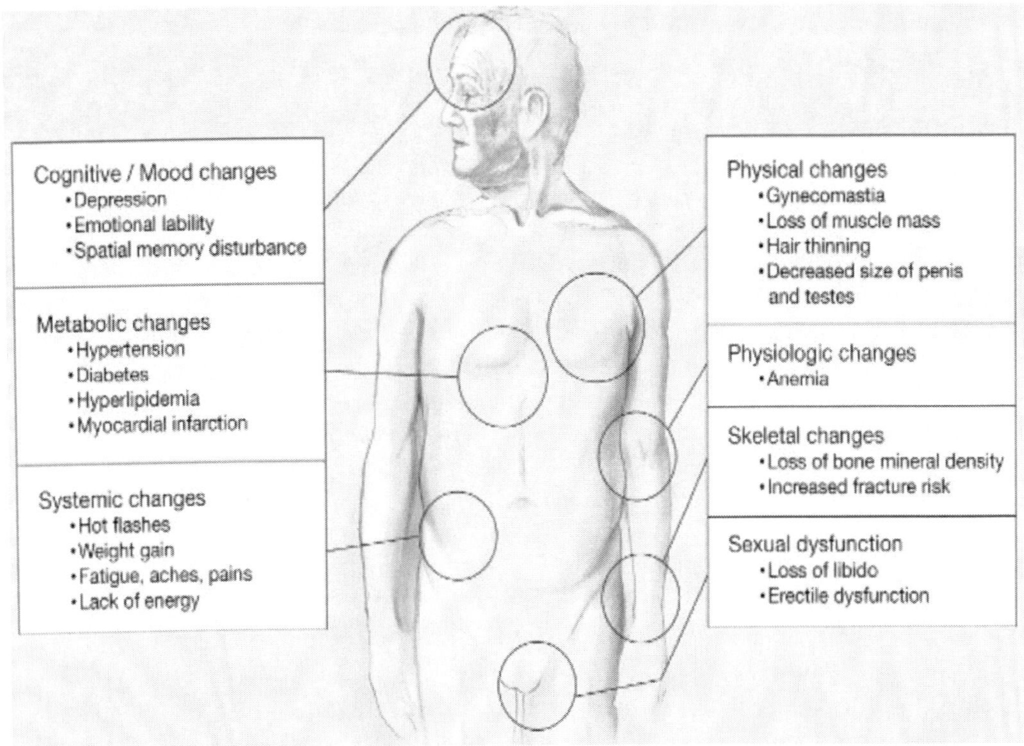

Figure 2. Adverse effects of ADT.

Men treated with orchiectomy alone were more likely to develop diabetes (adjusted HR = 1.34; $p < 0.001$) but not coronary heart disease, myocardial infarction, or sudden cardiac death (all $p > 0.20$) [217]. Taken together, these data suggest that metabolic syndrome may promote the development of cardiovascular risk factors. Given the risks and uncertain benefits of ADT for patients with biochemical recurrence, use of this therapy in this patient population should be carefully considered.

Sexual Dysfunction

Testosterone plays an important role in sexual function and the vast majority of men treated with ADT develop some form of sexual dysfunction [218]. Loss of libido usually develops first, followed by erectile dysfunction. An early cross-sectional study examined the effects of at least 12 months of ADT for prostate cancer on different parameters, including sexual function, lean body mass, muscle strength, BMD, and quality of life. On the Watts Sexual Function Questionnaire, patients receiving ADT had lower overall scores than either patients who had not undergone ADT or healthy age-matched controls ($p = 0.0001$), particularly in the areas of arousal, desire, and frequency of spontaneous early morning erections [219]. A study of 431 patients treated with GnRH-A and orchiectomy in the Prostate Cancer Outcomes Study of the SEER program found similar effects on sexual function [220]. Therapeutic efforts to prevent or ameliorate the sexual side effects of ADT have included intermittent ADT and AA monotherapy [95, 97, 98]. Other available options include vacuum devices, carnitine, and intracavernosal injections of prostaglandin [221, 222].

Emotional and Cognitive Side Effects

Men with prostate cancer often experience depression, especially if they have a history of depressive episodes. In men surviving at least 5 years after diagnosis, the incidence of at least one diagnosis of depression or cognitive dysfunction was significantly higher compared to controls (31.3% vs. 23.7%; $p < 0.001$) [223]. However, after adjusting for age, comorbidity, and tumor characteristics, the risks for patients on ADT were substantially reduced or eliminated. There are conflicting reports concerning the effects of ADT on cognitive function. Some data suggest a decline in attention and memory [224], as well as verbal fluency and visual memory [225], even in short-term ADT of 6 to 12 months [226]. Other data suggest a noted improvement in object recall among men undergoing ADT [227].

Systemic Effects

Up to 90% of patients receiving ADT experience normocytic normochromic anemia [228]. Although other etiologies should be ruled out, the anemia is typically temporary and related to androgen blockade. Treatment with erythropoietin is effective, though rarely necessary since the condition usually resolves when ADT is discontinued. Patients may experience fatigue that is independent of either anemia or psychological issues [218]. Gynecomastia resulting from ADT has been treated prophylactically with breast irradiation or with surgery and tamoxifen [100, 101]. Other systemic effects include dry eyes, thinning or loss of body hair, and vertigo [221].

Biology of Castrate-Resistant Prostate Cancer

While clinical studies of various forms of ADT have proven its clinical efficacy in several disease states and select prostate cancer patient populations, there is controversy regarding the use of such strategies in patients with biochemical failure. Current standard clinical practice dictates continued testosterone suppression even for CRPC, based on recent data that suggest that androgens and AR continue to play an important role in the biology of CRPC.

Patients with metastatic prostate cancer typically respond to ADT for an average of 14 to 20 months [83, 87], at which time CRPC develops. Once this occurs, survival ranges from 12 to 46 months [229, 230]. CRPC is sometimes referred to as androgen-independent prostate cancer, but "androgen-independent" is not a fully descriptive term since laboratory evidence suggests that castrate resistance is at least partially caused by sensitization to low (or castrate) levels of androgens. Other terms used in reference to castrate resistance, such as androgen-resistant, androgen-refractory, and hormone-refractory, are indistinct and vague, and are not appropriately descriptive of this disease state [231]. Androgen depletion-independent prostate cancer has been suggested as an alternative term [232] because it reflects the persistence or progression of prostate cancer in the presence of low, but not absent, levels of testosterone. However, castrate-resistant is the preferred term because it accurately describes the biology of the underlying mechanisms.

When ADT is instituted, a majority of prostate cancer cells undergo apoptosis, but some persist in the G1 phase of the cell cycle [233]. Clinical progression occurs with the proliferation of cells resistant to androgen ablation. In the last few years, compelling research has indicated that the biology and survival of CRPC is largely dependent on reactivation of AR, despite castrate levels of testosterone [234, 235]. Evidence suggesting different mechanisms of AR reactivation include the fact that CRPC manifests as progression that is accompanied by a rise in PSA that may still be sensitive to secondary treatments [235], and often responds to second-line hormonal therapy such as AR antagonists and adrenal ablation [236], indicating that CRPC is still dependent on AR. However, lack of response to secondary hormonal therapy does not necessarily rule out AR dependence. In a study of patients with CRPC treated with flutamide, men who previously had responded well to castration responded better to AR antagonists as secondary hormonal therapy. Men who had an incomplete response to castration did not respond as well to secondary hormonal therapy [237]. It may be that the men who responded well to ADT and secondary hormonal therapy had tumors whose biology was more reliant on AR.

Other evidence downstream of AR suggests that AR is reactivated and that the signal is intact. DNA microarray studies of human prostate cancer cell line xenografts that examined initial tumor growth, tumor response in the initial castrate setting, and castrate-resistant tumor growth showed that a subset of AR-responsive genes that are down-regulated upon castration are re-expressed in the castrate-resistant tumor [238]. DNA microarray studies from prostate cancer specimens collected from patients in similar states also suggest AR reactivation in castrate resistance. In fact, the gene expression patterns of castrate-resistant tumors were more similar to androgen-dependent tumors growing with normal levels of testosterone than to androgen-dependent tumors after initial exposure to castration [239].

Reactivation of AR function after castration occurs in part through increased expression of this gene. Gene amplification, or an increase in the genomic copy number of AR, occurs infrequently before ADT. However, AR gene amplification occurs in over 20% of CRPC [240, 241]. The mechanism of increased AR expression may not necessarily be limited to gene amplification. A DNA microarray study of mouse xenograft tumors comparing androgen-dependent and castrate-resistant tumors suggests that an increase in AR mRNA, by whatever mechanism, is common among CRPCs. Of all the genes analyzed in 7 matched sets of isogenic androgen-dependent and castrate-resistant tumors, AR was the only consistent transcript that increased in the transition to CRPC [242].

Several other mechanisms have been postulated for gain-of-function changes in AR signaling that lead to persistent activation despite castration. One mechanism involves changes in coregulatory molecules, such as overexpression of steroid receptor coactivators, which are proteins required for the full function of steroid receptors such as AR. This is perhaps best illustrated by mouse knockout models, such as that of steroid receptor coactivator-1 (SRC-1), in which organs responsive to sex steroids, such as the prostate, uterus, and mammary gland, failed to fully develop [243]. While a loss of steroid receptor coactivator expression may decrease AR function, increased expression may augment androgen signaling and therefore account for AR signaling in the castrate setting. While some findings suggest that this occurs, other studies on steroid receptor coactivator expression changes are not always consistent with this hypothesis [234].

Another mechanism of castrate resistance involves changes in the level of cytokines, growth factors, and membrane receptors in tumor tissue. Investigators have described a growing list of soluble factors that play a role in AR activation. Interleukin-6 (IL-6) is one of the most widely studied of these agents and has been shown to activate AR in a STAT3-dependent manner [244]. Furthermore, administration of an anti-IL-6 antibody inhibits the conversion of androgen-dependent to castrate-resistant phenotype in mice [245]. Insulin-like growth factor-1 has also been shown to activate AR and may be implicated in CRPC [246]. Her2/neu, a member of the epidermal growth factor receptor family, has a prominent role in the biology and treatment of breast cancer [247], but there is evidence that Her2/neu also plays a role in AR activation, in part by stabilizing AR protein and augmenting DNA binding activity [248]. The long list of cytokines and growth factors purported to activate AR is reviewed in detail elsewhere [249, 250]. It may be that these various growth factors all converge on a common signaling intermediary to activate AR, perhaps explaining the wide variety of growth factors on this list [251].

There is also evidence that AR activity may be up-regulated in the CRPC tumor microenvironment by an increase in the synthesis and availability of androgens. One study examined tissue concentrations of testosterone and DHT in benign prostate of men who had no hormonal treatment and compared it to androgen levels in locally recurrent CRPC [252]. Although DHT levels were over 90% lower in locally recurrent CRPC, testosterone levels were not significantly different between the 2 groups, suggesting that there may be local androgen production. This may be explained in part by an increase in genes, such as aldo-keto reductase family 1, member C3, that are responsible for converting adrenal androgens to testosterone in CRPC [253].

Data on the mechanisms involved in the development of castrate resistance suggest that various stimuli converge on AR as a nodal point, perhaps redundantly, and stimulate the activation of a subset of androgen-responsive genes. In addition, the majority of prostate cancers develop a translocation between the upstream, androgen-responsive TMPRSS2 gene and erythroblast transformation-specific (ETS) family oncogenes [254]. AR-mediated ETS activation may further hasten tumor growth upon reactivation of AR in the castrate-resistant state. These new data on this class of translocations further strengthen the argument for developing novel drugs to target AR in CRPC.

Potentially Therapeutic Approaches Targeting AR

There are several potential strategies for directing a novel therapy at AR in CRPC. The first would be to target the growth factors and receptors responsible for activating AR. However, there are several such growth factors and there is likely some redundancy to growth-factor activation of AR. Therefore, targeting only one growth factor may not be sufficient to prevent AR activation. It has been proposed that the intracellular signaling proteins at a point of convergence downstream of these growth factors would serve as a more attractive target [251]. Drugs fitting this description would mainly be kinase inhibitors.

At the level of AR itself, the interaction between AR and a required steroid receptor coactivator could, in principle, be inhibited. The observation that one patient with complete androgen insufficiency syndrome and complete loss of AR function has a wild-type AR and likely a mutation in a steroid receptor coactivator, suggests that at least one coactivator is absolutely required for AR function [255]. However, developing small-molecule inhibitors of protein-protein interactions is problematic. "Clefts" or "pockets" in which small molecules readily bind are typically not as obvious in the interface of 2 proteins as they are in catalytic domains of kinases, although these problems may not be insurmountable. Peptides that bind AR in steroid receptor coactivator binding domains have been developed, along with crystal structures of the interaction between these peptides and AR [256]. Such studies may pave the way to developing small molecules that antagonize coactivator function.

The interaction between the AR zinc-finger DNA binding domain and the androgen response element DNA motif to which it binds may also be interrupted. A compound that antagonizes the interaction between the estrogen receptor zinc-finger domain and the estrogen response element has been developed. This compound inhibits estrogen-induced transcription and cell proliferation. Furthermore, it reduces tumor growth in a mouse xenograft model of breast cancer [257]. In principle, such a strategy is also possible with AR. Since this strategy involves a drug that does not bind the classical steroid receptor ligand binding domain, it should also be effective in cells that have become resistant to classical antagonists.

Heat-shock proteins are required for normal protein folding, processing, and turnover [258]. Normal AR activity and processing are also dependent on heat-shock protein 90 (hsp90). Inhibitors of hsp90, such as geldanamycin and its analogs, disrupt AR activity and AR-mediated transcription [259]. However, inhibitors of hsp90 are not specific for AR and, like conventional cytotoxic chemotherapy, their development for the treatment of CRPC will depend on their therapeutic index [260]. Proteolysis targeting chimeric molecules (PROTACS), a strategy for selectively targeting AR, has been developed. This system uses an androgen ligand linked to a moiety that targets bound proteins to the ubiquitin-proteosome pathway, thereby specifically targeting and degrading AR [261]. Such strategies for developing specific drugs to target AR are promising for the development and treatment of CRPC.

Given the multiple levels of AR activation in CRPC, some have argued for a multifaceted approach to targeting AR for inhibition [262]. Strategies that combine newer targeted agents against AR could be undertaken as early as initiation of ADT to minimize the opportunity for castrate-resistant cells to arise. A similar strategy is being adopted, for instance, in the setting of chronic myelogenous leukemia, for which there are now multiple drugs, such as imatinib and dasatinib, to target a central, critical protein (BCR-ABL) for tumor survival [263].

Finally, there are new data on the existence of prostate cancer stem cells [264, 265]. Cancer stem cells are thought to represent a small subset of the total tumor-cell population, yet they have the potential for continuous replication. If prostate cancer arises from bona fide stem cells, then CRPC must originate from these stem cells as well. However, evidence suggests that neither normal prostate stem cells nor prostate cancer stem cells express AR. The castrate-resistance conundrum is that CRPC leads to gain-of-function changes in AR, yet it arises from AR-negative prostate cancer stem cells, which by definition can neither directly respond to ADT nor undergo selection for changes in AR function [266]. Resolution of the castrate-resistance conundrum will aid in understanding the development of CRPC and identifying new targets for therapy.

Conclusion

ADT is being employed more frequently for prostate cancer patients with symptomatic and asymptomatic metastatic disease, patients with biochemical recurrence, and about 14% of newly diagnosed patients [218]. ADT offers palliation of bone pain and, in some patient populations, a survival benefit. Unfortunately, nearly all men progress to castrate-resistant phenotype within 2 years, and overall survival after the start of ADT is 28 to 63 months [82, 229, 267]. For some disease states and patient populations, the issue of who should receive ADT, and when and how they should receive it, remains controversial. Data suggest that ADT is most strongly indicated for prostate cancer patients with metastatic disease, as an adjunct to RT for high-risk patients (Gleason score $\geq$ 8, stage $\geq$ T2c), for patients with microscopic lymph node disease after RP, and as primary therapy for patients who are not candidates for local therapy (figure 3). Of the available options for testosterone-lowering therapy, orchiectomy and GnRH-A have similar efficacy and safety. The choice to undergo one or the other should therefore be a matter of patient preference after factors such as comorbidities or side effects of each procedure have been thoroughly discussed. Whether CAB is better than ADT monotherapy is still unclear. Although several trials showed a small survival benefit (1% to 5%) within 5 to 10 years of analysis, the large United States intergroup study (INT-0105), which was negative, is considered by many to be the definitive trial comparing CAB to ADT monotherapy because of its size and prospective evaluation of quality-of-life benefits in men with minimal disease and good performance status. Given the considerable cost, toxicity, and minimal survival benefit, the consensus has not been in favor of routine use of CAB over ADT monotherapy.

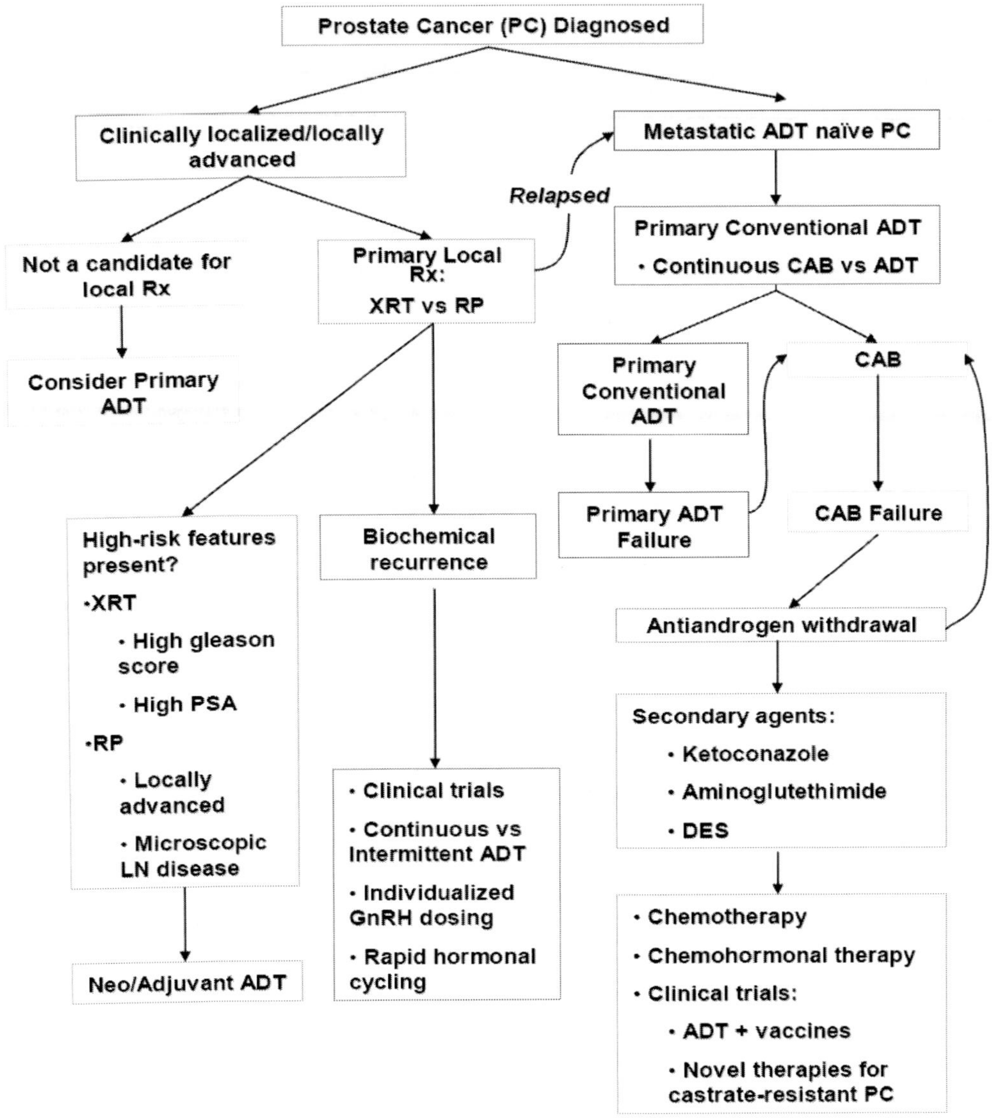

Figure 3. Algorithm for Clinical Application of ADT.

The timing of hormonal therapy is also a subject of ongoing debate. The benefits of early treatment—a trend towards survival benefit, reduced disease progression, and lower morbidity from the complications of urethral obstruction, pathologic fractures, spinal cord compression, ureteral obstruction, and extraskeletal metastases—must be weighed against the potential side effects of prolonged ADT. The side effects of continuous ADT treatment have not raised specific concerns for patients with metastatic disease since, on average, these men do not survive long enough for the side effects to become clinically significant. However, because improved diagnostic methods have resulted in earlier diagnosis of prostate cancer and earlier detection of biochemical recurrence, ADT is being instituted in men with earlier stages of disease, whose life expectancy could well be over 10 years. For these men, systemic side effects can be a substantial burden. This is especially true for patients presenting with biochemical recurrence, where no data from prospective trials are currently available to specifically address a possible overall survival benefit. However, given the benefits seen in patients with locally advanced or high-grade disease, and the risk of earlier metastasis in patients with aggressive high-risk features, there is a trend toward initiating ADT earlier in the course of disease. Alternative methods of delivering ADT have been studied, including the use of intermittent ADT as opposed to continuous ADT, and although its survival benefit and impact on disease control have not been fully ascertained, the strategy is gaining favor as a way to improve quality of life and lower health care costs for prostate cancer patients. When disease progresses after first-line ADT, secondary ADT should be considered. In addition, there is a rationale for combining ADT with other therapies, and studies are ongoing to determine the clinical benefit of this approach. Eventually, most patients with metastatic prostate cancer treated with ADT will develop CRPC. Recent insights into the biology of CRPC have led to the search for novel therapies targeting the AR pathway. This may eventually lead to another class of drugs that could lengthen the period when prostate cancer is responsive to hormonal therapy.

References

[1] Huggins C, Hodges C. Studies on prostatic cancer: I. The effect of estrogen and of androgen injection on serum phosphatases in metastatic carcinoma of the prostate. *Cancer Res.* 1941;1:293-7.

[2] Gelmann EP. Molecular biology of the androgen receptor. *J. Clin. Oncol.* 2002 Jul 1;20(13):3001-15.

[3] Tyagi RK, Lavrovsky Y, Ahn SC, Song CS, Chatterjee B, Roy AK. Dynamics of intracellular movement and nucleocytoplasmic recycling of the ligand-activated androgen receptor in living cells. *Mol. Endocrinol.* 2000 Aug;14(8):1162-74.

[4] DePrimo SE, Diehn M, Nelson JB, Reiter RE, Matese J, Fero M, et al. Transcriptional programs activated by exposure of human prostate cancer cells to androgen. *Genome Biol.* 2002 Jun 14;3(7):RESEARCH0032.

[5] Zhou ZX, Lane MV, Kemppainen JA, French FS, Wilson EM. Specificity of ligand-dependent androgen receptor stabilization: receptor domain interactions influence ligand dissociation and receptor stability. *Mol. Endocrinol.* 1995 Feb;9(2):208-18.

[6] Gottlieb B, Pinsky L, Beitel LK, Trifiro M. Androgen insensitivity. *Am. J. Med. Genet.* 1999 Dec 29;89(4):210-7.

[7] Treatment and survival of patients with cancer of the prostate. The Veterans Administration Co-operative Urological Research Group. *Surg. Gynecol. Obstet.* 1967 May;124(5):1011-7.

[8] Loblaw DA, Mendelson DS, Talcott JA, Virgo KS, Somerfield MR, Ben-Josef E, et al. American Society of Clinical Oncology recommendations for the initial hormonal management of androgen-sensitive metastatic, recurrent, or progressive prostate cancer. *J. Clin. Oncol.* 2004 Jul 15;22(14):2927-41.

[9] McLeod DG. Hormonal therapy: historical perspective to future directions. *Urology.* 2003 Feb;61(2 Suppl 1):3-7.

[10] Grayhack JT, Keeler TC, Kozlowski JM. Carcinoma of the prostate. Hormonal therapy. *Cancer.* 1987 Aug 1;60(3 Suppl):589-601.

[11] de Voogt HJ, Smith PH, Pavone-Macaluso M, de Pauw M, Suciu S. Cardiovascular side effects of diethylstilbestrol, cyproterone acetate, medroxyprogesterone acetate and estramustine phosphate used for the treatment of advanced prostatic cancer: results from European Organization for Research on Treatment of Cancer trials 30761 and 30762. *J. Urol.* 1986 Feb;135(2):303-7.

[12] Matsuo H, Baba Y, Nair RM, Arimura A, Schally AV. Structure of the porcine LH- and FSH-releasing hormone. I. The proposed amino acid sequence. *Biochem. Biophys. Res. Commun.* 1971 Jun 18;43(6):1334-9.

[13] Schally AV, Arimura A, Kastin AJ, Matsuo H, Baba Y, Redding TW, et al. Gonadotropin-releasing hormone: one polypeptide regulates secretion of luteinizing and follicle-stimulating hormones. *Science.* 1971 Sep 10;173(4001):1036-8.

[14] Schally AV. Luteinizing hormone-releasing hormone analogs: their impact on the control of tumorigenesis. *Peptides.* 1999;20(10):1247-62.

[15] Casper RF. Clinical uses of gonadotropin-releasing hormone analogues. *CMAJ.* 1991 Jan 15;144(2):153-8.

[16] Filicori M. Gonadotropin-releasing hormone agonists. A guide to use and selection. *Drugs.* 1994 Jul;48(1):41-58.

[17] Kraus S, Naor Z, Seger R. Gonadotropin-releasing hormone in apoptosis of prostate cancer cells. *Cancer Lett.* 2006 Mar 28;234(2):109-23.

[18] Kuhn JM, Billebaud T, Navratil H, Moulonguet A, Fiet J, Grise P, et al. Prevention of the transient adverse effects of a gonadotropin-releasing hormone analogue (buserelin) in metastatic prostatic carcinoma by administration of an antiandrogen (nilutamide). *N. Engl. J. Med.* 1989 Aug 17;321(7):413-8.

[19] Scardino PT. National Comprehensive Cancer Network (NCCN) Practice Guidelines version 2. 2005 [cited; Available from: http://www.nccn.org

[20] Mongiat-Artus P, Teillac P. Abarelix: the first gonadotrophin-releasing hormone antagonist for the treatment of prostate cancer. *Expert Opin. Pharmacother.* 2004 Oct;5(10):2171-9.

[21] Abarelix package insert. [cited 2006 September 2]; Available from: http://www.fda.gov/CDER/foi/label/2003/021320lbl.pdf

[22] Seidenfeld J, Samson DJ, Hasselblad V, Aronson N, Albertsen PC, Bennett CL, et al. Single-therapy androgen suppression in men with advanced prostate cancer: a systematic review and meta-analysis. *Ann. Intern. Med.* 2000 Apr 4;132(7):566-77.

[23] Cassileth BR, Soloway MS, Vogelzang NJ, Schellhammer PS, Seidmon EJ, Hait HI, et al. Patients' choice of treatment in stage D prostate cancer. *Urology.* 1989 May;33(5 Suppl):57-62.

[24] Thompson IM, Pauler DK, Goodman PJ, Tangen CM, Lucia MS, Parnes HL, et al. Prevalence of prostate cancer among men with a prostate-specific antigen level < or =4.0 ng per milliliter. *N. Engl. J. Med.* 2004 May 27;350(22):2239-46.

[25] Bubley GJ, Carducci M, Dahut W, Dawson N, Daliani D, Eisenberger M, et al. Eligibility and response guidelines for phase II clinical trials in androgen-independent prostate cancer: recommendations from the Prostate-Specific Antigen Working Group. *J. Clin. Oncol.* 1999 Nov;17(11):3461-7.

[26] Oefelein MG, Feng A, Scolieri MJ, Ricchiutti D, Resnick MI. Reassessment of the definition of castrate levels of testosterone: implications for clinical decision making. *Urology.* 2000 Dec 20;56(6):1021-4.

[27] Russell DW, Wilson JD. Steroid 5 alpha-reductase: two genes/two enzymes. *Annu. Rev. Biochem.* 1994;63:25-61.

[28] Titus MA, Gregory CW, Ford OH, 3rd, Schell MJ, Maygarden SJ, Mohler JL. Steroid 5alpha-reductase isozymes I and II in recurrent prostate cancer. *Clin. Cancer Res.* 2005 Jun 15;11(12):4365-71.

[29] McConnell JD, Wilson JD, George FW, Geller J, Pappas F, Stoner E. Finasteride, an inhibitor of 5 alpha-reductase, suppresses prostatic dihydrotestosterone in men with benign prostatic hyperplasia. *J. Clin. Endocrinol. Metab.* 1992 Mar;74(3):505-8.

[30] Thompson IM, Goodman PJ, Tangen CM, Lucia MS, Miller GJ, Ford LG, et al. The influence of finasteride on the development of prostate cancer. *N. Engl. J. Med.* 2003 Jul 17;349(3):215-24.

[31] Thompson IM, Chi C, Ankerst DP, Goodman PJ, Tangen CM, Lippman SM, et al. Effect of finasteride on the sensitivity of PSA for detecting prostate cancer. *J. Natl. Cancer Inst.* 2006 Aug 16;98(16):1128-33.

[32] Bramson HN, Hermann D, Batchelor KW, Lee FW, James MK, Frye SV. Unique preclinical characteristics of GG745, a potent dual inhibitor of 5AR. *J. Pharmacol. Exp. Ther.* 1997 Sep;282(3):1496-502.

[33] Roehrborn CG, Boyle P, Nickel JC, Hoefner K, Andriole G. Efficacy and safety of a dual inhibitor of 5-alpha-reductase types 1 and 2 (dutasteride) in men with benign prostatic hyperplasia. *Urology.* 2002 Sep;60(3):434-41.

[34] Gomella LG. Chemoprevention using dutasteride: the REDUCE trial. *Curr. Opin. Urol.* 2005 Jan;15(1):29-32.

[35] SELECT Trial SWOG page. [cited 2006 September 12]; Available from: http://www.crab.org/select/

[36] Virtamo J, Pietinen P, Huttunen JK, Korhonen P, Malila N, Virtanen MJ, et al. Incidence of cancer and mortality following alpha-tocopherol and beta-carotene supplementation: a postintervention follow-up. *JAMA.* 2003 Jul 23;290(4):476-85.

[37] Duffield-Lillico AJ, Dalkin BL, Reid ME, Turnbull BW, Slate EH, Jacobs ET, et al. Selenium supplementation, baseline plasma selenium status and incidence of prostate cancer: an analysis of the complete treatment period of the Nutritional Prevention of Cancer Trial. *BJU Int.* 2003 May; 91(7):608-12.

[38] Pilepich MV, Krall JM, John MJ, Rubin P, Porter AT, Marcial VA, et al. Hormonal cytoreduction in locally advanced carcinoma of the prostate treated with definitive radiotherapy: preliminary results of RTOG 83-07. *Int. J. Radiat. Oncol. Biol. Phys.* 1989 Mar;16(3):813-7.

[39] Bolla M, Gonzalez D, Warde P, Dubois JB, Mirimanoff RO, Storme G, et al. Improved survival in patients with locally advanced prostate cancer treated with radiotherapy and goserelin. *N. Engl. J. Med.* 1997 Jul 31;337(5):295-300.

[40] Bolla M, Collette L, Blank L, Warde P, Dubois JB, Mirimanoff RO, et al. Long-term results with immediate androgen suppression and external irradiation in patients with locally advanced prostate cancer (an EORTC study): a phase III randomised trial. *Lancet.* 2002 Jul 13;360(9327):103-6.

[41] Pilepich MV, Caplan R, Byhardt RW, Lawton CA, Gallagher MJ, Mesic JB, et al. Phase III trial of androgen suppression using goserelin in unfavorable-prognosis carcinoma of the prostate treated with definitive radiotherapy: report of Radiation Therapy Oncology Group Protocol 85-31. *J. Clin. Oncol.* 1997 Mar;15(3):1013-21.

[42] Lawton CA, Winter K, Murray K, Machtay M, Mesic JB, Hanks GE, et al. Updated results of the phase III Radiation Therapy Oncology Group (RTOG) trial 85-31 evaluating the potential benefit of androgen suppression following standard radiation therapy for unfavorable prognosis carcinoma of the prostate. *Int. J. Radiat. Oncol. Biol. Phys.* 2001 Mar 15;49(4):937-46.

[43] Pilepich M, Winter K, Lawton C, Do D, Re R, Mi M, et al. Phase III trial of androgen suppression adjuvant to definitive radiotherapy. Long term results of RTOG 85-31 [abstract]. *Proc. Am. Soc. Clin. Oncol.* 2003;22:A1530.

[44] Pilepich MV, Winter K, Lawton CA, Krisch RE, Wolkov HB, Movsas B, et al. Androgen suppression adjuvant to definitive radiotherapy in prostate carcinoma--long-term results of phase III RTOG 85-31. *Int. J. Radiat. Oncol. Biol. Phys.* 2005 Apr 1;61(5):1285-90.

[45] Lawton CA, Winter K, Grignon D, Pilepich MV. Androgen suppression plus radiation versus radiation alone for patients with stage D1/pathologic node-positive adenocarcinoma of the prostate: updated results based on national prospective randomized trial Radiation Therapy Oncology Group 85-31. *J. Clin. Oncol.* 2005 Feb 1;23(4):800-7.

[46] D'Amico AV, Manola J, Loffredo M, Renshaw AA, DellaCroce A, Kantoff PW. 6-month androgen suppression plus radiation therapy vs radiation therapy alone for patients with clinically localized prostate cancer: a randomized controlled trial. *JAMA.* 2004 Aug 18;292(7):821-7.

[47] D'Amico AV, Loffredo M, Renshaw AA, Loffredo B, Chen MH. Six-month androgen suppression plus radiation therapy compared with radiation therapy alone for men with prostate cancer and a rapidly increasing pretreatment prostate-specific antigen level. *J. Clin. Oncol.* 2006 Sep 1;24(25):4190-5.

[48] Pilepich MV, Krall JM, al-Sarraf M, John MJ, Doggett RL, Sause WT, et al. Androgen deprivation with radiation therapy compared with radiation therapy alone for locally advanced prostatic carcinoma: a randomized comparative trial of the Radiation Therapy Oncology Group. *Urology.* 1995 Apr;45(4):616-23.

[49] Pilepich MV, Winter K, John MJ, Mesic JB, Sause W, Rubin P, et al. Phase III radiation therapy oncology group (RTOG) trial 86-10 of androgen deprivation adjuvant to definitive radiotherapy in locally advanced carcinoma of the prostate. *Int. J. Radiat. Oncol. Biol. Phys.* 2001 Aug 1;50(5):1243-52.

[50] Shipley WU, Desilvio M, Pilepich MV, Roach M, 3rd, Wolkov HB, Sause WT, et al. Early initiation of salvage hormone therapy influences survival in patients who failed initial radiation for locally advanced prostate cancer: A secondary analysis of RTOG protocol 86-10. *Int. J. Radiat. Oncol. Biol. Phys.* 2006 Mar 15;64(4):1162-7.

[51] Hellerstedt BA, Pienta KJ. The truth is out there: an overall perspective on androgen deprivation. *Urol. Oncol.* 2003 Jul-Aug;21(4):272-81.

[52] Soloway MS, Pareek K, Sharifi R, Wajsman Z, McLeod D, Wood DP, Jr., et al. Neoadjuvant androgen ablation before radical prostatectomy in cT2bNxMo prostate cancer: 5-year results. *J. Urol.* 2002 Jan;167(1):112-6.

[53] Gleave ME, La Bianca SE, Goldenberg SL, Jones EC, Bruchovsky N, Sullivan LD. Long-term neoadjuvant hormone therapy prior to radical prostatectomy: evaluation of risk for biochemical recurrence at 5-year follow-up. *Urology.* 2000 Aug 1;56(2):289-94.

[54] Gleave ME, Goldenberg SL, Chin JL, Warner J, Saad F, Klotz LH, et al. Randomized comparative study of 3 versus 8-month neoadjuvant hormonal therapy before radical prostatectomy: biochemical and pathological effects. *J. Urol.* 2001 Aug;166(2):500-6; discussion 6-7.

[55] Gerber GS, Thisted RA, Chodak GW, Schroder FH, Frohmuller HG, Scardino PT, et al. Results of radical prostatectomy in men with locally advanced prostate cancer: multi-institutional pooled analysis. *Eur. Urol.* 1997;32(4):385-90.

[56] Messing EM, Manola J, Sarosdy M, Wilding G, Crawford ED, Trump D. Immediate hormonal therapy compared with observation after radical prostatectomy and pelvic lymphadenectomy in men with node-positive prostate cancer. *N. Engl. J. Med.* 1999 Dec 9;341(24):1781-8.

[57] Cheng L, Zincke H, Blute ML, Bergstralh EJ, Scherer B, Bostwick DG. Risk of prostate carcinoma death in patients with lymph node metastasis. *Cancer.* 2001 Jan 1;91(1):66-73.

[58] Studer UE, Hauri D, Hanselmann S, Chollet D, Leisinger HJ, Gasser T, et al. Immediate versus deferred hormonal treatment for patients with prostate cancer who are not suitable for curative local treatment: results of the randomized trial SAKK 08/88. *J. Clin. Oncol.* 2004 Oct 15;22(20):4109-18.

[59] Studer UE, Whelan P, Albrecht W, Casselman J, de Reijke T, Hauri D, et al. Immediate or deferred androgen deprivation for patients with prostate cancer not suitable for local treatment with curative intent: European Organisation for Research and Treatment of Cancer (EORTC) Trial 30891. *J. Clin. Oncol.* 2006 Apr 20;24(12):1868-76.

[60] Sorensen HT, Mellemkjaer L, Steffensen FH, Olsen JH, Nielsen GL. The risk of a diagnosis of cancer after primary deep venous thrombosis or pulmonary embolism. *N. Engl. J. Med.* 1998 Apr 23;338(17):1169-73.

[61] Gulley J, Parnes H, Wright J, Dahut W. Early treatment gets the benefit [letter]. *J. Clin. Oncol.* 2006;24(32):in press.

[62] Ravery V. The significance of recurrent PSA after radical prostatectomy: benign versus malignant sources. *Semin. Urol. Oncol.* 1999 Aug;17(3):127-9.

[63] Ciezki JP, Reddy CA, Garcia J, Angermeier K, Ulchaker J, Mahadevan A, et al. PSA kinetics after prostate brachytherapy: PSA bounce phenomenon and its implications for PSA doubling time. *Int. J. Radiat. Oncol. Biol. Phys.* 2006 Feb 1;64(2):512-7.

[64] Jhaveri FM, Zippe CD, Klein EA, Kupelian PA. Biochemical failure does not predict overall survival after radical prostatectomy for localized prostate cancer: 10-year results. *Urology.* 1999 Nov;54(5):884-90.

[65] Pound CR, Partin AW, Eisenberger MA, Chan DW, Pearson JD, Walsh PC. Natural history of progression after PSA elevation following radical prostatectomy. *Jama.* 1999 May 5;281(17):1591-7.

[66] Freedland SJ, Humphreys EB, Mangold LA, Eisenberger M, Dorey FJ, Walsh PC, et al. Risk of prostate cancer-specific mortality following biochemical recurrence after radical prostatectomy. *Jama.* 2005 Jul 27;294(4):433-9.

[67] Cannon GM, Jr., Walsh PC, Partin AW, Pound CR. Prostate-specific antigen doubling time in the identification of patients at risk for progression after treatment and biochemical recurrence for prostate cancer. *Urology*. 2003 Dec 29;62 Suppl 1:2-8.

[68] Partin AW, Pearson JD, Landis PK, Carter HB, Pound CR, Clemens JQ, et al. Evaluation of serum prostate-specific antigen velocity after radical prostatectomy to distinguish local recurrence from distant metastases. *Urology*. 1994 May;43(5):649-59.

[69] Zhou P, Chen MH, McLeod D, Carroll PR, Moul JW, D'Amico AV. Predictors of prostate cancer-specific mortality after radical prostatectomy or radiation therapy. *J. Clin. Oncol.* 2005 Oct 1;23(28):6992-8.

[70] D'Amico AV, Moul JW, Carroll PR, Sun L, Lubeck D, Chen MH. Surrogate end point for prostate cancer-specific mortality after radical prostatectomy or radiation therapy. *J. Natl. Cancer Inst.* 2003 Sep 17;95(18):1376-83.

[71] Isaacs JT, Coffey DS. Adaptation versus selection as the mechanism responsible for the relapse of prostatic cancer to androgen ablation therapy as studied in the Dunning R-3327-H adenocarcinoma. *Cancer Res.* 1981 Dec;41(12 Pt 1):5070-5.

[72] Zincke H, Lau W, Bergstralh E, Blute ML. Role of early adjuvant hormonal therapy after radical prostatectomy for prostate cancer. *J. Urol.* 2001 Dec;166(6):2208-15.

[73] Moul JW, Wu H, Sun L, McLeod DG, Amling C, Donahue T, et al. Early versus delayed hormonal therapy for prostate specific antigen only recurrence of prostate cancer after radical prostatectomy. *J. Urol.* 2004 Mar;171(3):1141-7.

[74] Immediate versus deferred treatment for advanced prostatic cancer: initial results of the Medical Research Council Trial. The Medical Research Council Prostate Cancer Working Party Investigators Group. *Br. J. Urol.* 1997 Feb;79(2):235-46.

[75] Kirk D. Immediate vs deferred hormone treatment for prostate cancer. How safe is androgen deprivation? *Br. J. Urol.* 2000;86(Suppl 3):S220.

[76] Nair B, Wilt T, MacDonald R, Rutks I. Early versus deferred androgen suppression in the treatment of advanced prostatic cancer. *Cochrane Database Syst. Rev.* 2002(1): CD003506.

[77] Picus J, Halabi S, Small E, Do D, Re R, Mi M, et al. Long term efficacy of peripheral androgen blockade on prostate cancer: CALGB 9782 [abstract]. *J. Clin. Oncol.* 2006; 24(18S Pt 1):A4573.

[78] Huggins C, Stevens R, Hodges C. Studies in prostatic cancer. II. The effects of castration on advanced cancer of the prostate gland. *Arch. Surg*. 1941;43:209-23.

[79] Ferrari P, Castagnetti G, Ferrari G, Pollastri CA, Tavoni F, Dotti A. Combination treatment in M1 prostate cancer. *Cancer*. 1993 Dec 15;72(12 Suppl):3880-5.

[80] Robinson MR, Smith PH, Richards B, Newling DW, de Pauw M, Sylvester R. The final analysis of the EORTC Genito-Urinary Tract Cancer Co-Operative Group phase III clinical trial (protocol 30805) comparing orchidectomy, orchidectomy plus cyproterone acetate and low dose stilboestrol in the management of metastatic carcinoma of the prostate. *Eur. Urol.* 1995;28(4):273-83.

[81] Labrie F, Dupont A, Belanger A, Lacoursiere Y, Raynaud JP, Husson JM, et al. New approach in the treatment of prostate cancer: complete instead of partial withdrawal of androgens. *Prostate.* 1983;4(6):579-94.

[82] Denis L, Murphy GP. Overview of phase III trials on combined androgen treatment in patients with metastatic prostate cancer. *Cancer*. 1993 Dec 15;72(12 Suppl):3888-95.

[83] Crawford ED, Eisenberger MA, McLeod DG, Spaulding JT, Benson R, Dorr FA, et al. A controlled trial of leuprolide with and without flutamide in prostatic carcinoma. *N. Engl. J. Med.* 1989 Aug 17;321(7):419-24.

[84] Denis LJ, Carneiro de Moura JL, Bono A, Sylvester R, Whelan P, Newling D, et al. Goserelin acetate and flutamide versus bilateral orchiectomy: a phase III EORTC trial (30853). EORTC GU Group and EORTC Data Center. *Urology.* 1993 Aug;42(2):119-29; discussion 29-30.

[85] Janknegt RA, Abbou CC, Bartoletti R, Bernstein-Hahn L, Bracken B, Brisset JM, et al. Orchiectomy and nilutamide or placebo as treatment of metastatic prostatic cancer in a multinational double-blind randomized trial. *J. Urol.* 1993 Jan;149(1):77-82; discussion 3.

[86] Dijkman GA, Janknegt RA, De Reijke TM, Debruyne FM. Long-term efficacy and safety of nilutamide plus castration in advanced prostate cancer, and the significance of early prostate specific antigen normalization. International Anandron Study Group. *J. Urol.* 1997 Jul;158(1):160-3.

[87] Eisenberger MA, Blumenstein BA, Crawford ED, Miller G, McLeod DG, Loehrer PJ, et al. Bilateral orchiectomy with or without flutamide for metastatic prostate cancer. *N. Engl. J. Med.* 1998 Oct 8;339(15):1036-42.

[88] Maximum androgen blockade in advanced prostate cancer: an overview of 22 randomised trials with 3283 deaths in 5710 patients. Prostate Cancer Trialists' Collaborative Group. *Lancet.* 1995 Jul 29;346(8970):265-9.

[89] Samson DJ, Seidenfeld J, Schmitt B, Hasselblad V, Albertsen PC, Bennett CL, et al. Systematic review and meta-analysis of monotherapy compared with combined androgen blockade for patients with advanced prostate carcinoma. *Cancer.* 2002 Jul 15;95(2):361-76.

[90] Verhelst J, Denis L, Van Vliet P, Van Poppel H, Braeckman J, Van Cangh P, et al. Endocrine profiles during administration of the new non-steroidal anti-androgen Casodex in prostate cancer. *Clin. Endocrinol.* (Oxf). 1994 Oct;41(4):525-30.

[91] Boccon-Gibod L, Fournier G, Bottet P, Marechal JM, Guiter J, Rischman P, et al. Flutamide versus orchidectomy in the treatment of metastatic prostate carcinoma. *Eur. Urol.* 1997;32(4):391-5; discussion 5-6.

[92] Pavone-Macaluso M. Flutamide monotherapy versus combined androgen blockade in advanced prostate cancer. Interim report of an Italian multicenter randomized study [abstract]. Proceedings of the 23rd Congress of the Societe Internationale d'Urologie, Sydney, Australia. 1994:A354.

[93] Boccardo F, Rubagotti A, Barichello M, Battaglia M, Carmignani G, Comeri G, et al. Bicalutamide monotherapy versus flutamide plus goserelin in prostate cancer patients: results of an Italian Prostate Cancer Project study. *J. Clin. Oncol.* 1999 Jul;17(7):2027-38.

[94] Kolvenbag GJ, Nash A. Bicalutamide dosages used in the treatment of prostate cancer. *Prostate.* 1999 Apr 1;39(1):47-53.

[95] Tyrrell CJ, Kaisary AV, Iversen P, Anderson JB, Baert L, Tammela T, et al. A randomised comparison of 'Casodex' (bicalutamide) 150 mg monotherapy versus castration in the treatment of metastatic and locally advanced prostate cancer. *Eur. Urol.* 1998;33(5):447-56.

[96] Kaisary AV, Iversen P, Tyrrell CJ, Carroll K, Morris T. Is there a role for antiandrogen monotherapy in patients with metastatic prostate cancer? *Prostate Cancer Prostatic Dis.* 2001;4(4):196-203.

[97] Iversen P, Tyrrell CJ, Kaisary AV, Anderson JB, Van Poppel H, Tammela TL, et al. Bicalutamide monotherapy compared with castration in patients with nonmetastatic locally advanced prostate cancer: 6.3 years of followup. *J. Urol.* 2000 Nov; 164(5): 1579-82.

[98] Kolvenbag GJ, Iversen P, Newling DW. Antiandrogen monotherapy: a new form of treatment for patients with prostate cancer. *Urology.* 2001 Aug;58(2 Suppl 1):16-23.

[99] McLeod DG, Iversen P, See WA, Morris T, Armstrong J, Wirth MP. Bicalutamide 150 mg plus standard care vs standard care alone for early prostate cancer. *BJU Int.* 2006 Feb; 97(2):247-54.

[100] Widmark A, Damber L, Fossa S, Do D, Re R, Mi M, et al. Does prophylactic radiotherapy prevent antiandrogen induced gynecomastia? Evaluation of patients in the randomized Scandinavian trial SPCG7/SFUO-3. *Int. J. Radiat. Oncol. Biol. Phys.* 1999;45(Suppl):262-3.

[101] McLeod DG, Iversen P. Gynecomastia in patients with prostate cancer: a review of treatment options. *Urology.* 2000 Nov 1;56(5):713-20.

[102] See WA, Wirth MP, McLeod DG, Iversen P, Klimberg I, Gleason D, et al. Bicalutamide as immediate therapy either alone or as adjuvant to standard care of patients with localized or locally advanced prostate cancer: first analysis of the early prostate cancer program. *J. Urol.* 2002 Aug;168(2):429-35.

[103] Wirth MP, See WA, McLeod DG, Iversen P, Morris T, Carroll K. Bicalutamide 150 mg in addition to standard care in patients with localized or locally advanced prostate cancer: results from the second analysis of the early prostate cancer program at median followup of 5.4 years. *J. Urol.* 2004 Nov;172(5 Pt 1):1865-70.

[104] McLeod DG, See WA, Klimberg I, Gleason D, Chodak G, Montie J, et al. The bicalutamide 150 mg early prostate cancer program: findings of the North American trial at 7.7-year median followup. *J. Urol.* 2006 Jul;176(1):75-80.

[105] EORTC Protocol 30991. 2003 [cited September 11, 2006]; Available from: http://www.eortc.be/protoc/Details.asp?Protocol=30991

[106] Kelly WK, Scher HI. Prostate specific antigen decline after antiandrogen withdrawal: the flutamide withdrawal syndrome. *J. Urol.* 1993 Mar;149(3):607-9.

[107] Gleave M, Small E. Androgen deprivation therapy for prostate cancer. In: Carroll P, Nelson W, eds. *Report to the Nation on Prostate Cancer 2004*. Elmwood Park: Medscape 2004:27-36.

[108] Dupont A, Gomez JL, Cusan L, Koutsilieris M, Labrie F. Response to flutamide withdrawal in advanced prostate cancer in progression under combination therapy. *J. Urol.* 1993 Sep;150(3):908-13.

[109] Scher HI, Kelly WK. Flutamide withdrawal syndrome: its impact on clinical trials in hormone-refractory prostate cancer. *J. Clin. Oncol.* 1993 Aug;11(8):1566-72.

[110] Veldscholte J, Ris-Stalpers C, Kuiper GG, Jenster G, Berrevoets C, Claassen E, et al. A mutation in the ligand binding domain of the androgen receptor of human LNCaP cells affects steroid binding characteristics and response to anti-androgens. *Biochem. Biophys. Res. Commun.* 1990 Dec 14;173(2):534-40.

[111] Young CY, Montgomery BT, Andrews PE, Qui SD, Bilhartz DL, Tindall DJ. Hormonal regulation of prostate-specific antigen messenger RNA in human prostatic adenocarcinoma cell line LNCaP. *Cancer Res.* 1991 Jul 15;51(14):3748-52.

[112] Schuurmans AL, Bolt J, Voorhorst MM, Blankenstein RA, Mulder E. Regulation of growth and epidermal growth factor receptor levels of LNCaP prostate tumor cells by different steroids. *Int. J. Cancer.* 1988 Dec 15;42(6):917-22.

[113] Nieh PT. Withdrawal phenomenon with the antiandrogen casodex. *J. Urol.* 1995 Mar;153(3 Pt 2):1070-2; discussion 2-3.

[114] Feldman BJ, Feldman D. The development of androgen-independent prostate cancer. *Nat. Rev. Cancer.* 2001 Oct;1(1):34-45.

[115] Akakura K, Bruchovsky N, Goldenberg SL, Rennie PS, Buckley AR, Sullivan LD. Effects of intermittent androgen suppression on androgen-dependent tumors. Apoptosis and serum prostate-specific antigen. *Cancer.* 1993 May 1;71(9):2782-90.

[116] Gleave M, Miyake H, Chi K. Beyond simple castration: targeting the molecular basis of treatment resistance in advanced prostate cancer. *Cancer Chemother. Pharmacol.* 2005 Nov;56 Suppl 1:47-57.

[117] Gulley J, Figg W, Dahut W. Even more cost savings? Letter to the Editor. *J. Oncol. Pract.* 2006;2(4):202.

[118] Tunn U, Kurek R, Kienle E, Do D, Re R, Mi M, et al. Intermittent is as effective as continuous androgen deprivation in patients with PSA relapse after radical prostatectomy. *J. Urol.* 2004;171:A1458.

[119] Gulley JL, Figg WD, Steinberg SM, Carter J, Sartor O, Higano CS, et al. A prospective analysis of the time to normalization of serum androgens following 6 months of androgen deprivation therapy in patients on a randomized phase III clinical trial using limited hormonal therapy. *J. Urol.* 2005 May;173(5):1567-71.

[120] Bruchovsky N, Klotz L, Crook J, Malone S, Ludgate C, Morris WJ, et al. Final results of the Canadian prospective phase II trial of intermittent androgen suppression for men in biochemical recurrence after radiotherapy for locally advanced prostate cancer: clinical parameters. *Cancer.* 2006 Jul 15;107(2):389-95.

[121] Klotz LH, Herr HW, Morse MJ, Whitmore WF, Jr. Intermittent endocrine therapy for advanced prostate cancer. *Cancer.* 1986 Dec 1;58(11):2546-50.

[122] Goldenberg SL, Bruchovsky N, Gleave ME, Sullivan LD, Akakura K. Intermittent androgen suppression in the treatment of prostate cancer: a preliminary report. *Urology.* 1995 May;45(5):839-44; discussion 44-5.

[123] Strum SB, Scholz MC, McDermed JE. Intermittent androgen deprivation in prostate cancer patients: factors predictive of prolonged time off therapy. *Oncologist.* 2000; 5(1):45-52.

[124] Calais Da Silva F, Calais Da Silva F, Bono A, Do D, Re R, Mi M, et al. Phase III intermittent MAB vs continuous MAB [abstract]. *J. Clin. Oncol.* 2006;24(18S Pt 1):A4513.

[125] Bruchovsky N, Klotz L, Crook J, Do D, Re R, Mi M, et al. A phase II study of intermittent androgen suppression (IAS) in men with a rising serum PSA after radiation for localized prostate cancer. *J. Urol.* 1998;159(5):1287.

[126] Higano CS, Ellis W, Russell K, Lange PH. Intermittent androgen suppression with leuprolide and flutamide for prostate cancer: a pilot study. *Urology.* 1996 Nov; 48(5):800-4.

[127] Rashid MH, Chaudhary UB. Intermittent androgen deprivation therapy for prostate cancer. *Oncologist.* 2004;9(3):295-301.

[128] Salonen A, Viitanen J, Ala-Opas M, Do D, Re R, Mi M, et al. Finnish multicenter study to compare intermittent and continuous androgen deprivation in patients with advanced prostate cancer [abstract]. *J. Urol.* 2006;175(4 Suppl):1198.

[129] Bhandari MS, Crook J, Hussain M. Should intermittent androgen deprivation be used in routine clinical practice? *J. Clin. Oncol.* 2005 Nov 10;23(32):8212-8.

[130] Hussain M, Tangen CM, Higano C, Schelhammer PF, Faulkner J, Crawford ED, et al. Absolute prostate-specific antigen value after androgen deprivation is a strong independent predictor of survival in new metastatic prostate cancer: data from Southwest Oncology Group Trial 9346 (INT-0162). *J. Clin. Oncol.* 2006 Aug 20; 24(24):3984-90.

[131] Wagmiller J, Griggs J, Dick A, Sahasrabudhe D. Individualized strategy for dosing luteinizing hormone-releasing hormone agonists for androgen-independent prostate cancer: identification of outcomes and costs. *J. Oncol. Pract.* 2006;2(2):57-66.

[132] Oefelein MG. Health related quality of life using serum testosterone as the trigger to re-dose long acting depot luteinizing hormone-releasing hormone agonists in patients with prostate cancer. *J. Urol.* 2003 Jan;169(1):251-5.

[133] Pedraza R, Kwart AM. Hormonal therapy for patients with advanced adenocarcinoma of the prostate: is there a role for discontinuing treatment after prolonged androgen suppression? *Urology.* 2003 Apr;61(4):770-3.

[134] Oefelein MG. Serum testosterone-based luteinizing hormone-releasing hormone agonist redosing schedule for chronic androgen ablation: a phase I assessment. *Urology.* 1999 Oct;54(4):694-9.

[135] Nordquist L, Morris M, Sauter N, Do D, Re R, Mi M, et al. Rapid hormone cycling for prostate cancer (PC) patients: the MENS cycle [abstract]. *Proc. Am. Soc. Clin. Oncol.* 2003;22:A1669.

[136] Nordquist L, Sauter N, Morris M, Do D, Re R, Mi M, et al. Hormonal induction followed by rapid hormonal cycling for prostate cancer (PC): the MENS cycle [abstract]. *Proc. Am. Soc. Clin. Oncol.* 2004;22(14S):A4609.

[137] Beekman K, Kelly W, Carducci M, Do D, Re R, Mi M, et al. Rapid hormonal cycling with docetaxel for patients with rising PSA after definitive local therapy for prostate cancer [abstract]. Multidisciplinary Prostate Cancer Symposium: Current Status and Future Directions for Prevention and Management Orlando, FL. 2005:A145.

[138] Clinical Trials Webpage. [cited 2006 September 27]; Available from: http://clinicaltrials.gov/show/NCT00070369

[139] Trachtenberg J, Halpern N, Pont A. Ketoconazole: a novel and rapid treatment for advanced prostatic cancer. *J. Urol.* 1983 Jul;130(1):152-3.

[140] Trachtenberg J. High-dose ketoconazole in the treatment of metastatic prostatic cancer. *Prostate.* 1985;6(4):455.

[141] Harris KA, Weinberg V, Bok RA, Kakefuda M, Small EJ. Low dose ketoconazole with replacement doses of hydrocortisone in patients with progressive androgen independent prostate cancer. *J. Urol.* 2002 Aug;168(2):542-5.

[142] Eichenberger T, Trachtenberg J, Toor P, Keating A. Ketoconazole: a possible direct cytotoxic effect on prostate carcinoma cells. *J. Urol.* 1989 Jan;141(1):190-1.

[143] Small EJ, Baron A, Bok R. Simultaneous antiandrogen withdrawal and treatment with ketoconazole and hydrocortisone in patients with advanced prostate carcinoma. *Cancer.* 1997 Nov 1;80(9):1755-9.

[144] Lowe FC, Somers WJ. The use of ketoconazole in the emergency management of disseminated intravascular coagulation due to metastatic prostatic cancer. *J. Urol.* 1987 May;137(5):1000-2.

[145] Litt MR, Bell WR, Lepor HA. Disseminated intravascular coagulation in prostatic carcinoma reversed by ketoconazole. *JAMA.* 1987 Sep 11;258(10):1361-2.

[146] Johnson DE, Babaian RJ, von Eschenbach AC, Wishnow KI, Tenney D. Ketoconazole therapy for hormonally refractive metastatic prostate cancer. *Urology.* 1988 Feb;31(2):132-4.

[147] Kruit WH, Stoter G, Klijn JG. Effect of combination therapy with aminoglutethimide and hydrocortisone on prostate-specific antigen response in metastatic prostate cancer refractory to standard endocrine therapy. *Anticancer Drugs.* 2004 Oct;15(9):843-7.

[148] Elomaa I, Taube T, Blomqvist C, Rissanen P, Rannikko S, Alfthan O. Aminoglutethimide for advanced prostatic cancer resistant to conventional hormonal therapy. *Eur. Urol.* 1988;14(2):104-6.

[149] Chang AY, Bennett JM, Pandya KJ, Asbury R, McCune C. A study of aminoglutethemide and hydrocortisone in patients with advanced and refractory prostate carcinoma. *Am. J. Clin. Oncol.* 1989 Aug;12(4):358-60.

[150] Lam JS, Leppert JT, Vemulapalli SN, Shvarts O, Belldegrun AS. Secondary hormonal therapy for advanced prostate cancer. *J. Urol.* 2006 Jan;175(1):27-34.

[151] O'Donnell A, Judson I, Dowsett M, Raynaud F, Dearnaley D, Mason M, et al. Hormonal impact of the 17alpha-hydroxylase/C(17,20)-lyase inhibitor abiraterone acetate (CB7630) in patients with prostate cancer. *Br. J. Cancer.* 2004 Jun 14;90(12): 2317-25.

[152] de Bono J, O'Donnell A, Attard G, Do D, Re R, Mi M, et al. Clinical and endocrine evaluation of abiraterone acetate (AA), a rationally designed small molecule inhibitor of androgen synthesis targeting 17a hydroxylase (17OH)/17,20 lyase, in patients with hormone refractory prostate cancer [abstract]. *ASCO 2005 Prostate Cancer Symposium.* 2005:A290.

[153] Gulley J, Dahut WL. Chemotherapy for prostate cancer: finally an advance! *Am. J. Ther.* 2004 Jul-Aug;11(4):288-94.

[154] Isaacs JT, Heston WD, Weissman RM, Coffey DS. Animal models of the hormone-sensitive and -insensitive prostatic adenocarcinomas, Dunning R-3327-H, R-3327-HI, and R-3327-AT. *Cancer Res.* 1978 Nov;38(11 Pt 2):4353-9.

[155] Murphy GP, Beckley S, Brady MF, Chu TM, deKernion JB, Dhabuwala C, et al. Treatment of newly diagnosed metastatic prostate cancer patients with chemotherapy agents in combination with hormones versus hormones alone. *Cancer.* 1983 Apr 1;51(7):1264-72.

[156] Murphy GP, Huben RP, Priore R. Results of another trial of chemotherapy with and without hormones in patients with newly diagnosed metastatic prostate cancer. *Urology.* 1986 Jul;28(1):36-40.

[157] Wang J, Halford S, Rigg A, Roylance R, Lynch M, Waxman J. Adjuvant mitozantrone chemotherapy in advanced prostate cancer. *BJU Int.* 2000 Oct;86(6):675-80.

[158] Hudes G. Estramustine-based chemotherapy. *Semin. Urol. Oncol.* 1997 Feb;15(1):13-9.

[159] Benson R, Hartley-Asp B. Mechanisms of action and clinical uses of estramustine. *Cancer Invest.* 1990;8(3-4):375-80.

[160] Clark PE, Peereboom DM, Dreicer R, Levin HS, Clark SB, Klein EA. Phase II trial of neoadjuvant estramustine and etoposide plus radical prostatectomy for locally advanced prostate cancer. *Urology.* 2001 Feb;57(2):281-5.

[161] Zelefsky MJ, Kelly WK, Scher HI, Lee H, Smart T, Metz E, et al. Results of a phase II study using estramustine phosphate and vinblastine in combination with high-dose three-dimensional conformal radiotherapy for patients with locally advanced prostate cancer. *J. Clin. Oncol.* 2000 May;18(9):1936-41.

[162] Khil MS, Kim JH, Bricker LJ, Cerny JC. Tumor control of locally advanced prostate cancer following combined estramustine, vinblastine, and radiation therapy. *Cancer J. Sci. Am.* 1997 Sep-Oct;3(5):289-96.

[163] Ben-Josef E, Porter AT, Han S, Mertens W, Chuba P, Fontana J, et al. Neoadjuvant estramustine and etoposide followed by concurrent estramustine and definitive radiotherapy for locally advanced prostate cancer: feasibility and preliminary results. *Int. J. Radiat. Oncol. Biol. Phys.* 2001 Mar 1;49(3):699-703.

[164] Pettaway CA, Pisters LL, Troncoso P, Slaton J, Finn L, Kamoi K, et al. Neoadjuvant chemotherapy and hormonal therapy followed by radical prostatectomy: feasibility and preliminary results. *J. Clin. Oncol.* 2000 Mar;18(5):1050-7.

[165] Hussain M, Smith DC, El-Rayes BF, Du W, Vaishampayan U, Fontana J, et al. Neoadjuvant docetaxel and estramustine chemotherapy in high-risk/locallyadvanced prostate cancer. *Urology.* 2003 Apr;61(4):774-80.

[166] Petrylak D, Shelton G, England-Owen C, Do D, Re R, Mi M, et al. Response and preliminary survival results of a phase II study of docetaxel (D) + estramustine (E) in patients with androgen-independent prostate cancer (AIPC) [abstract]. *Proc. Am. Soc. Clin. Oncol.* 2000;19:A334.

[167] Kent EC, Hussain MH. The rationale for adjuvant chemotherapy for high-risk prostate cancer. *Curr. Opin. Urol.* 2003 Mar;13(2):123-31.

[168] Hudes G, Einhorn L, Ross E, Balsham A, Loehrer P, Ramsey H, et al. Vinblastine versus vinblastine plus oral estramustine phosphate for patients with hormone-refractory prostate cancer: a Hoosier Oncology Group and Fox Chase Network phase III trial. *J. Clin. Oncol.* 1999 Oct;17(10):3160-6.

[169] Petrylak DP, Tangen CM, Hussain MH, Lara PN, Jr., Jones JA, Taplin ME, et al. Docetaxel and estramustine compared with mitoxantrone and prednisone for advanced refractory prostate cancer. *N. Engl. J. Med.* 2004 Oct 7;351(15):1513-20.

[170] Berry DL, Moinpour CM, Jiang CS, Ankerst DP, Petrylak DP, Vinson LV, et al. Quality of life and pain in advanced stage prostate cancer: results of a Southwest Oncology Group randomized trial comparing docetaxel and estramustine to mitoxantrone and prednisone. *J. Clin. Oncol.* 2006 Jun 20;24(18):2828-35.

[171] Tannock IF, de Wit R, Berry WR, Horti J, Pluzanska A, Chi KN, et al. Docetaxel plus prednisone or mitoxantrone plus prednisone for advanced prostate cancer. *N. Engl. J. Med.* 2004 Oct 7;351(15):1502-12.

[172] Tang Y, Khan MA, Goloubeva O, Lee DI, Jelovac D, Brodie AM, et al. Docetaxel followed by castration improves outcomes in LNCaP prostate cancer-bearing severe combined immunodeficient mice. *Clin. Cancer Res.* 2006 Jan 1;12(1):169-74.

[173] Perlman H, Zhang X, Chen MW, Walsh K, Buttyan R. An elevated bax/bcl-2 ratio corresponds with the onset of prostate epithelial cell apoptosis. *Cell Death Differ.* 1999 Jan;6(1):48-54.

[174] Hussain A, Dawson N, Amin P, Engstrom C, Dorsey B, Siegel E, et al. Docetaxel followed by hormone therapy in men experiencing increasing prostate-specific antigen after primary local treatments for prostate cancer. *J. Clin. Oncol.* 2005 Apr 20;23(12): 2789-96.

[175] Olsen NJ, Zhou P, Ong H, Kovacs WJ. Testosterone induces expression of transforming growth factor-beta 1 in the murine thymus. *J. Steroid Biochem. Mol. Biol.* 1993 May;45(5):327-32.

[176] Kim IY, Kim JH, Zelner DJ, Ahn HJ, Sensibar JA, Lee C. Transforming growth factor-beta1 is a mediator of androgen-regulated growth arrest in an androgen-responsive prostatic cancer cell line, LNCaP. *Endocrinology.* 1996 Mar;137(3):991-9.

[177] Zatelli MC, Rossi R, degli Uberti EC. Androgen influences transforming growth factor-beta1 gene expression in human adrenocortical cells. *J. Clin. Endocrinol. Metab.* 2000 Feb;85(2):847-52.

[178] Zatelli MC, Rossi R, del Senno L, degli Uberti EC. Role of transforming growth factor beta 1 (TGF beta 1) in mediating androgen-induced growth inhibition in human adrenal cortex in vitro. *Steroids.* 1998 May-Jun;63(5-6):243-5.

[179] Salm SN, Koikawa Y, Ogilvie V, Tsujimura A, Coetzee S, Moscatelli D, et al. Transforming growth factor-beta is an autocrine mitogen for a novel androgen-responsive murine prostatic smooth muscle cell line, PSMC1. *J. Cell Physiol.* 2000 Dec;185(3):416-24.

[180] Dammann CE, Ramadurai SM, McCants DD, Pham LD, Nielsen HC. Androgen regulation of signaling pathways in late fetal mouse lung development. *Endocrinology.* 2000 Aug;141(8):2923-9.

[181] Inui S, Fukuzato Y, Nakajima T, Yoshikawa K, Itami S. Androgen-inducible TGF-beta1 from balding dermal papilla cells inhibits epithelial cell growth: a clue to understand paradoxical effects of androgen on human hair growth. *FASEB J.* 2002 Dec;16(14):1967-9.

[182] Roden AC, Moser MT, Tri SD, Mercader M, Kuntz SM, Dong H, et al. Augmentation of T cell levels and responses induced by androgen deprivation. *J. Immunol.* 2004 Nov 15;173(10):6098-108.

[183] Hsueh EC, Gupta RK, Lefor A, Reyzin G, Ye W, Morton DL. Androgen blockade enhances response to melanoma vaccine. *J. Surg. Res.* 2003 Apr;110(2):393-8.

[184] Drake CG, Doody AD, Mihalyo MA, Huang CT, Kelleher E, Ravi S, et al. Androgen ablation mitigates tolerance to a prostate/prostate cancer-restricted antigen. *Cancer Cell.* 2005 Mar;7(3):239-49.

[185] Olsen NJ, Kovacs WJ. Gonadal steroids and immunity. *Endocr. Rev.* 1996 Aug;17(4):369-84.

[186] Mercader M, Bodner BK, Moser MT, Kwon PS, Park ES, Manecke RG, et al. T cell infiltration of the prostate induced by androgen withdrawal in patients with prostate cancer. *Proc. Natl. Acad. Sci. U. S. A.* 2001 Dec 4;98(25):14565-70.

[187] Sutherland JS, Goldberg GL, Hammett MV, Uldrich AP, Berzins SP, Heng TS, et al. Activation of thymic regeneration in mice and humans following androgen blockade. *J. Immunol.* 2005 Aug 15;175(4):2741-53.

[188] Schwarer A, Sutherland J, Muirhead J, Prince H. Sex steroid suppression with luteinizing hormone-releasing hormone agonist (LHRH-A) allows thymic reactivation and improved immunological recovery after allogencic hematopoietic stem cell transplantation (alloHSCT). American Society of Hematology Annual Meeting [abstract]. *Blood.* 2003;102(11):A260.

[189] Sanda MG, Smith DC, Charles LG, Hwang C, Pienta KJ, Schlom J, et al. Recombinant vaccinia-PSA (PROSTVAC) can induce a prostate-specific immune response in androgen-modulated human prostate cancer. *Urology.* 1999 Feb;53(2):260-6.

[190] Eder JP, Kantoff PW, Roper K, Xu GX, Bubley GJ, Boyden J, et al. A phase I trial of a recombinant vaccinia virus expressing prostate-specific antigen in advanced prostate cancer. *Clin. Cancer Res.* 2000 May;6(5):1632-8.

[191] Gulley J, Chen AP, Dahut W, Arlen PM, Bastian A, Steinberg SM, et al. Phase I study of a vaccine using recombinant vaccinia virus expressing PSA (rV-PSA) in patients with metastatic androgen-independent prostate cancer. *Prostate.* 2002 Oct 1;53(2):109-17.

[192] Arlen PM, Gulley JL, Todd N, Lieberman R, Steinberg SM, Morin S, et al. Antiandrogen, vaccine and combination therapy in patients with nonmetastatic hormone refractory prostate cancer. *J. Urol.* 2005 Aug;174(2):539-46.

[193] Kassouf W, Tanguay S, Aprikian AG. Nilutamide as second line hormone therapy for prostate cancer after androgen ablation fails. *J. Urol.* 2003 May;169(5):1742-4.

[194] Desai A, Stadler WM, Vogelzang NJ. Nilutamide: possible utility as a second-line hormonal agent. *Urology.* 2001 Dec;58(6):1016-20.

[195] Daniell HW, Dunn SR, Ferguson DW, Lomas G, Niazi Z, Stratte PT. Progressive osteoporosis during androgen deprivation therapy for prostate cancer. *J. Urol.* 2000 Jan;163(1):181-6.

[196] Stoch SA, Parker RA, Chen L, Bubley G, Ko YJ, Vincelette A, et al. Bone loss in men with prostate cancer treated with gonadotropin-releasing hormone agonists. *J. Clin. Endocrinol.* Metab. 2001 Jun;86(6):2787-91.

[197] Preston DM, Torrens JI, Harding P, Howard RS, Duncan WE, McLeod DG. Androgen deprivation in men with prostate cancer is associated with an increased rate of bone loss. *Prostate Cancer Prostatic Dis.* 2002;5(4):304-10.

[198] Oefelein MG, Ricchuiti V, Conrad W, Seftel A, Bodner D, Goldman H, et al. Skeletal fracture associated with androgen suppression induced osteoporosis: the clinical incidence and risk factors for patients with prostate cancer. *J. Urol.* 2001 Nov;166(5): 1724-8.

[199] Diamond TH, Bucci J, Kersley JH, Aslan P, Lynch WB, Bryant C. Osteoporosis and spinal fractures in men with prostate cancer: risk factors and effects of androgen deprivation therapy. *J. Urol.* 2004 Aug;172(2):529-32.

[200] Shahinian VB, Kuo YF, Freeman JL, Goodwin JS. Risk of fracture after androgen deprivation for prostate cancer. *N. Engl. J. Med.* 2005 Jan 13;352(2):154-64.

[201] Smith MR, McGovern FJ, Zietman AL, Fallon MA, Hayden DL, Schoenfeld DA, et al. Pamidronate to prevent bone loss during androgen-deprivation therapy for prostate cancer. *N. Engl. J. Med.* 2001 Sep 27;345(13):948-55.

[202] Smith MR, Eastham J, Gleason DM, Shasha D, Tchekmedyian S, Zinner N. Randomized controlled trial of zoledronic acid to prevent bone loss in men receiving

androgen deprivation therapy for nonmetastatic prostate cancer. *J. Urol.* 2003 Jun;169(6):2008-12.

[203] Michaelson M, Lee H, Kaufman D, Kantoff P, Finkelstein J, Smith M. Annual zoledronic acid to prevent gonadotropin-releasing hormone agonist-induced bone loss in men with prostate cancer: a randomized placebo-controlled trial [abstract]. *J. Clin. Oncol.* 2006;24(18S Pt I):4515.

[204] Alibhai SM, Rahman S, Warde PR, Jewett MA, Jaffer T, Cheung AM. Prevention and management of osteoporosis in men receiving androgen deprivation therapy: a survey of urologists and radiation oncologists. *Urology.* 2006 Jul;68(1):126-31.

[205] Higano CS. Management of bone loss in men with prostate cancer. *J. Urol.* 2003 Dec;170(6 Pt 2):S59-63; discussion S4.

[206] Holzbeierlein JM, McLaughlin MD, Thrasher JB. Complications of androgen deprivation therapy for prostate cancer. *Curr. Opin. Urol.* 2004 May;14(3):177-83.

[207] Loprinzi CL, Michalak JC, Quella SK, O'Fallon JR, Hatfield AK, Nelimark RA, et al. Megestrol acetate for the prevention of hot flashes. *N. Engl. J. Med.* 1994 Aug 11;331(6):347-52.

[208] Sartor O, Eastham JA. Progressive prostate cancer associated with use of megestrol acetate administered for control of hot flashes. *South Med. J.* 1999 Apr;92(4):415-6.

[209] Quella SK, Loprinzi CL, Sloan J, Novotny P, Perez EA, Burch PA, et al. Pilot evaluation of venlafaxine for the treatment of hot flashes in men undergoing androgen ablation therapy for prostate cancer. *J. Urol.* 1999 Jul;162(1):98-102.

[210] Loprinzi CL, Barton DL, Carpenter LA, Sloan JA, Novotny PJ, Gettman MT, et al. Pilot evaluation of paroxetine for treating hot flashes in men. *Mayo Clin. Proc.* 2004 Oct;79(10):1247-51.

[211] Sharifi N, Gulley JL, Dahut WL. Androgen deprivation therapy for prostate cancer. *JAMA.* 2005 Jul 13;294(2):238-44.

[212] Muller M, Grobbee DE, den Tonkelaar I, Lamberts SW, van der Schouw YT. Endogenous sex hormones and metabolic syndrome in aging men. *J. Clin. Endocrinol. Metab.* 2005 May;90(5):2618-23.

[213] Smith MR, Finkelstein JS, McGovern FJ, Zietman AL, Fallon MA, Schoenfeld DA, et al. Changes in body composition during androgen deprivation therapy for prostate cancer. *J. Clin. Endocrinol. Metab.* 2002 Feb;87(2):599-603.

[214] Berruti A, Dogliotti L, Terrone C, Cerutti S, Isaia G, Tarabuzzi R, et al. Changes in bone mineral density, lean body mass and fat content as measured by dual energy x-ray absorptiometry in patients with prostate cancer without apparent bone metastases given androgen deprivation therapy. *J. Urol.* 2002 Jun;167(6):2361-7; discussion 7.

[215] Nishiyama T, Ishizaki F, Anraku T, Shimura H, Takahashi K. The influence of androgen deprivation therapy on metabolism in patients with prostate cancer. *J. Clin. Endocrinol. Metab.* 2005 Feb;90(2):657-60.

[216] Braga-Basaria M, Dobs AS, Muller DC, Carducci MA, John M, Egan J, et al. Metabolic syndrome in men with prostate cancer undergoing long-term androgen-deprivation therapy. *J. Clin. Oncol.* 2006 Aug 20;24(24):3979-83.

[217] Keating NL, O'Malley AJ, Smith MR. Diabetes and cardiovascular disease during androgen deprivation therapy for prostate cancer. *J. Clin. Oncol.* 2006 Sep 20;24(27): 4448-56.

[218] Managing the side effects of androgen deprivation therapy. *J. Support Oncol.* 2006 Feb; 4(2):97-8.

[219] Basaria S, Lieb J, 2nd, Tang AM, DeWeese T, Carducci M, Eisenberger M, et al. Long-term effects of androgen deprivation therapy in prostate cancer patients. *Clin. Endocrinol.* (Oxf). 2002 Jun;56(6):779-86.

[220] Potosky AL, Knopf K, Clegg LX, Albertsen PC, Stanford JL, Hamilton AS, et al. Quality-of-life outcomes after primary androgen deprivation therapy: results from the Prostate Cancer Outcomes Study. *J. Clin. Oncol.* 2001 Sep 1;19(17):3750-7.

[221] Chen AC, Petrylak DP. Complications of androgen deprivation therapy in men with prostate cancer. *Curr. Oncol. Rep.* 2004 May;6(3):209-15.

[222] Cavallini G, Caracciolo S, Vitali G, Modenini F, Biagiotti G. Carnitine versus androgen administration in the treatment of sexual dysfunction, depressed mood, and fatigue associated with male aging. *Urology.* 2004 Apr;63(4):641-6.

[223] Shahinian VB, Kuo YF, Freeman JL, Goodwin JS. Risk of the "androgen deprivation syndrome" in men receiving androgen deprivation for prostate cancer. *Arch. Intern. Med.* 2006 Feb 27;166(4):465-71.

[224] Green HJ, Pakenham KI, Headley BC, Yaxley J, Nicol DL, Mactaggart PN, et al. Altered cognitive function in men treated for prostate cancer with luteinizing hormone-releasing hormone analogues and cyproterone acetate: a randomized controlled trial. *BJU Int.* 2002 Sep;90(4):427-32.

[225] Salminen EK, Portin RI, Koskinen AI, Helenius HY, Nurmi MJ. Estradiol and cognition during androgen deprivation in men with prostate carcinoma. *Cancer.* 2005 Apr 1;103(7):1381-7.

[226] Jenkins VA, Bloomfield DJ, Shilling VM, Edginton TL. Does neoadjuvant hormone therapy for early prostate cancer affect cognition? Results from a pilot study. *BJU Int.* 2005 Jul;96(1):48-53.

[227] Salminen E, Portin R, Korpela J, Backman H, Parvinen LM, Helenius H, et al. Androgen deprivation and cognition in prostate cancer. *Br. J. Cancer.* 2003 Sep 15;89 (6):971-6.

[228] Strum SB, McDermed JE, Scholz MC, Johnson H, Tisman G. Anaemia associated with androgen deprivation in patients with prostate cancer receiving combined hormone blockade. *Br. J. Urol.* 1997 Jun;79(6):933-41.

[229] Sharifi N, Dahut WL, Steinberg SM, Figg WD, Tarassoff C, Arlen P, et al. A retrospective study of the time to clinical endpoints for advanced prostate cancer. *BJU Int.* 2005 Nov;96(7):985-9.

[230] Oefelein MG, Agarwal PK, Resnick MI. Survival of patients with hormone refractory prostate cancer in the prostate specific antigen era. *J. Urol.* 2004 Apr;171(4):1525-8.

[231] Gregory CW, Johnson RT, Jr., Mohler JL, French FS, Wilson EM. Androgen receptor stabilization in recurrent prostate cancer is associated with hypersensitivity to low androgen. *Cancer Res.* 2001 Apr 1;61(7):2892-8.

[232] Roy-Burman P, Tindall DJ, Robins DM, Greenberg NM, Hendrix MJ, Mohla S, et al. Androgens and prostate cancer: are the descriptors valid? *Cancer Biol. Ther.* 2005 Jan;4(1):4-5.

[233] Agus DB, Cordon-Cardo C, Fox W, Drobnjak M, Koff A, Golde DW, et al. Prostate cancer cell cycle regulators: response to androgen withdrawal and development of androgen independence. *J. Natl. Cancer Inst.* 1999 Nov 3;91(21):1869-76.

[234] Sharifi N, Farrar WL. Androgen receptor as a therapeutic target for androgen independent prostate cancer. *Am. J. Ther.* 2006 Mar-Apr;13(2):166-70.

[235] Scher HI, Sawyers CL. Biology of progressive, castration-resistant prostate cancer: directed therapies targeting the androgen-receptor signaling axis. *J. Clin. Oncol.* 2005 Nov 10;23(32):8253-61.

[236] Figg WD, Liu Y, Arlen P, Gulley J, Steinberg SM, Liewehr DJ, et al. A randomized, phase II trial of ketoconazole plus alendronate versus ketoconazole alone in patients with androgen independent prostate cancer and bone metastases. *J. Urol.* 2005 Mar;173(3):790-6.

[237] Fujikawa K, Matsui Y, Fukuzawa S, Takeuchi H. Prostate-specific antigen levels and clinical response to flutamide as the second hormone therapy for hormone-refractory prostate carcinoma. *Eur. Urol.* 2000 Feb;37(2):218-22.

[238] Mousses S, Wagner U, Chen Y, Kim JW, Bubendorf L, Bittner M, et al. Failure of hormone therapy in prostate cancer involves systematic restoration of androgen responsive genes and activation of rapamycin sensitive signaling. *Oncogene.* 2001 Oct 11;20(46):6718-23.

[239] Holzbeierlein J, Lal P, LaTulippe E, Smith A, Satagopan J, Zhang L, et al. Gene expression analysis of human prostate carcinoma during hormonal therapy identifies androgen-responsive genes and mechanisms of therapy resistance. *Am. J. Pathol.* 2004 Jan;164(1):217-27.

[240] Edwards J, Krishna NS, Grigor KM, Bartlett JM. Androgen receptor gene amplification and protein expression in hormone refractory prostate cancer. *Br. J. Cancer.* 2003 Aug 4;89(3):552-6.

[241] Bubendorf L, Kononen J, Koivisto P, Schraml P, Moch H, Gasser TC, et al. Survey of gene amplifications during prostate cancer progression by high-throughput fluorescence in situ hybridization on tissue microarrays. *Cancer Res.* 1999 Feb 15;59(4):803-6.

[242] Chen CD, Welsbie DS, Tran C, Baek SH, Chen R, Vessella R, et al. Molecular determinants of resistance to antiandrogen therapy. *Nat. Med.* 2004 Jan;10(1):33-9.

[243] Xu J, Qiu Y, DeMayo FJ, Tsai SY, Tsai MJ, O'Malley BW. Partial hormone resistance in mice with disruption of the steroid receptor coactivator-1 (SRC-1) gene. *Science.* 1998 Mar 20;279(5358):1922-5.

[244] Chen T, Wang LH, Farrar WL. Interleukin 6 activates androgen receptor-mediated gene expression through a signal transducer and activator of transcription 3-dependent pathway in LNCaP prostate cancer cells. *Cancer Res.* 2000 Apr 15;60(8):2132-5.

[245] Wallner L, Dai J, Escara-Wilke J, Zhang J, Yao Z, Lu Y, et al. Inhibition of interleukin-6 with CNTO328, an anti-interleukin-6 monoclonal antibody, inhibits conversion of androgen-dependent prostate cancer to an androgen-independent phenotype in orchiectomized mice. *Cancer Res.* 2006 Mar 15;66(6):3087-95.

[246] Papatsoris AG, Karamouzis MV, Papavassiliou AG. Novel insights into the implication of the IGF-1 network in prostate cancer. *Trends Mol. Med.* 2005 Feb;11(2):52-5.

[247] Finn RS, Slamon DJ. Monoclonal antibody therapy for breast cancer: herceptin. *Cancer Chemother. Biol. Response Modif.* 2003;21:223-33.

[248] Mellinghoff IK, Vivanco I, Kwon A, Tran C, Wongvipat J, Sawyers CL. HER2/neu kinase-dependent modulation of androgen receptor function through effects on DNA binding and stability. *Cancer Cell.* 2004 Nov;6(5):517-27.

[249] Gioeli D. Signal transduction in prostate cancer progression. *Clin. Sci.* (Lond). 2005 Apr;108(4):293-308.

[250] Heinlein CA, Chang C. Androgen receptor in prostate cancer. *Endocr. Rev.* 2004 Apr;25(2):276-308.

[251] Weber MJ, Gioeli D. Ras signaling in prostate cancer progression. *J. Cell Biochem.* 2004 Jan 1;91(1):13-25.

[252] Titus MA, Schell MJ, Lih FB, Tomer KB, Mohler JL. Testosterone and dihydrotestosterone tissue levels in recurrent prostate cancer. *Clin. Cancer Res.* 2005 Jul 1;11(13):4653-7.

[253] Stanbrough M, Bubley GJ, Ross K, Golub TR, Rubin MA, Penning TM, et al. Increased expression of genes converting adrenal androgens to testosterone in androgen-independent prostate cancer. *Cancer Res.* 2006 Mar 1;66(5):2815-25.

[254] Tomlins SA, Rhodes DR, Perner S, Dhanasekaran SM, Mehra R, Sun XW, et al. Recurrent fusion of TMPRSS2 and ETS transcription factor genes in prostate cancer. *Science.* 2005 Oct 28;310(5748):644-8.

[255] Adachi M, Takayanagi R, Tomura A, Imasaki K, Kato S, Goto K, et al. Androgen-insensitivity syndrome as a possible coactivator disease. N Engl J Med. 2000 Sep 21;343(12):856-62.

[256] Hur E, Pfaff SJ, Payne ES, Gron H, Buehrer BM, Fletterick RJ. Recognition and accommodation at the androgen receptor coactivator binding interface. *PLoS Biol.* 2004 Sep;2(9):E274.

[257] Wang LH, Yang XY, Zhang X, Mihalic K, Fan YX, Xiao W, et al. Suppression of breast cancer by chemical modulation of vulnerable zinc fingers in estrogen receptor. *Nat. Med.* 2004 Jan;10(1):40-7.

[258] Whitesell L, Lindquist SL. HSP90 and the chaperoning of cancer. *Nat. Rev. Cancer.* 2005 Oct;5(10):761-72.

[259] Georget V, Terouanne B, Nicolas JC, Sultan C. Mechanism of antiandrogen action: key role of hsp90 in conformational change and transcriptional activity of the androgen receptor. *Biochemistry.* 2002 Oct 1;41(39):11824-31.

[260] Drysdale MJ, Brough PA, Massey A, Jensen MR, Schoepfer J. Targeting Hsp90 for the treatment of cancer. *Curr. Opin. Drug Discov. Devel.* 2006 Jul;9(4):483-95.

[261] Schneekloth JS, Jr., Fonseca FN, Koldobskiy M, Mandal A, Deshaies R, Sakamoto K, et al. Chemical genetic control of protein levels: selective in vivo targeted degradation. *J. Am. Chem. Soc.* 2004 Mar 31;126(12):3748-54.

[262] Singh P, Uzgare A, Litvinov I, Denmeade SR, Isaacs JT. Combinatorial androgen receptor targeted therapy for prostate cancer. *Endocr. Relat. Cancer.* 2006 Sep;13(3):653-66.

[263] Talpaz M, Shah NP, Kantarjian H, Donato N, Nicoll J, Paquette R, et al. Dasatinib in imatinib-resistant Philadelphia chromosome-positive leukemias. *N. Engl. J. Med.* 2006 Jun 15;354(24):2531-41.

[264] ollins AT, Berry PA, Hyde C, Stower MJ, Maitland NJ. Prospective identification of tumorigenic prostate cancer stem cells. *Cancer Res.* 2005 Dec 1;65(23):10946-51.

[265] Patrawala L, Calhoun T, Schneider-Broussard R, Li H, Bhatia B, Tang S, et al. Highly purified CD44+ prostate cancer cells from xenograft human tumors are enriched in tumorigenic and metastatic progenitor cells. *Oncogene.* 2006 Mar 16;25(12):1696-708.

[266] Sharifi N, Kawasaki BT, Hurt EM, Farrar WL. Stem cells in prostate cancer: resolving the castrate-resistant conundrum and implications for hormonal therapy. *Cancer Biol. Ther.* 2006 Aug 28;5(8):[Epub ahead of print].

[267] Smaletz O, Scher HI, Small EJ, Verbel DA, McMillan A, Regan K, et al. Nomogram for overall survival of patients with progressive metastatic prostate cancer after castration. *J. Clin. Oncol.* 2002 Oct 1;20(19):3972-82.

[268] Hanks G, Lu J, Machtay M, Do D, Re R, Mi M, et al. RTOG Protocol 92-02: a phase III trial of the use of long term androgen suppression following neoadjuvant hormonal cytoreduction and radiotherapy in locally advanced carcinoma of the prostate [abstract]. *Proc. Am. Soc. Clin. Oncol.* 2000;19:1264.

[269] D'Amico AV, Schultz D, Loffredo M, Dugal R, Hurwitz M, Kaplan I, et al. Biochemical outcome following external beam radiation therapy with or without androgen suppression therapy for clinically localized prostate cancer. *Jama.* 2000 Sep 13;284(10):1280-3.

[270] Schulman CC, Debruyne FM, Forster G, Selvaggi FP, Zlotta AR, Witjes WP. 4-Year follow-up results of a European prospective randomized study on neoadjuvant hormonal therapy prior to radical prostatectomy in T2-3N0M0 prostate cancer. European Study Group on Neoadjuvant Treatment of Prostate Cancer. *Eur. Urol.* 2000 Dec;38(6):706-13.

[271] Aus G, Abrahamsson PA, Ahlgren G, Hugosson J, Lundberg S, Schain M, et al. Three-month neoadjuvant hormonal therapy before radical prostatectomy: a 7-year follow-up of a randomized controlled trial. *BJU Int.* 2002 Oct;90(6):561-6.

[272] Denis LJ, Keuppens F, Smith PH, Whelan P, de Moura JL, Newling D, et al. Maximal androgen blockade: final analysis of EORTC phase III trial 30853. EORTC Genito-Urinary Tract Cancer Cooperative Group and the EORTC Data Center. *Eur. Urol.* 1998;33(2):144-51.

[273] Tyrrell CJ, Altwein JE, Klippel F, Varenhorst E, Lunglmayr G, Boccardo F, et al. Multicenter randomized trial comparing Zoladex with Zoladex plus flutamide in the treatment of advanced prostate cancer. Survival update. International Prostate Cancer Study Group. *Cancer.* 1993 Dec 15;72(12 Suppl):3878-9.

[274] Iversen P, Rasmussen F, Klarskov P, Christensen IJ. Long-term results of Danish Prostatic Cancer Group trial 86. Goserelin acetate plus flutamide versus orchiectomy in advanced prostate cancer. *Cancer.* 1993 Dec 15;72(12 Suppl):3851-4.

[275] Boccardo F, Decensi A, Guarneri D, Rubagotti A, Oneto F, Martorana G, et al. Zoladex with or without flutamide in the treatment of locally advanced or metastatic prostate cancer: interim analysis of an ongoing PONCAP study. Italian Prostatic Cancer Project (PONCAP). *Eur. Urol.* 1990;18 Suppl 3:48-53.

[276] Bracarda S, Rosi P, Pizzirusso G, Do D, Re R, Mi M, et al. Intermittent androgen suppression (IAS) in advanced or inoperable prostate cancer. *Ann. Oncol.* 1998;9:272O.

[277] Theyer G, Hamilton G. Current status of intermittent androgen suppression in the treatment of prostate cancer. *Urology.* 1998 Sep;52(3):353-9.

[278] Gleave M, Bruchovsky N, Goldenberg SL, Rennie P. Intermittent androgen suppression for prostate cancer: rationale and clinical experience. *Eur. Urol.* 1998;34 Suppl 3:37-41.

[279] Crook JM, Szumacher E, Malone S, Huan S, Segal R. Intermittent androgen suppression in the management of prostate cancer. *Urology.* 1999 Mar;53(3):530-4.

[280] Grossfeld GD, Chaudhary UB, Reese DM, Carroll PR, Small EJ. Intermittent androgen deprivation: update of cycling characteristics in patients without clinically apparent metastatic prostate cancer. *Urology.* 2001 Aug;58(2):240-5.

[281] Prapotnich D, Fizazi K, Escudier B, Mombet A, Cathala N, Vallancien G. A 10-year clinical experience with intermittent hormonal therapy for prostate cancer. *Eur. Urol.* 2003 Mar;43(3):233-39; discussion 9-40.

[282] Tangen C, Hussain M, Wilding G, Do D, Re R, Mi M, et al. Determinants of prostate specific antigen normalization in prostate cancer patients treated with androgen deprivation on Southwest Oncology Group Study 9346 (INT-0162) [abstract]. *Proc. Am. Soc. Clin. Oncol.* 2003;22:A1591.

Index

A

B

C

E

F

G

H

I

J

K

L

M

N

O

P

Q

R

S

T

U

V